TYRANNOSAURUS REX, THE TYRANT KING

Life of the Past

James O. Farlow, editor

TYRANNOSAURUS REX, THE TYRANT KING

Edited by Peter Larson and Kenneth Carpenter

Indiana University Press

This book is a publication of

Indiana University Press
Office of Scholarly Publishing
Herman B Wells Library 350
1320 East 10th Street
Bloomington, Indiana 47405 USA

iupress.indiana.edu

© 2008 by Indiana University Press

All rights reserved

No part of this book may be reproduced or utilized in any form or by any means, electronic or mechanical, including photocopying and recording, or by any information storage and retrieval system, without permission in writing from the publisher.

The paper used in this publication meets the minimum requirements of the American National Standard for Information Sciences— Permanence of Paper for Printed Library Materials, ANSI Z39.48-1992.

Manufactured in the United States of America

Library of Congress Cataloging-in-Publication Data

Tyrannosaurus rex, the tyrant king / edited by
  Peter Larson and Kenneth Carpenter.
    p. cm. — (Life of the past) Includes index.
  ISBN-13: 978-0-253-35087-9 (cloth : alk. paper) 1. Tyrannosaurus
    rex—Research. I. Larson, Peter L. II. Carpenter, Kenneth, date
  QE862.S3T977 2008
  567.912'9—dc22
                              2007045376

2 3 4 5 6 24 23 22 21 20 19

CONTENTS

ix   ONLINE MATERIALS

xi   CONTRIBUTORS

xiii   PREFACE

xv   INSTITUTIONAL ABBREVIATIONS

1 One Hundred Years of *Tyrannosaurus rex*: The Skeletons

1   NEAL L. LARSON

2 Wyoming's *Dynamosaurus imperiosus* and Other Early Discoveries of *Tyrannosaurus rex* in the Rocky Mountain West

57   BRENT H. BREITHAUPT, ELIZABETH H. SOUTHWELL, AND NEFFRA A. MATTHEWS

3 How Old Is *T. rex*? Challenges with the Dating of Terrestrial Strata Deposited during the Maastrichtian Stage of the Cretaceous Period

63   KIRK JOHNSON

4 Preliminary Account of the Tyrannosaurid Pete from the Lance Formation of Wyoming

67   KRAIG DERSTLER AND JOHN M. MYERS

5 Taphonomy of the *Tyrannosaurus rex* Peck's Rex from the Hell Creek Formation of Montana

75   KRAIG DERSTLER AND JOHN M. MYERS

6 Taphonomy and Environment of Deposition of a Juvenile Tyrannosaurid Skeleton from the Hell Creek Formation (Latest Maastrichtian) of Southeastern Montana

83 MICHAEL D. HENDERSON AND WILLIAM H. HARRISON

7 One Pretty Amazing *T. rex*

93 MARY HIGBY SCHWEITZER, JENNIFER L. WITTMEYER, AND JOHN R. HORNER

8 Variation and Sexual Dimorphism in *Tyrannosaurus rex*

103 PETER LARSON

9 Why *Tyrannosaurus rex* Had Puny Arms: An Integral Morphodynamic Solution to a Simple Puzzle in Theropod Paleobiology

131 MARTIN LOCKLEY, REIJI KUKIHARA, AND LAURA MITCHELL

10 Looking Again at the Forelimb of *Tyrannosaurus rex*

167 CHRISTINE LIPKIN AND KENNETH CARPENTER

11 Rex, Sit: Digital Modeling of *Tyrannosaurus rex* at Rest

193 KENT A. STEVENS, PETER LARSON, ERIC D. WILLS, AND ART ANDERSON

12 *T. rex* Speed Trap

205 PHILLIP L. MANNING

13 Atlas of the Skull Bones of *Tyrannosaurus rex*

233 PETER LARSON

14 Palatal Kinesis of *Tyrannosaurus rex*

245 HANS C. E. LARSSON

15 Reconstruction of the Jaw Musculature of *Tyrannosaurus rex*

255 RALPH E. MOLNAR

16 Vestigialism in a Dinosaur

283 WILLIAM L. ABLER

17 Tyrannosaurid Pathologies as Clues to Nature and Nurture in the Cretaceous

287 BRUCE M. ROTHSCHILD AND RALPH E. MOLNAR

18 The Extreme Lifestyles and Habits of the Gigantic Tyrannosaurid Superpredators of the Late Cretaceous of North America and Asia

307 GREGORY S. PAUL

19 An Analysis of Predator-Prey Behavior in a Head-to-Head Encounter between *Tyrannosaurus rex* and *Triceratops*

355 JOHN HAPP

20 A Critical Reappraisal of the Obligate Scavenging Hypothesis for *Tyrannosaurus rex* and Other Tyrant Dinosaurs

371 THOMAS R. HOLTZ JR.

21 *Tyrannosaurus rex:* A Century of Celebrity

399 DONALD F. GLUT

429 INDEX

ONLINE MATERIALS

https://www.iupress.indiana.edu/media/tyrannosaurusrex/ contains material relating to three of the chapters in the volume.

Chapter 7

Color figures 7.1, 7.2, 7.3 (figures in black and white reproduction in book chapter)

Chapter 11

Rex, Sit: Modeling Tyrannosaurid Postures

Chapter 13

Atlas of the Skull Bones of *Tyrannosaurus rex*

CONTRIBUTORS

William L. Abler, 1200 Warren Creek Rd., Arcata, CA 95521, USA.

Art Anderson, Virtual Surfaces Inc., 832 E Rand Rd., Suite 16, Mt. Prospect, IL 60056, USA.

Brent H. Breithaupt, Geological Museum, University of Wyoming, Laramie, WY 82071, USA.

Kenneth Carpenter, Department of Earth Sciences, Denver Museum of Nature and Science, 2001 Colorado Blvd., Denver, CO 80206, USA.

Kraig Derstler, Department of Earth and Environmental Sciences, University of New Orleans, New Orleans, LA 70148, USA.

Donald F. Glut, 2805 N Keystone St., Burbank, CA 91504-1604, USA.

John Happ, 3889 Chestnut Hill Rd., Harpers Ferry, WV 25425, USA.

William H. Harrison, Department of Geology and Environmental Geosciences, Northern Illinois University, Dekalb, IL 60115, USA.

Michael D. Henderson, Burpee Museum of Natural History, 737 N Main St., Rockford, IL 61103, USA.

Thomas R. Holtz Jr., Department of Geology, University of Maryland, College Park, MD 20742, USA.

John R. Horner, Museum of the Rockies, Montana State University, Bozeman, MT 59717, USA.

Kirk Johnson, Denver Museum of Nature and Science, 2001 Colorado Blvd., Denver, CO 80206, USA.

Reiji Kukihara, Dinosaur Tracks Museum, University of Colorado at Denver, P.O. Box 173364, Denver, CO 80217, USA.

Neal L. Larson, Black Hills Institute of Geological Research Inc., P.O. Box 643, Hill City, SD 57745, USA.

Peter Larson, Black Hills Institute of Geological Research Inc., P.O. Box 643, Hill City, SD 57745, USA.

Hans C. E. Larsson, Redpath Museum, McGill University, 859 Sherbrooke St. W, Montreal, Quebec H3A 2K6, Canada.

Christine Lipkin, University of Chicago, Chicago, IL 60637, USA.

Martin Lockley, Dinosaur Tracks Museum, University of Colorado at Denver, P.O. Box 173364, Denver, CO 80217, USA.

Phillip L. Manning, The Manchester Museum, University of Manchester, Oxford Road, Manchester, M13 9PL UK.

Neffra A. Matthews, Geological Museum, University of Wyoming, Laramie, WY 82071, USA.

Laura Mitchell, Dinosaur Tracks Museum, University of Colorado at Denver, P.O. Box 173364, Denver, CO 80217, USA.

Ralph E. Molnar, Museum of Northern Arizona, 3101 N Fort Valley Rd., Flagstaff, AZ 86001, USA.

John M. Myers, Department of Geology, 108 Thompson Hall, Kansas State University, Manhattan, KS 66506, USA.

Gregory S. Paul, 3109 N Calvert St., Side, Baltimore MD 21218, USA.

Bruce M. Rothschild, Arthritis Center of Northeast Ohio, 5500 Market, Youngstown, OH 44512, USA.

Elizabeth H. Southwell, Geological Museum, University of Wyoming, Laramie, WY, 82071, USA.

Mary Higby Schweitzer, Department of Marine, Earth and Atmospheric Sciences, North Carolina State University, Raleigh NC 27695, USA.

Kent A. Stevens, Department of Computer and Information Science, University of Oregon, Eugene, OR 97403, USA.

Eric D. Wills, Department of Computer and Information Science, University of Oregon, Eugene, OR 97403, USA.

Jennifer L. Wittmeyer, North Carolina Museum of Natural Science, Raleigh, NC 27695, USA.

PREFACE

The archetypal dinosaur, *Tyrannosaurus rex*, celebrated its 100th anniversary in 2005. This occasion was observed by a conference, 100 Years of *Tyrannosaurus rex*: A Symposium, hosted at the Black Hills Institute of Geological Research in Hill City, SD, June 10–11, 2005. The symposium brought together an international cast of dinosaur paleontologists who presented the results of their research to the general public. The paleontologists all agree, however, that the most enjoyable and intellectually stimulating time was spent in the basement storage room of the Black Hills Institute, surrounded by bones and casts of *Tyrannosaurus*. There, individual bones could be examined during the many long hours of discussion and debate about *Tyrannosaurus*. These debates undoubtedly had an influence on the chapters comprising the resulting volume. The editors, Peter Larson and Kenneth Carpenter, thank the authors for their contributions. To fill out this volume, a few additional chapters were solicited, and we thank these authors as well.

We also thank the many individuals of the Black Hills Institute of Geological Research, the Mammoth Site of Hot Springs, the Hill City Public Schools, the Journey Museum, the Kirby Science Center, and numerous businesses and the local residents for their assistance in making the symposium possible.

Thanks to Robert Sloan, editor at Indiana University Press for his support of this volume, and to James O. Farlow, editor of the Life of the Past series for the press. Marion Zenker (BHI) and Deborah Longhofer (DMNS) provided clerical support.

INSTITUTIONAL ABBREVIATIONS

AMNH	*American Museum of Natural History, New York, NY*
BHI	*Black Hills Institute of Geological Research, Hill City, SD*
BMNH	*Natural History Museum, London, England*
BMR	*Burpee Museum of Natural History, Rockford, IL*
CM	*Carnegie Museum of Natural History, Pittsburg, PA*
CMNH	*Cleveland Museum of Natural History, Cleveland, OH*
DMNH	*Denver Museum of Nature & Science, Denver, CO*
FMNH	*Field Museum of Natural History, Chicago IL*
IVPP	*Institute of Vertebrate Paleontology and Palaeoanthropology, Beijing, China*
LACM	*Natural History Museum of Los Angeles County, Los Angeles, CA*
LDP	*Lance Dinosaur Project, New Orleans, LA*
LL	*The Manchester Museum, Manchester, England*
MMS	*The Science Museum of Minnesota, St. Paul, MN*
MOR	*Museum of the Rockies, Montana State University, Bozeman, MT*
NCSM	*North Carolina State Museum of Natural Sciences, Raleigh, NC*
NMMNH	*New Mexico Museum of Natural History, Albuquerque, NM*
PIN	*Paleontological Institute, Russian Academy of Sciences, Moscow, Russia*
RMM	*Red Mountian Museum, Birmingham, AL*
ROM	*Royal Ontario Museum, Toronto, Ontario, Canada*
RSM	*Royal Saskatchewan Museum, Regina, Saskatchewan, Canada*
RTMP	*Royal Tyrrell Museum of Palaeontology, Drumheller, Alberta, Canada*
SDSM	*Museum of Geology, South Dakota School of Mines and Technology, Rapid City, SD*
SUP	*Shenandoah University, Winchester, VA*
TCM	*The Children's Museum, Indianapolis, IN*
UCMP	*University of California, Museum of Paleontology, Berkeley, CA*

UCRC	*University of Chicago Research Collection, Chicago, IL*
UMNH	*Utah Museum of Natural History, University of Utah, Salt Lake City, UT*
USNM	*National Museum of Natural History, Smithsonian Institution, Washington, DC*
UUVP	*University of Utah Vertebrate Paleontology, Salt Lake City, UT*
UWGM	*University of Wisconsin Geology Museum, Madison, WI*
YPM	*Yale Peabody Museum of Natural History, New Haven, CT*
ZPAL	*Institute of Palaeobiology of the Polish Acadamy of Sciences, Warsaw, Poland*

TYRANNOSAURUS REX, THE TYRANT KING

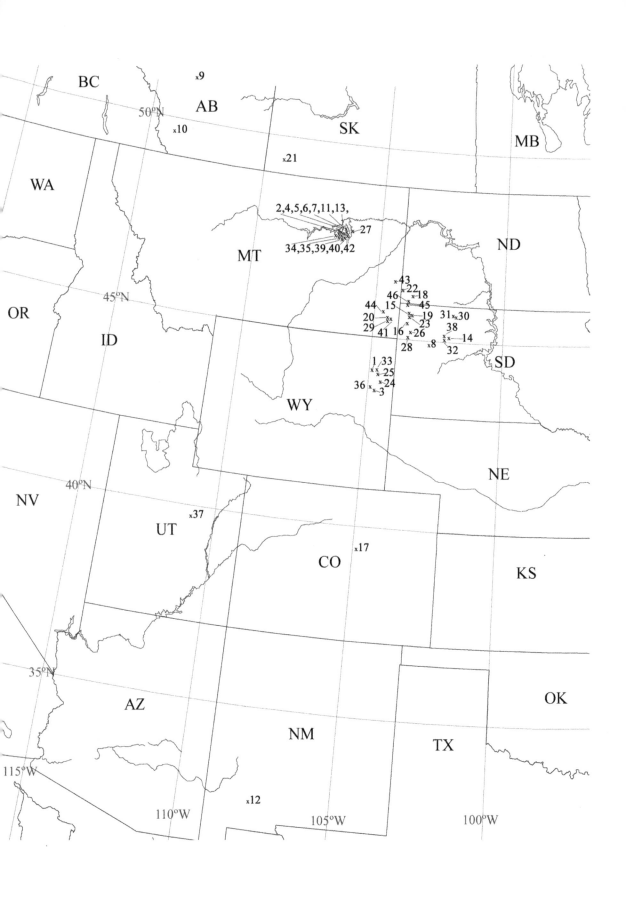

ONE HUNDRED YEARS OF *TYRANNOSAURUS REX:* THE SKELETONS

Neal L. Larson

Introduction

In 1905, Henry Fairfield Osborn, of the American Museum of Natural History, published the name *Tyrannosaurus rex* for a large theropod discovered by Barnum Brown within the Upper Maastrichtian Hell Creek Formation of Montana. Later, *Dynamosaurus imperiosus* Osborn 1905 and 2 cervical vertebrae described as *Manospondylus gigas* (Cope 1892) were included in the species (Osborn 1916). The most complete *T. rex* skeleton, AMNH 5027, was also excavated by Barnum Brown from the Hell Creek Formation of Montana in 1908. Because there were so few specimens, most people believed that *T. rex* was rare.

A dinosaur renaissance beginning in the 1960s inspired many paleontologists to seek out new and more complete specimens. Since 1965, 42 skeletons (5% to 80% complete by bone count) of *T. rex* have been excavated from the Western Interior of North America. To date, *T. rex* has been found in the uppermost Cretaceous terrestrial rocks from central Alberta to southern New Mexico, and from north-central South Dakota to central Utah. These skeletons, along with innumerable numbers of bones and teeth, have been discovered, excavated, displayed, and written about in numerous scientific and popular books and publications.

The discovery of the *Tyrannosaurus* specimen known as Sue and its subsequent publicity (Glut 2000; Fiffer 2000; Larson and Donnan 2002) reignited this renaissance. The rumors of Sue's value lured dozens of amateur paleontologists and untrained fortune hunters into the field. Although most knew nothing about collecting or caring for fossils once they were discovered, people were looking for dinosaurs in areas that no one had ever searched. Within the last 15 years, most exposures of the Hell Creek and Lance Formations have been explored, resulting in many new specimens of *Tyrannosaurus* and other dinosaurs being discovered. These new *Tyrannosaurus* specimens have yielded more information about *T. rex* than was once thought possible, as evidenced by this volume. Much has now been learned about the respiratory, digestive, and reproductive systems, feeding habits, range, sex, lifestyles, injuries, and diseases of the tyrant-lizard king. These new specimens have provided data on arm and jaw strength, speed, and growth rate, and findings have bolstered the finding of a close relationship of bird and theropods. Because of these new discoveries, the once-rare *T. rex* is now one of the most abundant large theropods in many museum collections. An annotated catalog of these specimens is presented below.

Figure 1.1. Site map for **Tyrannosaurus rex** discoveries. See text and Table 1.1 for number identification.

Tyrannosaurus rex Specimens

Brief descriptions follow each of the more complete *Tyrannosaurus rex* specimens recovered to date. Each listed specimen has a minimum of 10 associated skeletal bones from several parts of the body or a fairly complete skull. The list excludes specimens consisting of only a braincase and/or a few skull bones, or only foot bones, or only caudal bones, as well as the countless specimens of *T. rex* teeth and isolated bones. Although there may be some omissions, the total number of skeletons is particularly impressive when considering that most of these specimens have been found in the last 15 years. Considering the wide geographic range of the specimens, and assuming there is only one North American species, the paleogeographic range of *T. rex* was immense—possibly larger than that of most other large theropods (Fig. 1.1).

Although some authors have considered *Nanotyrannus lancensis* (Bakker et al. 1988) synonymous with *T. rex* (Carr 1999; Carr and Williamson 2004; Holtz 2004; Glut 1997, 2000, 2003, 2006), others (Bakker et al. 1988; Currie 2003; Larson this volume) do not agree. Tooth count, bone shape, and foramen placement and size, along with many other skeletal differences, seem to clearly separate the 2 genera. Because there is so much evidence separating *Nanotyrannus* from *T. rex*, I have excluded *Nanotyrannus* from the following list of *T. rex* specimens. *Aublysodon molnari* Paul, 1988 (=*Stygivenator molnari* Olshevsky, Ford, and Yamamoto, 1995), was excluded for similar reasons.

The following *Tyrannosaurus rex* skeletons have been assigned a percentage of completeness on the basis of the number of bones found with each skeleton. The total number of bones in a skeleton of *T. rex* is approximately 300 (Appendix). All identifiable bones, whether complete or incomplete, are counted as bones in the totals. For many skeletons, some bones were fragmented or eroded, and others were fragmented before burial. In most instances, if the incomplete bones can be positively identified, they are treated as complete bones for the purpose of calculating the percentage of completeness for each skeleton. Every effort was made to avoid inflating the count (for instance, rib heads were counted but rib shafts were not, unless they could be proved to be from separate ribs already counted). The completeness of a skeleton was then derived by dividing the number of bones found with each specimen by the total number of bones in a skeleton.

I attempted to locate and list all *Tyrannosaurus* specimens collected from the upper Maastrichtian terrestrial rocks of North America. All specimens are listed, whether they are in private hands or public institutions. Any errors or omissions are solely my responsibility. All known institutions, repositories, and private collectors that may have *Tyrannosaurus rex* skeletal elements in their collections were contacted for data, and most were cooperative. The following specimens are listed chronologically and alphabetically by their catalog number and/or their nickname or moniker. The year of excavation is given rather than the year of discovery because some of them were not excavated until many decades after they were initially found (Table 1.1).

Table 1.1. Summary of Tyrannousaurus Rex Skeletons by Year of Excavation (as of August 2006)

Location	Year Excavated	Specimen No./ Name	Discoverer, Year	State/ Province	No. of Elements	% of Skeleton	Skull/ Parts	Sex*	Current Location
1	1900	BMNH R7994	Brown, 1900	WY	40	13	Y	--	Natural History Museum, London, England
2	1902–1905	CM 9380	Brown, 1902	MT	34	11	Y	F	Carnegie Museum of Natural History, Pittsburgh, PA
3	1902	CM 1400	Peterson, 1902	WY	29	10	Y	?	Carnegie Museum of Natural History, Pittsburgh, PA
4	1908	AMNH 5027	Brown, 1908	MT	143	48	Y	M?	American Museum of Natural History, NY, NY
5	1967	MOR 008	MacMannis, 1967	MT	46	15	Y	F?	Museum of the Rockies, Bozeman, MT
6	1967–1969	LACM 23844	Garbani, 1966	MT	74	25	Y	M	Natural History Museum of LA County, LA, CA
7	1967	LACM 23845	Garbani, 1969	MT	37	12	Y	?	Natural History Museum of LA County, LA, CA
8	1981	SDSM 12047	Floden, 1980	SD	82	27	Y	M?	Museum of Geology, South Dakota School of Mines and Technology, Rapid City, SD
9	1981	RTMP 81.12.1	Sternberg, 1946	AB	49	16	Y	M	Royal Tyrrell Museum of Palaentology, Drumheller, AB
10	1981	RTMP 81.6.1 Black Beauty	Baker, 1980	AB	85	28	Y	F	Royal Tyrrell Museum of Palaentology, Drumheller, AB
11	1981	MOR 009 Hager rex	Hager, 1981	MT	58	19	N	?	Museum of the Rockies, Bozeman, MT
12	1983, 2003	NMMNH PR1081	Staton and LaPoint, 1982	NM	10	3	Y	?	New Mexico Museum of Natural History and Science, Albuquerque, NM
13	1990	MOR 555 Wankel T. rex	Wankel, 1988	MT	146	49	Y	M	Museum of the Rockies, Bozeman, MT
14	1990	FMNH PR2081 Sue	Herdrickson 1990	SD	219	73	Y	f	Field Museum, Chicago, IL

Table 1.1 continued

Location	Year Excavated	Specimen No./Name	Discoverer, Year	State/Province	No. of Elements	% of Skeleton	Skull/Parts	Sex*	Current Location
15	1992, 1993, 2003	BHI 3033 Stan	Sacrison, 1987	SD	190	63	Y	M	Black Hills Institute of Geological Research, Hill City, SD
16	1992	Samson	Zimmerscheid, 1987	SD	121	40	Y	F	Private, Carnegie Museum of Natural History, Pittsburgh, PA
17	1992	DMNH 2827	Fickle, 1992	CO	10	3	N	?	Denver Museum of Nature & Science, Denver, CO
18	1993	Bowman	Pearson, 1992	ND	45+	15+	N	?	Pioneer Trails Regional Museum, Bowman, ND
19	1993–1996, 2006	BHI 4100 Duffy	Sacrison, 1993	SD	79	26%	Y	?	Black Hills Institute of Geological Research, Hill City, SD
20	1993	UWGM 181	Pallen, 1993	MT	20	7	Y	?	Geological Museum, University of Wisconsin, Madison, WI
21	1994–2001	RSM 2523.8 Scotty	Gebhardt, 1991	SK	120+	40+	Y	F	T. rex Discovery Centre, Eastend, SK
22	1994–1995	BHI 6219007	Garstka, 1994	ND	12	2	Y	?	Various, incl. Black Hills Inst. Geol. Res., Hill City, SD
23	1995	BHI 6239 Steven	Sacrison, 1995	SD	15	5	N	F?	Black Hills Inst. Geol. Res., Hill City, SD
24	1995	LDP 977-2 Pete	Patchus, 1995	WY	35+	12+	N	?	Univ. New Orleans, New Orleans, LA
25	1995–1996	Barnum	Theisen, 1995	WY	47	16	N	?	Private
26	1996–1998	BHI 4182 Fox	Fox, 1994	SD	29	10	Y	F?	Black Hills Inst. Geol. Res., Hill City, SD
27	1997–2004	MOR 980 Pecks Rex	Tremblay, 1997	MT	120+	40+	Y	M	Ft. Peck Paleo. Station, Ft. Peck, MT
28	1997	Tinker	Eatman, 1997	SD	73	24	Y	?	Private
29	1998–1999	Ollie	Pfister, 1998	MT	124	41	Y	?	Great Plains Paleo., Madison, WI
30	1998	Rex B	Alley, 1998	SD	24	8	Y	M	Black Hills Inst. Geol. Res., Hill City, SD
31	1999	Rex-C	Alley, 1999	SD	18	6	Y	F	Private, SD

Table 1.1 continued

Location	Year Excavated	Specimen No./Name	Discoverer, Year	State/Province	No. of Elements	% of Skeleton	Skull/Parts	Sex*	Current Location
32	2000	BHI 6248 E. D. Cope	Derlinger, 1999	SD	30	10	Y	?	Black Hills Inst. Geol. Res., Hill City, SD
33	2000	Monty	Landowner, 1999	WY	53	18	Y	?	Babiarz Inst. Paleon. Studies, Mesa, AZ
34	2000–2003	MOR 1125 B-rex	Harmon, 2000	MT	111	37	Y	F	Museum of the Rockies, Bozeman, MT
35	2000–2001	MOR 1126 C-rex	Horner, 2000	MT	26	9	Y	?	Museum of the Rockies, Bozeman, MT
36	2001	UCRC PV1	Zerbst, before 1950	WY	60+	20+	N	?	University of Chicago, Chicago, IL
37	2001	UMNH 110000	Difley and Saharatian, 2001	UT	26	9	Y	?	Utah Museum of Natural History, Salt Lake City, UT
38	2001–2002	TCM 2001.90.1 Bucky	Derflinger, 1998	SD	101	34	N	F	The Children's Museum, Indianapolis, IN
39	2001	MOR 1128 G-rex	Wilson, 2001	MT	23	7	N	F	Museum of the Rockies, Bozeman, MT
40	2001	MOR 1152 F-rex	Stewart, 2001	MT	25?	8?	N	?	Museum of the Rockies, Bozeman, MT
41	2001–2002	Otto	Pfister, 2001	MT	32	11	N	?	Great Plains Paleontology, Madison, WI
42	2002–2003	MOR/USNM N-rex	Myrhvold, 2001	MT	40	13	Y	?	National Museum of Natural History, Washington, DC
43	2002–2004	BHI 6230 Wyrex	Wells and Wyrick, 2002	MT	114	38	Y	M	Black Hills Institute of Geological Research, Hill City, SD
44	2003–2005	LACM 7509/10167 Thomas	Curry, 2003	MT	110+	37+	Y	?	Natural History Museum of Los Angles County, LA, CA
45	2004–2005	Wayne	Olson, 2004	ND	24	8	N	?	Private, Fargo, ND
	2005	Ivan	Olson, 2005	SD	116	39	N	?	Private, Fargo, ND

Note—See text and Figure 1.1.
* Sex: M, gracile morph; F, robust morph. See P. Larson (this volume).

1900–1909

The years 1900 through 1909 marked the dawn of our understanding of *Tyrannosaurus rex*. This decade includes the initial discovery and description of *Tyrannosaurus rex* Osborn 1905. Four incomplete skeletons of *T. rex* were unearthed during this time, and each one would contribute tremendously to the knowledge of this magnificent dinosaur (Osborn 1905, 1906, 1912, 1916). Nearly 60 years would pass before any other skeletons were collected, and it would be nearly 80 more years before ideas of how they lived, acted, and walked would change.

BMNH R7994
(Originally AMNH 5866)
(Holotype of Dynamosaurus imperiosus Osborn 1905)

DISCOVERED: The year 1900, Barnum Brown, a professional collector employed by AMNH.
LOCATION: Seven Mile Creek, Weston County, WY (Fig. 1.1, site 1).
FORMATION: Lance Formation.
EXCAVATED: American Museum expedition under Barnum Brown, 1900.
REPOSITORY: Natural History Museum, London, England, UK.
ACQUISITION: Purchased from the American Museum of Natural History, 1960.
DESCRIBED: Osborn (1905, 1906); Newman (1970); Carpenter (1990); Glut (1997).
SKELETAL REMAINS: The skull consists of both palatines and both dentaries. The postcranial skeleton has all 10 cervical vertebrae, plus 9 left and 4 right cervical ribs. The first 5 dorsal and the 5 sacral vertebrae are also present, along with the right ilium, the left ischium, and the right femur.
COMPLETENESS: Forty bones, or 13% of a skeleton by bone count.
ON DISPLAY? Yes, according to Phillip Manning (personal communication 2005); some parts, the dentary, and maybe another bone or two, are displayed at the Natural History Museum, London, England, UK.
COMMENTS: This *Tyrannosaurus rex* specimen had the first articulated neck with cervical ribs. It was discovered with numerous scutes of what is now known to be from an *Ankylosaurus* (Carpenter 2004). The species was synonymized with *Tyrannosaurus rex* by Osborn (1906) in his second contribution on *Tyrannosaurus*.

CM 9380
(Originally AMNH 973)
(Holotype of Tyrannosaurus rex Osborn 1905)

DISCOVERED: The year 1902, Barnum Brown, a professional collector employed by AMNH.
LOCATION: From Quarry No. 1, near Jordan, Garfield County, MT (Osborn 1905) (Fig. 1.1, site 2).
FORMATION: Hell Creek Formation, 220 feet above the Bearpaw Shale.

Figure 1.2. *CMNH 9380, skeleton on display. Photo by Peter Larson.*

EXCAVATED: American Museum expedition, under Barnum Brown, 1902–1905.

REPOSITORY: Carnegie Museum of Natural History, Pittsburg, PA.

ACQUISITION: Purchased from the American Museum of Natural History, 1941.

DESCRIBED: Osborn (1905, 1906, 1912, 1916); Carpenter (1990); Glut (1997).

SKELETAL REMAINS: The partial skeleton consists of the right maxilla, both lacrimals, left squamosal and ectopterygoid, both dentaries, and the left surangular. It also has 1 cervical, 7 dorsals, and 5 sacral vertebrae; 3 gastralia; right scapula and left humerus; both ilia, pubes, and ischia; the left femur and part of the right tibia; and 3 metatarsals (McIntosh 1981).

Figure 1.3. *AMNH 5027*, skeleton on display. Photo courtesy AMNH library.

COMPLETENESS: Thirty-four bones, or 11% of a skeleton by bone count.

ON DISPLAY? Was on display at the Carnegie Museum of Natural History, Pittsburg, PA, since the early 1940s. In 2005, the skeleton was dismantled, but it is scheduled to be freshly restored, remounted, and back on display sometime late 2007.

COMMENTS: The American Museum of Natural History sold the skeleton soon after the beginning of World War II. It has often been stated that it was necessary to ensure that a *Tyrannosaurus rex* would survive in the event of a bombing. Because the skeletal pose had not been redone since it went on display in the 1940s, the Carnegie Museum of Natural History is now in the process of conserving and remounting the original skeleton.

CM 1400

DISCOVERED: The year 1902, Olaf Peterson, a professional collector employed by Carnegie Museum.
LOCATION: Snyder Creek, Niobrara County, WY (Fig. 1.1, site 3).
FORMATION: Lance Formation.
EXCAVATED: Olaf Peterson and Carnegie Museum expedition, 1902.
REPOSITORY: Carnegie Museum of Natural History, Pittsburg, PA.
DESCRIBED: Partially described by McIntosh (1981).
SKELETAL REMAINS: McIntosh (1981) noted that CM 1400 contains the left maxilla, premaxilla, and pterygoid, the nasals, and the braincase. The postcranial skeleton consists of 2 cervical ribs, 1 dorsal vertebrae, 1 dorsal rib, 3 chevrons, the left ischium, and both pubes (incomplete).
COMPLETENESS: Twenty-nine bones, or 10% of a skeleton by bone count.
ON DISPLAY? No.
COMMENTS: Because the Carnegie was in possession of this large theropod specimen, it encouraged Osborn (1905) to publish on *Tyrannosaurus rex* earlier than he originally intended (http://paleo.amnh.org/projects/t-rex/index.html).

AMNH 5027

DISCOVERED: The year 1908, Barnum Brown, a professional collector employed by AMNH.
LOCATION: Near Dry Creek, McCone County, MT (Fig. 1.1, site 4).
FORMATION: Hell Creek Formation, 220 feet above the Bearpaw Shale.
EXCAVATED: American Museum field crew under Barnum Brown, 1908.
REPOSITORY: American Museum of Natural History, New York.
DESCRIBED: Osborn (1912, 1916).
SKELETAL REMAINS: The specimen boasts the first complete skull, all of the cervical, dorsal, and sacral vertebrae, plus 18 caudal vertebrae and 7 chevrons; 9 cervical ribs from the right side; 20 dorsal ribs; both ilia, ischia, and pubes.
COMPLETENESS: A total of 143 bones, or 48%, of a skeleton by bone count.
ON DISPLAY? Yes, at the American Museum of Natural History, New York.
COMMENTS: This specimen of *Tyrannosaurus rex* was the classic one, the best *T. rex* skeleton on view anywhere until Sue and Stan were prepared in 2000 and 1995, respectively. The skull and much of the skeleton were articulated, but the skeleton lacked legs, feet, forelimbs, and the distal end of the tail. With 143 bones, this was the most complete *T. rex* skeleton until 1990 with the excavation of MOR 555 (146 bones) and a *T. rex* named Sue (219 bones). In 1996, the skeleton, which had been mounted in an upright position, was remounted into a more natural pose. There are casts of the skull and of the skeleton in museum collections throughout the world.

Figure 1.4. *MOR 008, cast of skull. Photo by Ed Gerken.*

1910–1959

Wars, rumors of wars, and the Great Depression kept many paleontologists out of the field and in the lab from the 1910s through the 1950s. Although there was only some dinosaur collecting from Upper Cretaceous formations undertaken during this period in the United States, there was substantial collecting in Canada. Some paleontologists, such as the Sternberg family and Barnum Brown, did extensive collecting in Alberta, helping to build the collections and displays in the major museums of Canada, Europe, and the United States. It was also during this time that Roy Chapman Andrews discovered an entirely new Late Cretaceous dinosaur fauna in Mongolia, which would later include the Asian tyrannosaurid, *Tarbosaurus*.

1960–1979

Beginning in the mid 1960s, new dinosaur finds once again began to lure more people into the field, and many amateurs would make important discoveries. New discoveries in Montana would begin to establish the Museum of the Rockies and Los Angeles County Museum as major repositories for dinosaurs. For the first time, non–East Coast museums would possess original specimens of the tyrant-lizard king.

Discoveries of *Deinonychus* and the resultant revolutionary concepts about dinosaurs by John Ostrom (Yale University) would help to change the way we looked at dinosaurs. Many modern researchers would use Ostrom's brilliant research as a springboard to present *Tyrannosaurus* and other dinosaurs as active, warm-blooded, birdlike animals instead of cold-blooded, lizardlike creatures (e.g., Bakker 1986; Paul this volume).

MOR 008

DISCOVERED: In 1967, by Dr. William MacMannis, an archeologist from Montana State University (Larson and Donnan 2002).
LOCATION: Garfield County, MT (Fig. 1.1, site 5).
FORMATION: Hell Creek Formation.
EXCAVATED: A team from the Museum of the Rockies, 1967.
REPOSITORY: Museum of the Rockies, Bozeman, MT.
DESCRIBED: Partial description in Molnar (1991).
SKELETAL REMAINS: The skull is missing only the left premaxilla, the right palatine, the right epipterygoid, and the vomer. The lower jaw is missing the splenials, the coronoids, the right dentary, and the left prearticular. An atlas is also present.
COMPLETENESS: Forty-six bones, or 15% of a skeleton by bone count.
ON DISPLAY? A cast of the skull is on display at the Black Hills Museum of Natural History, Hill City, SD.
COMMENTS: The specimen consists of only an articulated skull and the atlas of a very large (Sue size), robust adult. It was on display at the Museum of the Rockies until 1990, when it was moved to the collections. Portions of this skull were molded and cast to supplement the missing portions of MOR 555 (a gracile *Tyrannosaurus rex*).

LACM 23844

DISCOVERED: The year 1966, Harley Garbani, plumber and amateur paleontologist (Dingus 2004).
LOCATION: L. D. Engdahl Ranch, Garfield County, MT (Glut 2002) (Fig. 1.1, site 6).
FORMATION: Hell Creek Formation.
EXCAVATED: J. R. McDonald and LACM field crew, 1967–1969 (Glut 2002).
REPOSITORY: Natural History Museum of Los Angeles County, Los Angeles, CA.
DESCRIBED: Partial description of the skull in Molnar (1991); see also Molnar (this volume).
SKELETAL REMAINS: The skull consists of the right premaxilla, maxilla, postorbital, squamosal, and pterygoid; both jugals, both lacrimals, the left quadrate and quadratojugal; nasals; frontals and the occipital; both surangulars, dentaries, articulars, and angulars; the left splenial; and prearticular. The postcranial skeleton consists of 2 cervical, 7 dorsal, and 4 caudal vertebrae; 5 ribs; 10 chevrons; the right scapula, femur, and astragalus; and both ischia, left tibia, 4 metatarsals, and 10 pes phalanges. Many of the bones are incomplete (Glut 2002).
COMPLETENESS: Seventy-four bones, or 25% of a skeleton by bone count.

Figure 1.5. *LACM 23844, skull.* Photo by Dick Meier, courtesy Natural History Museum of Los Angeles County.

ON DISPLAY? Yes, at the Natural History Museum of Los Angeles County, Los Angeles, CA.

COMMENTS: The skull and partial skeleton were disarticulated and provided the first detail for many of the bones (Molnar 1991). The specimen was excavated over the span of several field seasons. The rancher assisted by bulldozing the overburden away (Dingus 2004). There are several casts of this skull on display throughout the world.

LACM 23845
(Holotype of Albertosaurus megragracilis Paul 1988 and of Dinotyrannus megragracilis Olshevsky et al. 1995)

DISCOVERED: The year 1969, Harley Garbani, plumber and amateur paleontologist (Dingus 2004).

LOCATION: L. D. Engdahl Ranch, Garfield County, MT (Glut 2002) (Fig. 1.1, site 7).

FORMATION: Hell Creek Formation.

EXCAVATED: Harley Garbani and the LACM field crew, 1967 (Dingus 2004).

REPOSITORY: Natural History Museum of Los Angeles County, Los Angeles, CA.

DESCRIBED: Molnar (1980); Paul (1988); Olshevsky et al. (1995); Larson and Donnan (2002); Carr and Williamson (2004).

SKELETAL REMAINS: The skull consists of the right maxilla, right lacrimal, frontals, nasals, parietals, both dentaries, right coronoid, and right surangular. The postcranial skeleton has the right scapula, coracoid, and ulna; both ischia; the right femur, tibia, and astragalus; a nearly complete right foot (missing metatarsals I and IV and the phalanges from the first toe); and there are also 2 phalanges from the left foot (R. Farrar, personal communication 2005).

COMPLETENESS: Thirty-seven bones, or 12% of a skeleton by bone count.

ON DISPLAY? No.

COMMENTS: This specimen is a juvenile *Tyrannosaurus rex* (P. Larson, personal communication 2005; Carr and Williamson 2004). It was discovered about 2 feet above LACM 23844, in scrap piles that were created while bulldozing the overburden away from the lower *T. rex*. Because of bulldozing, much of the skeleton was fragmentary (Dingus 2004).

The 1980s

Several new and important skeletons were found during this decade. The first *Tyrannosaurus* skeletons from South Dakota, New Mexico, and Alberta increased the known range of *T. rex* by hundreds of miles in all directions. (*Manospondylus gigas* Cope, 1892, was collected from South Dakota, but that specimen consisted of only 2 cervical vertebrae.) Toward the end of this decade, one of the Alberta *T. rex* skeletons would become the first original *T. rex* skeletons to tour the world.

SDSM 12047
(Mud Butte T. rex)

DISCOVERED: The year 1980, Jennings Floden, rancher and amateur fossil collector (Larson and Donnan 2002).

LOCATION: Jennings Floden Ranch, near Zeona, Butte County, SD (Fig. 1.1, site 8).

FORMATION: Hell Creek Formation.

EXCAVATED: Phil Bjork, Jennings Floden, and neighbors 1981; Floden and neighbors 1983.

REPOSITORY: Museum of Geology, South Dakota School of Mines and Technology, Rapid City.

DESCRIBED: Partial description by Bjork (1982); Carpenter (1990).

SKELETAL REMAINS: Nearly complete skull, missing only the premaxillae. The lower jaw has both dentaries, the right angular, and the right coronoid. The postcranial skeleton includes parts of 3 ribs and

Figure 1.6. *SDSM 12047, skull on display as Museum of Geology. Photo by Peter Larson.*

has a complete section of tail from the 15th to the 34th caudal (including chevrons).

COMPLETENESS: Eighty-one bones, or 27% of a skeleton by bone count.

ON DISPLAY? Yes, the skull is on display in the Museum of Geology, Rapid City, SD.

COMMENTS: After the initial digging in 1981, the South Dakota School of Mines had found the skull, some vertebrae, and some ribs. Floden believed that more of the skeleton was still in the ground and recruited neighbors to reopen the dig. They uncovered the skull within a few days. Two years later, they discovered articulated tail vertebrae and the second lower jaw (Smith-Hill 1983). Associated with the skeleton were elements from a gar (*Lepisosteus*), 2 turtles (including a baenid), 2 crocodiles (*Branchychampsa* and *Leidyosuchus*), *Champsosaurus*, and the teeth from the theropod *Nanotyrannus* (Bjork 1982). The skull and skeleton are currently being prepared, and a description of the specimen is also planned.

RTMP.81.12.1
(Huxley *T. rex*)

DISCOVERED: The year 1946, Charles M. Sternberg, a professional paleontologist.

LOCATION: Near the town of Huxley, along the Red Deer River, Alberta (Fig. 1.1, site 9).

FORMATION: Scollard Formation.

EXCAVATED: Phil Currie with the Provincial Museum of Alberta, Edmonton, in 1981.
REPOSITORY: Royal Tyrrell Museum of Palaeontology, Drumheller, Alberta.
ACQUISITION: Transferred from the Provincial Museum of Alberta in Edmonton.
DESCRIBED: Partial description by Currie (1993).
SKELETAL REMAINS: From the skull, only a right postorbital exists. The postcranial skeleton consists of 8 anterior caudals, 5 anterior chevrons, part of the sacrum, 7 anterior dorsal vertebrae, 1 rib, both ilia, left ischium, left pubis, both femora, both tibiae, both fibulae, right astragalus and calcaneum, the 2 right tarsals, 1 metatarsal, and 7 pes phalanges from the right foot (A. Neuman, written communication February 1994).
COMPLETENESS: Forty-nine bones, or 16% of a skeleton by bone count.
ON DISPLAY? Yes, at the Royal Tyrrell Museum of Palaeontology, Drumheller, Alberta.
COMMENTS: Sternberg found the specimen halfway down a steep cutbank on the Red Deer River in Alberta. He believed that the specimen was mostly gone, so he did not attempt to excavate it. Currie, while investigating some of Sternberg's earlier finds, determined that it would be worth excavating to see whether there was more of the skeleton (Currie 1993). At 52° north latitude, this is the northernmost *Tyrannosaurus rex*.

Figure 1.7. *Huxley-rex*, RTMP 81.6.1, mixture of real bones and cast, with original pelvis in foreground. Photo by Peter Larson.

Figure 1.8. *Black Beauty RTMP 81.6.1, cast skull.* Photo by Ed Gerken.

RTMP 81.6.1
(*Black Beauty; Cowley T. rex*)

DISCOVERED: The year 1980, Jeff Baker, a high school student (Currie 1993).

LOCATION: Near the confluence of the Crowsnest and Willow Rivers (Crowsnest Pass) in southwestern Alberta (Fig. 1.1, site 10).

FORMATION: Willow Creek Formation.

EXCAVATED: Phil Currie with the Provincial Museum of Alberta, Edmonton, in 1981.

REPOSITORY: Royal Tyrrell Museum of Palaeontology, Drumheller, Alberta.

ACQUISITION: Transferred from the Provincial Museum of Alberta in Edmonton.

DESCRIBED: Partial description by Carpenter (1990) and Currie (1993).

SKELETAL REMAINS: According to Currie (1993), the specimen consists of a nearly complete skull but only partial lower jaws (it has both dentaries, the left splenial, and the right angular). It also has 5 cervical and 7 dorsal vertebrae; 2 cervical and 8 dorsal ribs; a right humerus; a manus phalange; both femora; both tibiae; the left fibula, calcaneum, and astragalus; 4 metatarsals; and 5 pes phalanges (A. Neuman, written communication February 1994).

COMPLETENESS: Eighty-five bones, or 28% of a skeleton by bone count.

ON DISPLAY? Yes, the skull is on display at the Royal Tyrrell Museum of Palaeontology, Drumheller, Alberta. The skeleton has been part of an international traveling exhibit.

COMMENTS: Black Beauty got its name from its beautifully colored black bones. It was discovered when Jeff Baker took a break from fishing and went walking in the hills. This would be the first *T. rex* to receive a nickname, but it would not be the last. It has an articulated skull showing minor distortion. It became the first *Tyrannosaurus rex* skeleton to go on tour, spending considerable time traveling across Canada and Japan. This specimen is the westernmost *Tyrannosaurus rex*, found at 114° west longitude.

MOR 009
(Hager rex)

DISCOVERED: The year 1981, Mick Hager, former director of the Museum of the Rockies (Larson and Donnan 2002).

LOCATION: Garfield County, MT (Fig. 1.1, site 11).

FORMATION: Hell Creek Formation.

EXCAVATED: Museum of the Rockies field crew, 1981.

REPOSITORY: Museum of the Rockies, Bozeman, MT.

SKELETAL REMAINS: The skeleton consists of the right dentary, parts of 4 ribs, 22 caudal vertebrae, and 7 chevrons; the right ilia, both ischia, and both pubes; both femora (left incomplete), tibiae (incomplete), the right fibula and astragalus; and 4 metatarsals and 7 pes phalanges (P. Larson, personal communication 2005).

COMPLETENESS: Fifty-eight bones, or 19% of a skeleton by bone count.

ON DISPLAY? No.

COMMENTS: There is a good possibility that more of the specimen remains uncollected (Mick Hagar to P. Larson, personal communication 2005).

NMMNH P-1013-1

DISCOVERED: The year 1982, D. Staton and J. LaPoint, amateurs.

LOCATION: The east side of Elephant Butte Reservoir, near Truth or Consequences, NM (Fig. 1.1, site 12).

FORMATION: Hall Lake Member of the McRae Formation.

EXCAVATED: Gillette and staff from the New Mexico Museum of Natural History 1983; Tom Williamson, September 2003.

REPOSITORY: New Mexico Museum of Natural History, Albuquerque, NM.

DESCRIBED: Partial description by Gillette et al. (1986); Carr and Williamson (2000); Carr and Williamson (2004); and Williamson and Carr (2005).

SKELETAL REMAINS: The specimen consists of a left dentary, an incomplete palatine (originally identified as an articular), a right prearticular, some incomplete teeth, and a nearly complete chevron (Gillette et al. 1986); a right squamosal and postorbital; an articular, splenial, and at least 2 chevrons; and some tooth fragments (Tom Williamson, personal communication 2005).

COMPLETENESS: Ten bones, or 3% of a skeleton by bone count.

ON DISPLAY? Yes, the lower jaw is on display at the New Mexico Museum of Natural History, Albuquerque, NM.

COMMENTS: This *Tyrannosaurus rex* extended the known range in the northern Great Plains south 1300 km nearly to the Mexican border. The specimen was found in an area that is normally submerged beneath the Elephant Butte Reservoir. It was discovered while D. Staton and J. LaPoint were taking a break from a sailboat outing (Gillette et al. 1986). The low level of the reservoir in 2003 gave Williamson the opportunity to return to the site and collect more bones some 20 years after its initial excavation. This specimen is the southernmost *T. rex* at 33° north latitude.

The 1990s

The 1990s became the decade for the most incredible *Tyrannosaurus rex* discoveries, with nearly twice as many specimens found and excavated than in all of the years before. With the 3 most complete *T. rex* skeletons excavated in the first 2 years, who could have expected that there would be so many more specimens found? And who could have anticipated all of the publicity and controversies that some of these dinosaurs would generate (e.g., Davies 1997; Donnan and Counter 2000; Glut 2000, 2002; Fiffer 2000; Larson and Donnan 2002)?

MOR 555
(Wankel rex; Devil rex)

DISCOVERED: The year 1988, Cathy Wankel, novice.

LOCATION: From the south side of Fort Peck Lake, McCone County, MT (Fig. 1.1, site 13).

FORMATION: Hell Creek Formation.

EXCAVATED: Pat Leiggi and Museum of the Rockies field crew, 1990.

REPOSITORY: Museum of the Rockies, Bozeman, MT.

DESCRIBED: Partial description by Horner and Lessem (1993); forelimb described by Carpenter and Smith (2001).

SKELETAL REMAINS: The skull consists of a complete braincase, right dentary, left maxilla, nasals, vomer, both squamosals, both postorbitals, both quadrates and quadratojugals, both jugals, both lacrimals, both pterygoids, and both epipterygoids. The skeleton consists of both scapulae and coracoids, the left arm and most of the left hand (missing only 3 bones), complete legs and left foot (minus 1 phalange), plus the right II, III, and V metatarsals. Also collected were a complete pelvic girdle (including the sacrum), nearly all of the dorsal and cervical ver-

Figure 1.9. MOR 555, cast of skeleton. Photo by Ed Gerken.

tebrae (missing only the atlas), the first 14 caudal vertebrae plus 1 additional, 6 chevrons, 4 dorsal ribs, and 5 cervical ribs.
COMPLETENESS: Fourteen bones, or 49% of a skeleton by bone count.
ON DISPLAY? Yes, a portion of the skeleton went on display in 2005 at the Museum of the Rockies, Bozeman, MT.
COMMENTS: This dinosaur was discovered while Cathy Wankel was hiking on a butte in the Hell Creek Formation on the south side of Fort Peck Lake while the rest of her family was fishing (Horner and Lessem 1993). Some resin casts of this skeleton may be seen at the Dallas Museum of Natural History, Houston Museum of Nature and Science, University of California Museum of Paleontology, Museum of the Rockies, and Black Hills Museum of Natural History, among many others. A bronze casting of this dinosaur may also be viewed outside the Museum of the Rockies, Bozeman, MT.

FMNH PR2081
(Sue, formerly BHI 2033)

DISCOVERED: The year 1990, Susan Hendrickson, amateur archeologist and paleontologist.
LOCATION: Maurice Williams Ranch, north of the town of Faith, Ziebach County, SD (Fig. 1.1, site 14).
FORMATION: Hell Creek Formation, about 15 feet above the contact with the Fox Hills Formation (P. Larson, personal communication 2006).
EXCAVATED: Black Hills Institute field crew from August 14 to September 1, 1990.
REPOSITORY: Field Museum of Natural History, Chicago, IL.
ACQUISITION: Sotheby's Auction, 1997.
DESCRIBED: A full description of the skull and skeleton was published by Brochu (2003); it was also partially described by Larson (1994, 2000);

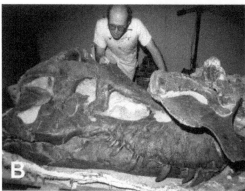

Figure 1.10. *Sue FMNH PR2081*, during excavation (A); skull during preparation with Terry Wentz (B); on display at the Field Museum (C). Photos: (A, C) Peter Larson; (B) Ed Gerken.

Donnan and Counter (2000); Carpenter and Smith (2001); Larson and Donnan (2002); and Larson and Rigby (2005).

WEB SITE: http://www.fieldmuseum.org/sue/ and http://www.bhigr.com/pages/info/info_sue.htm.

SKELETAL REMAINS: A complete skull, 61 vertebrae (9 cervical, 11 dorsal, 5 sacral, 36 caudal), 25 chevrons, 13 cervical ribs, both proatlas, 19 dorsal ribs. It also has both scapulae and coracoids, the furcula, the right arm and most of the right hand, a complete pelvic girdle, both legs, both calcanea and astragali, a tarsal, a nearly complete right foot (missing 2 bones), and a single pes phalange from the left foot (Brochu 2003).

COMPLETENESS: A total of 219 bones, or 73% of a skeleton by bone count.

ON DISPLAY? Yes, in 2000, Sue went on display at the Field Museum of Natural History, Chicago, IL.

COMMENTS: The dinosaur was named in honor of its discoverer, Sue Hendrickson. The skeleton was found along with the remains of 3 other theropods (a partial tibia and fibula from a *Tyrannosaurus rex* subadult, a frontal from a juvenile *T. rex*, and a lacrimal from *Nanotyrannus*), several skeletal elements of *Thescelosaurus*, stomach contents of *Edmontosaurus*, a turtle skull and scutes, crocodile teeth and parts, fish teeth and vertebrae, a varanid, abundant plants, and mollusks.

No other dinosaur has provoked so much attention as a dinosaur named Sue. Several groups, including Black Hills Institute and the

U.S. federal government, laid claim to the bones (see Davies 1997; Donnan and Counter 2000; Larson and Donnan 2002; Fiffer 2000). The Field Museum of Natural History in Chicago, aided by Walt Disney World, Ronald McDonald House, and other sources, acquired Sue from Sotheby's Auction in 1997. The rights to the Sue trademarked name were given to the Field Museum by BHI in 1999. On May 17, 2000, the Field Museum unveiled the prepared and mounted skeleton of Sue to the public. The Field Museum also developed 2 traveling exhibits around casts of the Sue skeleton.

BHI 2033 (Sue) and several other specimens were used by Stephan Pickering in an unpublished manuscript (Glut 1997) to establish a second species of *Tyrannosaurus*. Although there may be reasons to erect a second *Tyrannosaurus* species from the Late Maastrichtian of North American (Larson and Donnan 2002; Bakker, personal communication; Paul 1988; Larson this volume), both Sue and Stan compare favorably with the type of *T. rex*. As a side note, Pickering came to the BHI to study some of the specimens used to justify the new species.

BHI 3033
(Stan)

DISCOVERED: The year 1987, Stan Sacrison, amateur paleontologist.
LOCATION: The Niemi Ranch, near Buffalo, Harding County, SD (Fig. 1.1, site 15).
FORMATION: Hell Creek Formation, 16 m beneath the K-T boundary (K. Johnson, written communication 2000).
EXCAVATED: Black Hills Institute field crew from April 14 to May 7, 1992; a few more skeletal elements were collected in 1993 and 2003.
REPOSITORY: Black Hills Institute of Geological Research, Hill City, SD.
DESCRIBED: Partially described in Donnan and Counter (1999); Larson (2000); Larson and Donnan (2002); Hurum and Sabath (2003); P. L. Larson (this volume); Larsson (this volume); Stevens et al. (this volume).
WEB SITE: http://www.bhigr.com/pages/info/info_stan.htm.
SKELETAL REMAINS: The skeletal elements collected with Stan include: a nearly complete skull (missing the right articular and the left coronoid); 59 vertebrae (9 cervical, 14 dorsal, 5 sacral, 31 caudal); 24 chevrons; 14 cervical ribs (including the proatlas); 12 dorsal ribs; a nearly complete pelvic girdle (the distal ends of both ischia and pubes were weathered away); left femora; both tibiae; the left fibula; both calcanea and astragali; 3 left metatarsals; and 11 pes phalanges.
COMPLETENESS: A total of 190 bones, or 63% of a skeleton by bone count.
ON DISPLAY? Yes, since 1996 at the Black Hills Institute of Geological Research, Hill City, SD.
COMMENTS: Stan was the first of many *Tyrannosaurus rex* skeletons found in Harding County, SD. It was named in honor of Stan Sacrison, its discoverer. Soon after Stan found this dinosaur, he became discouraged because he was told that this skeleton was just another

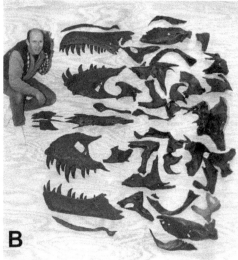

Figure 1.11. *Stan BHI 3033*, during excavation (A); disarticulated skull with Terry Wentz (B); on display at Black Hills Institute, with Brenda Larson (C). Photos: (A, B) Ed Gerken; (C) Larry Shaffer.

articulated *Triceratops*. The bones lay in the ground until 1992, when BHI became aware of the skeleton and began excavation. Stan was found with 2 non–*T. rex* bones: an *Edmontosaurus* vertebrae and an acid-etched Triceratops tibia with both ends missing, bitten (Larson and Donnan 2002), and innumerable well-preserved leaves (Johnson 1996).

Preparation of this skeleton was completed in May 1995, and Stan went on tour in Japan until June 1996 as part of the *T. rex* World Tour. Because Stan's skull was disarticulated and so well preserved, it provided much new information on the osteology and mechanics of tyrannosaurid skulls (Donnan and Counter 1999; Larson 2000; Larson and Donnan 2002; Hurum and Sabath 2003; P. Larson this volume; Larsson this volume).

More than 35 cast Stan skeletons and 50 Stan skulls are on display in public and private institutions worldwide, including: the Smithsonian Institute, Washington, DC; Oxford University, Oxford, England, UK; North American Museum of Ancient Life, Lehi, UT; Kenosha Public Museum, Kenosha, WI; and National Science Museum, Tokyo.

Samson
(Z-rex, Mr. Z, Mr. Zed)

DISCOVERED: The year 1987, Mike Zimmershied, landowner's son, amateur.
LOCATION: Donald Zimmershied Ranch, near the Jump-off, Harding County, SD (Fig. 1.1, site 16)
FORMATION: Hell Creek Formation.
EXCAVATED: Fred Nuss, Alan Deitrich, Steve and Stan Sacrison, 1992.
REPOSITORY: Private, Graham Lacey.
ACQUISITION: Purchased in 2004 from Fred Nuss and Alan Deitrich for an undisclosed amount.
DESCRIBED: Partially described in Glut (2002); Larson and Donnan (2002).
WEB SITE: http://www.carnegiemnh.org/ditw/paleolab/samson/index.htm.
SKELETAL REMAINS: According to Dale Russell (personal communication 1997), the skeleton has a nearly complete skull, 9 cervical and 7 dorsal vertebrae, 2 cervical and 10 dorsal ribs, 17 caudals, at least 4 chevrons, both femora, left fibula, abundant tibia pieces, 3 metatarsals, and 10 pes phalanges.
COMPLETENESS: A total of 121 bones, or 40% of a skeleton by bone count.
ON DISPLAY? Yes, currently at the Carnegie Museum Paleo Lab while undergoing preparation.
COMMENTS: The landowners knew that this was a of a *Tyrannosaurus rex* skeleton as early as 1987. In 1992, professional collectors Fred Nuss and Alan Deitrich, from Kansas, contracted with the landowner and proceeded to excavate the skeleton. They named it "Z-rex" for the first letter of the landowner's last name. The skeleton and skull were

Figure 1.12. *Samson (Z-rex); anterior view of skull block (A); side view of skull (B). Photos courtesy Dale Russell.*

then transported to Kansas for storage. It was repeatedly offered for sale over the next dozen years. Z-rex became the first *T. rex* offered for sale over the Internet and was offered for sale in the Wall Street Journal on April 23, 1999 (Glut 2002).

The specimen was acquired by international businessman, Graham Lacey, who changed its name from Z-rex to Samson and transported to the Carnegie Museum in Pittsburgh, where it is currently undergoing preparation. Samson's preparation can be observed in the Carnegie Museum Paleo Lab. More information on Samson can be obtained from Glut (2002, p. 578).

DMNH 2827

DISCOVERED: The year 1992, Charlie Fickle and his dog, both amateurs.
LOCATION: A housing development, Littleton, CO (Fig. 1.1, site 17).
FORMATION: Lower part of the Denver Formation.
EXCAVATED: Collected by the Denver Museum of Nature & Science under the direction of Kenneth Carpenter, 1992.
REPOSITORY: Denver Museum of Nature & Science, Denver, CO.
DESCRIBED: Carpenter and Young (2002).
SKELETAL REMAINS: Carpenter and Young (2002) reported that the partial skeleton consisted of a left femur, ilium, scapula, and coracoid; right tibia, fibula, and astragalus; a distal caudal vertebrae, ribs, and 3 teeth.
COMPLETENESS: Twelve bones, or 4% of a skeleton by bone count.
ON DISPLAY? Yes, some of this specimen can be seen at the Denver Museum of Nature & Science.
COMMENTS: This *Tyrannosaurus rex* has the distinction of being the only dinosaur discovered by a dog (K. Johnson, personal communication 2005). It is also the only *T. rex* found to date in Colorado (although teeth were previously known) and the only *T. rex* with a street address (K. Carpenter, personal communication 2005). DMNH 2827 was scattered before burial and was also damaged by earth-moving equipment before discovery (Carpenter and Young 2002).

Bowman

DISCOVERED: November 1992, Dean Pearson, amateur.
LOCATION: Near Rhame, Bowman County, ND (Fig. 1.1, site 18).
FORMATION: Hell Creek Formation, 32 m below the K-T boundary (Dean Pearson, notes with specimen).
EXCAVATED: Pioneer Trails Regional Museum, volunteers, 1992–1994.
REPOSITORY: Pioneer Trails Regional Museum, Bowman, ND.
SKELETAL REMAINS: According to Dean Pearson (notes with specimen), there were 45 bones of the skeleton collected. They consisted of ribs, vertebrae, both pubes, the distal end of the scapula, and gastralia (Don Wilkening, personal communication 2005).

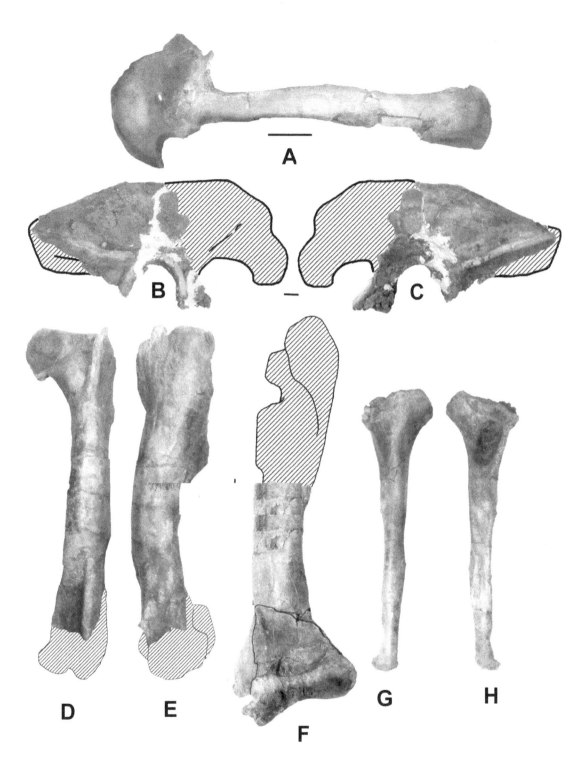

Figure 1.13. *DMNH 2827*, left scapula and coracoid (A); left ilium in medial (B) and lateral (C) views; left femur in anterior (D) and lateral (E) views; right tibia and astragalus in anterior view (F); right fibula in lateral (G) and medial (H) views.

COMPLETENESS: Less than 45 bones (because some are gastralia), or less than 15% of a skeleton by bone count.
ON DISPLAY? No.
COMMENTS: The museum has never given the specimen a collection or an acquisition number. The bones are still in plaster jackets, and they are also encased in a hard concretion that makes preparation difficult. Because of that, it is not known when the specimen will be prepared (Don Wilkening, personal communication 2005).

BHI 4100
(Duffy)

DISCOVERED: The year 1993, Stan Sacrison, amateur paleontologist.
LOCATION: John, Betty, and David Niemi Ranch, near Buffalo, Harding County, SD, about a quarter mile from the Stan excavation site (Fig. 1, site 19).
FORMATION: Hell Creek Formation, estimated to be about 16 m beneath the K-T boundary (Johnson 1996).
EXCAVATED: Black Hills Institute field crew, summers of 1993, 1994, 1996, and 2006.

Figure 1.14. *Duffy, BHI 4100, during excavation.* Photo by Ed Gerken.

REPOSITORY: Black Hills Institute of Geological Research, Hill City, SD.

DESCRIBED: Partially described by Counter (1996), Larson and Donnan (2002).

SKELETAL REMAINS: Most of the skull and a portion of the skeleton was recovered. The skull consists of: both maxillas, right premaxilla, nasals, both lacrimals, left postorbital, both quadrates, both jugals, left squamosal, both quadratojugals, left pterygoid, both palatines, left ectopterygoid, right epipterygoid, both dentaries, right surangular, right prearticular, the right splenial, right coronoid, a partial braincase, and 49 loose, rooted teeth. The skeleton has both scapulae and coracoids, 1 astragalus; the right ischium; 8 caudals; 13 dorsals and cervical vertebrae; 9 dorsal ribs; and 6 chevrons. About 50 loose teeth were found along with the specimen.

COMPLETENESS: Seventy-six bones, or 25% of a skeleton by bone count.

ON DISPLAY? Yes, some portions are on display at the Black Hills Institute of Geological Research, Hill City, SD, and a bronze cast of the left side of the skull is on the outside of their building. The skull bones are prepared, but most of the skeleton remains unprepared.

COMMENTS: Duffy was named for attorney Pat Duffy, who was the defense attorney for BHI in the Sue case (see Larson and Donnan 2004). The skull and partial skeleton of Duffy were disarticulated and scattered over a large area. This prompted Black Hills Institute to use a Bobcat to assist with the discovery of the bones. This ground-penetrating Bobcat was used to scrape thin layers of the quarry floor away during multiple passes, which led to the discovery of much of the specimen. The final dimensions of the quarry were approximately 55 feet by 70 feet. The Black Hills Institute originally excavated during 1993–1996, then returned in 2006 and removed the hills around the specimen. Within 3 days, they had excavated 3 additional skull bones and several other unidentifiable bones.

UWGM 181

DISCOVERED: Mike Pallett, University of Wisconsin geology student, 1993.

LOCATION: Near Ekalaka, Carter County, MT (Fig. 1.1, site 20).

FORMATION: Hell Creek Formation.

EXCAVATED: University of Wisconsin Field Crew, 1993.

REPOSITORY: University of Wisconsin, Geology Museum, Madison, WI.

SKELETAL REMAINS: A partial, fragmentary skull with 3 associated vertebrae. According to Richard Slaughter, director of the UW Museum of Geology, the following skull elements are represented: left dentary, right dentary, right surangular, splenial, right maxilla, left postorbital, right postorbital, right frontal (with attached prefrontal, parietal, and laterosphenoid), right jugal, right quadratojugal, right quadrate, and, tentatively, partial pterygoid, squamosal, and prearticular (not sided).

COMPLETENESS: Twenty bones, or 7% of a skeleton by bone count.

ON DISPLAY? No.

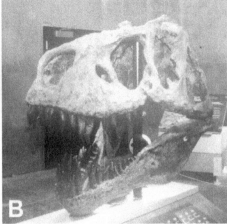

Figure 1.15. *Scotty, RSM 2523.8.*, during excavation (A); skull on display at the RSM Fossil Research Station (B). Photos courtesy Royal Saskatchewan Museum.

COMMENTS: "This specimen is really broken up, with close to 100 small pieces that have not been identified to element yet" (R. Slaughter, written communication 2005).

RSM 2523.8
(Scotty)

DISCOVERED: The year 1991, Robert Gebhart, schoolteacher.
LOCATION: Near Eastend, southern Saskatchewan (Fig. 1.1, site 21).
FORMATION: Frenchman Formation.
EXCAVATED: Royal Saskatchewan Museum field crew 1994 through 2001.
REPOSITORY: Royal Saskatchewan Museum.
SKELETAL REMAINS: The skull is nearly complete. It is missing both palatines, the epipterygoids, the splenials, the coronoids, the left angular, and about a third of the teeth. Most of the skull bones are incomplete. The postcranial skeleton is less known because most of it is still unprepared. There are more than 40 cervical, dorsal, and caudal (but no sacral) vertebrae, 16 dorsal ribs (at least), 1 scapula, 1 manus phalange, and the right femur, tibia, and fibula, along with both ilia, ischia, and pubes. There is also at least 1 metatarsal and several pes phalanges (Phil Currie, personal communication 2005).
COMPLETENESS: About 120 + bones to date, or at least 40% of a skeleton by bone count.
ON DISPLAY? Yes, the skull is on display and several portions of the skeleton can be seen undergoing preparation at the *T. rex* Discovery Centre in Eastend, Saskatchewan.
COMMENTS: This skeleton was discovered by Robert Gebhart as he accompanied Tim Tokaryk and John Storer on a field trip for the Royal Saskatchewan Museum. The specimen was nicknamed Scotty after a bottle of spirits consumed in the field by the discoverers.

Preparation is underway at the *T. rex* Discovery Centre in Eastend, Saskatchewan (owned and operated by the Royal Saskatchewan

Museum). The skull and skeleton were disarticulated and spread out over a large area (Glut 2003). Scotty is a large adult *Tyrannosaurus rex* skeleton with a femur length of 1290 mm (P. Currie, personal communication 2005). There are still years of work ahead to finish preparation because of the hard encasing rock, so eventually more of the skeleton may be discovered.

BHI 6219
(007, Double-O-Seven)

DISCOVERED: The year 1994, Bill Garstka, professional collector.
LOCATION: Near Marmouth, Slope County, ND (Fig. 1.1, site 22).
FORMATION: Hell Creek Formation.
EXCAVATED: Collected by Warfield Fossils field crew (led by Rick Hebdon and Bill Garstka) in 1994 and 1995.
REPOSITORY: The skull bones were sold to a private collector in New York, the foot bones to another collector, and the remaining portions were eventually purchased by Black Hills Institute in 2005.
ACQUISITION: Through purchase.
SKELETAL REMAINS: The skull consists of both maxillae, both premaxillae, and parts of the dentary. The skeleton consisted of 1 vertebrae, 1 dorsal rib, 1 metatarsal, 1 pes phalange, the distal end of left humerus, and parts of a tibia and fibula (Rick Hebdon, personal communication 2005).
COMPLETENESS: Twelve bones, or 3% of a skeleton by bone count.
ON DISPLAY? No.
COMMENTS: According to Rick Hebdon (personal communication 2005), most of the skeleton was found weathered out in an arroyo; the remaining portions were under a hard sandstone ledge. The field crew jackhammered on the sandstone ledge for 6 weeks in 1994 and came back in 1995 and dynamited the ledge. Not much of the skeleton was found, but the humerus and a large fragment of tooth indicate that it is probably a robust morphotype.

BHI 6249
(Steven)

DISCOVERED: The year 1995, Steve Sacrison, an amateur paleontologist.
LOCATION: John, Betty, and David Niemi Ranch, near Buffalo, Harding County, SD, about a quarter mile south of the Duffy excavation site (Fig. 1.1, site 23).
FORMATION: Hell Creek Formation, estimated to be 6 to 10 m beneath the K-T boundary (P. Larson, personal communication 2005).
EXCAVATED: Black Hills Institute field crew summer of 1995.
REPOSITORY: Black Hills Institute of Geological Research, Hill City, SD.
DESCRIBED: Partially described by Larson and Donnan (2002).
ACQUISITION: Collected by Black Hills Institute field crew.
SKELETAL REMAINS: Skeletal elements collected with Steven are a nearly

Figure 1.16 *Theropod eggshell (A) found with Steven, BHI 6249; close-up (B). Photos: (A) Ed Gerken.*

complete right femur, 6 dorsal vertebrae (mostly incomplete), 5 dorsal ribs, 1 phalange, and 2 incomplete skull bones.

COMPLETENESS: Fifteen bones, or 5% of a skeleton by bone count.

ON DISPLAY? No.

COMMENTS: Steven was named for Steve Sacrison, Stan Sacrison's twin brother, who found the specimen. The skeleton was scattered and most of its bones were incomplete. Some of them also appeared to have been bitten and chewed. This led P. Larson (1997) to first speculate cannibalism in *Tyrannosaurus* because they were the only animals large enough to have been able to bite through the large bones. Theropod eggshell was also discovered, but it will never be known whether the eggshell washed in or whether it came from this dinosaur.

LDP#977-2
(Pete)

DISCOVERED: The year 1995, Rob Patchus, a graduate student from University of New Orleans.

LOCATION: Swanson Ranch, Niobrara County, WY (Fig. 1.1, site 24).

FORMATION: Lance Formation, about 800 feet below the K-T contact.

EXCAVATED: Lance Dinosaur Project field crew in the summer of 1995.

REPOSITORY: Stored at the University of New Orleans, New Orleans, LA.

DESCRIBED: Derstler and Myers (2005, this volume).

SKELETAL REMAINS: According to Derstler and Myers (2005), the skeleton contains 2 short strings of articulated cervical and dorsal vertebrae along with some ribs and gastralia that are heavily weathered. There are also parts or fragments of the hind limbs, pelvis, ribs, and vertebrae weathered downslope from the specimen.

COMPLETENESS: Estimated to be between 10% to 15% of a skeleton by bone count.

ON DISPLAY? No.

COMMENTS: Pete was named in honor of Rob's grandfather, Pete Patchus, and for Pete Larson. Pete is undergoing preparation at UNO but is not yet on public display. Hurricane Katrina (August 2005) interrupted the work by putting the specimen off limits for more than 9 months. As a result, neither the records nor the specimen could be worked on, investigated, or inventoried (Derstler, personal communication 2005).

Barnum

DISCOVERED: The year 1995, Bruce Hamilton and Leon Theisen, professional collectors.

LOCATION: Near the headwaters of Seven Mile Creek, north of the Cheyenne River, Niobrara County, WY (Fig. 1.1, site 25).

FORMATION: Lance Formation.

EXCAVATED: Collected by Japh Boyce and R. J. B. Rockshop field crew in 1995 and 1996.

REPOSITORY: Status unknown.

WEB SITE: http://www.factmonster.com/spot/dino_bargaintrex.html.

ACQUISITION: Undisclosed investors purchased Barnum at a Bonhams and Butterfields auction in May 2004.

SKELETAL REMAINS: The skull consists of incomplete bones, including both maxillae, left jugal, left ectopterygoid, right squamosal, left dentary, surangular, and articular. The skeleton also has mostly incomplete bones consisting of 1 cervical, 4 dorsal, and 3 caudal vertebrae; 9 dorsal ribs, some gastralia; 6 metatarsals, 3 pes phalanges; left ilium, left ischium, both pubes (right incomplete), both femora (right incomplete), left tibia, fibula, astragalus, and calcaneum; partial left scapula, left humerus, and manus claw.

COMPLETENESS: Forty-seven bones, or 16% of a skeleton by bone count.

ON DISPLAY? No.

COMMENTS: According to Japh Boyce (personal communication 2005), he named the skeleton in honor of Barnum Brown, who had discovered the first *Tyrannosaurus rex* bones (described as *Dynamosaurus imperiosus* Osborn 1905) not far from this site. Boyce originally claimed that this specimen was the rest of the holotype skeleton (BMNH R7994, formerly AMNH 5866; see above), but according to Carpenter (personal communication 2006), if that were true, there would be 3 femurs for the type instead of 2.

According to Japh Boyce, he sold the specimen to a group of investors known as Tyrex in 2000. Soon after this, Tyrex brought a case against Boyce, claiming lack of clear title. After Boyce demonstrated a clear title, the specimen was auctioned off for $90,000 by

Figure 1.17. *Fox, BHI 4182, left dentary at the excavation, with Casey Smith. Photo by Ed Gerken.*

Bonhams and Butterfields, an auction house, in May 2004 to unknown investors (Japh Boyce, personal communication 2005).

BHI 4182
(Fox; Foxy Lady; County rex)

DISCOVERED: The year 1994, Lloyd Fox, rancher.
LOCATION: Land belonging to Harding County, located within the pasture of Lloyd, Eunice, and Russell Fox, near Redig, Short Pine Hills, Harding County, SD (Fig. 1.1, site 26)
FORMATION: Hell Creek Formation.
EXCAVATED: Black Hills Institute field crew, summers of 1996, 1997, and 1998.
DESCRIBED: Partially described by Larson and Donnan (2002).
REPOSITORY: Black Hills Institute of Geological Research, Hill City, SD.
SKELETAL REMAINS: The skull consists of the left postorbital, right quadratojugal, right ectopterygoid, both dentaries, both surangulars, both articulars, both prearticulars, both angulars, both splenials, both coronoids, and 43 loose, rooted teeth. The postcranial skeleton was found with 2 cervical, 1 dorsal, and 3 caudal vertebrae, 5 dorsal ribs, and 2 cervical ribs.
COMPLETENESS: To date, 29 bones, or 10% of a skeleton by bone count.
ON DISPLAY? Yes, the left dentary is on display at the Black Hills Museum of Natural History, Hill City, SD.
COMMENTS: Lloyd Fox discovered a *Tyrannosaurus rex*, later named Fox, in honor of Lloyd, in 1994 on a parcel of Harding County land. Black Hills Institute of Geological Research received permission from the

county commissioners to begin excavation in 1996. The specimen was widely scattered; more probably remains buried at the site. Portions of the skull and skeleton indicate that it is a robust morphotype.

MOR 980
(Peck's Rex)

DISCOVERED: The year 1997, Lou Tremblay, biology teacher.
LOCATION: McCone County, MT (Fig. 1.1, site 27).
FORMATION: Hell Creek Formation.
EXCAVATED: Collected from 1997 through 2004 by several different groups, including different parties led by J. Keith Rigby, Nate Murphy, and Kraig Derstler (see Derstler and Myers this volume).
REPOSITORY: Museum of the Rockies, Bozeman, MT, on loan to the Fort Peck Paleontology Field Station at Fort Peck, Montana.
DESCRIBED: Furcula by Larson and Rigby (2005); Lipkin and Carpenter this volume.
SKELETAL REMAINS: The skull consists of a braincase, both maxillae and premaxillae, incomplete nasals (most of the right), right postorbital, right jugal, both lacrimals, both quadrates, both quadratojugals, both dentaries, left splenial, left prearticular, right surangular, and the right articular. The forelimbs include the furcula, both scapulas, 1 coracoid, both humeri, metacarpal III, and 3 manus phalanges. Its hind limbs and pelvis include a complete right leg with calcaneum, astragalus, and 1 metatarsal; a complete left ilium, partial right ilium, both pubes, both ischia, and the sacral vertebrae. The skeleton also has some dorsal and cervical vertebrae, a string of 9 or 10 anterior caudal vertebrae (most with chevrons), many dorsal ribs, some cervical ribs, and quite a few gastralia (N. Murphy, personal communication 2005).
COMPLETENESS: There has not yet been a complete skeletal inventory. It is estimated to be about 120 bones, or 40% complete by bone count.
ON DISPLAY? Cast of the skeleton is on display at Fort Peck Interpretive Center, Fort Peck.

A   B

Figure 1.18. *Peck's Rex, MOR 980, cast of skeleton at the Fort Peck Interpretive Center (A); cast of skull (B). Photos courtesy Nate Murphy.*

COMMENTS: Peck's Rex was another *Tyrannosaurus rex* involved in controversy. The property was originally private, but the federal government foreclosed the land because of an unpaid loan. A California teacher discovered the *T. rex* while exploring for fossils as an Earth Watch volunteer, under the supervision of J. Keith Rigby, who began the excavation. After the Earth Watch group closed the quarry, later that year, the original landowners returned and tried to excavate the dinosaur themselves. The government seized the specimen, and eventually the specimen was put under the care of the Museum of the Rockies (http://www.nd.edu/~ndmag/au2003/rigbyrex.html).

The specimen received its nickname after nearby Fort Peck Lake. The excavation of this specimen was unusual because several different groups participated. J. Keith Rigby of Notre Dame University, along with Earth Watch volunteers, the original landowners, Nate Murphy from the Judith River Dinosaur Institute, and Kraig Derstler from the University of New Orleans were all involved with the recovery of the skeleton. For further discussion, see Derstler and Myers (this volume).

Tinker

DISCOVERED: The year 1997, Mark Eatman, professional collector.
LOCATION: Harding County Land, SD (Fig. 1.1, site 28).
FORMATION: Hell Creek Formation.
EXCAVATED: Mark Eatman in 1997, and Mike Farrell in 1998 (Glut 2002).
REPOSITORY: None, as of yet.
DESCRIBED: Partially described in Glut (2002).
SKELETAL REMAINS: According to Barry James (personal communication 2005), the specimen has most of a skull (although some bones are only pieces and fragments) consisting of the both premaxillae, left maxillae, both quadratojugals, right palatine, left jugal, quadrate, squamosal, pterygoid, a parietal, and part of 1 nasal. The lowers consist of left dentary, both articulars, both coronoids, right surangular, the left prearticular, angular, and splenial. The skeleton is made up of 20 partial caudal vertebrae and 12 chevrons; 5 dorsal and 2 cervical ribs and numerous rib fragments; both pubes, both ilia (incomplete), and the left ischia; parts of both scapula, the right coracoid, both humeri (left complete); and a manus and a pes claw.
COMPLETENESS: Seventy-three bones, or 24% of a skeleton by bone count.
ON DISPLAY? No.
COMMENTS: Because of the status of the land, its ownership has been tied up in litigation for several years. Bakker (personal communication 2002) has viewed this specimen, and is certain that it is a juvenile *T. rex*. He is also convinced that this specimen offers important information on growth rates for *T. rex* and will finally settle the differences between juvenile *Tyrannosaurus* and *Nanotyrannus*.

According to B. James (personal communication 2005), the

specimen was found with a right maxilla and jugal from a larger individual, along with a dentary from another theropod. He believes that there are perhaps 3 different individual *T. rex* specimens from the site. The specimen was heavily weathered and broken. The initial collecting was done without glue or plaster, so many of the bones were fragmented. In addition to *T. rex* bones, there were also a number of *Edmontosaurus* skeletal elements, plants, molluscs, and crocodile teeth found with the specimen (Glut 2002).

Ollie
(Rex A)

DISCOVERED: October 1998 by Craig Pfister, a professional collector.
LOCATION: Southeast of Ekalaka Carter County, MT (Fig. 1.1, site 29).
FORMATION: Hell Creek Formation.
EXCAVATED: Craig Pfister, 1998 and 1999.
REPOSITORY: Currently housed at Great Plains Paleontology, Madison, Wisconsin.
SKELETAL REMAINS: The skull consists of the left maxilla, both premaxillae, both postorbitals, both quadrates, left jugal, both pterygoids, and a partial braincase. The skeleton consists of both femora, both tibia, both fibula, both astragali and calcanea, 2 metatarsals, several phalanges, 1 ischia, the left pubis, right ilium, right scapula, both humeri, right radius, right ulna, several cervical vertebrae, multiple dorsal and caudal vertebrae, many cervical and dorsal ribs, and several chevrons (Craig Pfister written communication 2005).
COMPLETENESS: A total of 124 bones, or 41% of a skeleton to date (Craig Pfister, written communication 2005).
ON DISPLAY? No.
COMMENTS: The site measures 20 m by 17 m . Many of the bones show evidence of scavenging (Craig Pfister, written communication 2005).

Rex B
(Triceratops-Alley T. rex)

DISCOVERED: The year 1998 by Bill Alley, rancher and amateur collector.
LOCATION: Northeast of Isabel, Corson County, SD (Fig. 1.1, site 30).
FORMATION: Hell Creek Formation.
EXCAVATED: Bill Alley, 1998 through 2000.
REPOSITORY: Currently housed at Black Hills Institute of Geological Research, Hill City, SD, for cleaning, restoration, and casting.
SKELETAL REMAINS: The skull consists of the left maxilla, left premaxilla, left quadrate, left quadratojugal, both lacrimals, nasals, left ectopterygoid, and a braincase with disarticulated, nonfused frontals. The skeleton consists of only the right scapula, right coracoid, and 1 rib. There are also numerous fragmented bones with this specimen, and several other unprepared skull bones whose identification cannot yet be made.

Figure 1.19. *Rex C, maxilla and premaxilla (A), pes phalanges (B).*

COMPLETENESS: Twenty-four bones, or 8% of a skeleton to date.
ON DISPLAY? No.
COMMENTS: This specimen is of the gracile form nearly as large as Stan. The bones were scattered, with preservation ranging from good to poor. This is one of the most eastern *T. rex* specimens, at 101.5° west longitude.

Rex C

DISCOVERED: The year 1999 by Bill Alley, rancher and amateur collector.
LOCATION: Northeast of Isabel, Corson County, SD (Fig. 1.1, site 31).
FORMATION: Hell Creek Formation.
EXCAVATED: Bill Alley, 1999.
REPOSITORY: None.
SKELETAL REMAINS: The skull consists of the right maxilla, right premaxilla, right surangular, right articular and the right splenial. The skeleton consists of the right (?) tibia and astragalus (calcaneum ?), both fibula, 3 pes phalanges, right ischia, 1 cervical vertebrae, 1 dorsal vertebrae, 2 caudal vertebrae, 1 chevron, and several boxes of unidentified parts.
COMPLETENESS: Eighteen bones, or 6% of a skeleton to date.
ON DISPLAY? No.
COMMENTS: The maxilla and pes phalanges of this *Tyrannosaurus rex* are quite large, similar in size to Sue. Bone preservation is excellent. One of the pes phalanges has some severe pathologies, perhaps from a healed break. This is another easternmost *T. rex* specimen, at 101.5° west longitude.

2000–2005

The excavations of *Tyrannosaurus rex* continued into the 21st century. Although some of these discoveries were made in the 1990s, there were quite a number of *T. rex* skeletons found since 2000. There have been several television specials about some of these discoveries, such as Discovery Channel's *Valley of the T. rex*, which highlight many of the new discoveries made by the Museum of the Rockies near Hell Creek, south of Fort Peck Lake, MT.

Figure 1.20. *E. D. Cope, BHI 6248, excavation site. Photo by Dan Counter.*

BHI 6248
(*E. D. Cope*)

DISCOVERED: The year 1999, Bucky Derflinger, rancher, amateur fossil collector.

LOCATION: Wade Derflinger Ranch, near Usta, Perkins County, SD (Fig. 1.1, site 32).

FORMATION: Lower portion of the Hell Creek Formation.

EXCAVATED: Black Hills Institute field crew summer of 2000.

REPOSITORY: Currently stored at Black Hills Institute of Geological Research, Hill City, SD.

DESCRIBED: Partial description by Larson and Donnan (2002).

SKELETAL REMAINS: To date, E. D. Cope consists of a maxilla, dentary, ectopterygoid, angular, other undetermined skull bones, some vertebrae, ribs, and concretion-encased bones. Other than the left maxilla, the preparation of the specimen has not yet begun.

COMPLETENESS: About 10% of skeleton by bone count.

ON DISPLAY? The left maxilla is on display at the Black Hills Museum of Natural History, Hill City, SD.

COMMENTS: During the excavation, a number of centra were discovered piled on the surface. It appeared that some time ago, someone had intentionally piled these bones up. Larson and Donnan (2002) speculated that this could possibly be the site from which Edward Drinker Cope had collected 2 vertebrae from a partial *Tyrannosaurus rex* he described as *Manospondylus gigas*. Unfortunately, this question remains unsolved. Chemical analysis might answer the question.

This specimen was scattered over a large area like Duffy, Steven, and Fox. And like Steven, it appears that it was also cannibalized (Larson and Donnan 2002). Most of the skeletal elements are encased in sideritic and phosphatic concretions, making it difficult to prepare and recognize the bone elements. More of the skeleton could be buried at the site.

Monty

DISCOVERED: The year 1999, anonymous landowner.
LOCATION: Northern Niobrara County, WY (Fig. 1.1, site 33).
FORMATION: Lance Formation.
EXCAVATED: Collected by Craig Sundell, Fred Nuss, and crew in 2000.
REPOSITORY: Babiarz Institute of Paleontological Studies (BIOPSI), Mesa, AZ.
SKELETAL REMAINS: Babiarz (personal communication 2005) reports that the skull consists of a braincase; nasals; the right maxilla, lacrimal, postorbital, quadratojugal, and the squamosal; both quadrates, both pterygoids; the left premaxilla, jugal, and surangular. The skeleton consists of 4 cervicals, 2 dorsals, 3 caudals, 12 dorsal ribs, 4 gastralia, 1 pubis, a partial ilium, 1 pes phalange, an ulna(?), and several bones still in jackets.
COMPLETENESS: Fifty-three bones, or 18% (to date) of a skeleton by bone count.
ON DISPLAY? Yes, parts at Arizona State University.

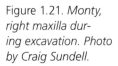

Figure 1.21. *Monty, right maxilla during excavation. Photo by Craig Sundell.*

Figure 1.22. *B-rex, MOR 1125, cast of skull on display at the Museum of the Rockies. Photo by Peter Larson.*

COMMENTS: The specimen was named after the landowner's first name, Monty. It has a fairly good braincase from which an endocast is being produced.

MOR 1125
(Bob; B-rex)

DISCOVERED: The year 2000 by Bob Harmon, a professional preparator for the Museum of the Rockies.
LOCATION: Near the Fort Peck Lake, Charles M. Russell National Wildlife Refuge, Garfield County, MT (Fig. 1.1, site 34).
FORMATION: Lower half of the Hell Creek Formation.
EXCAVATED: Bob Harmon, Nels Peterson, and a Museum of the Rockies field crew, 2001–2003.
REPOSITORY: Museum of the Rockies, Bozeman, MT.
DESCRIBED: Description of dinosaur skin tissue from the femur in Schwei-

tzer et al. (2004) and in the medullary bone from the same femur by Schweitzer et al. (2005). See also Schweitzer et al. (this volume).

SKELETAL REMAINS: The skeleton has a fairly complete yet disarticulated skull that is missing both premaxilla and a dentary, plus a few other skull bones. There are 3 cervical, 4 dorsal, 5 sacral, and 12 caudal vertebrae, along with 7 chevrons, 4 cervical, 13 dorsal ribs, left scapula and coracoid, the furcula, left ulna, both femora, tibiae, and fibulae, right calcaneum and astragalus, and 11 pes phalanges (P. Larson, personal communication 2005).

COMPLETENESS: A total of 111 bones, or 37% of a skeleton by bone count.

ON DISPLAY? Yes, the furcula, ulna, a portion of the femur (showing the structure within the bone) and a cast of the skull are on display at the Museum of the Rockies, Montana State University, Bozeman, MT.

COMMENTS: This specimen has made headlines with the discovery of medullary bone in the right femur indicating that it was an egg-producing female (see Schweitzer et al. 2005, this volume), proving that you can sex a rex, or at least some of them. It has also been referred to as Bob, after the discoverer, Bob Harmon, before it was determined to be female. It is of the robust morphotype.

MOR 1126
(C-rex)

DISCOVERED: The year 2000, Celeste Horner.

LOCATION: South of the Fort Peck Lake, Garfield County, MT (Fig. 1.1, site 35).

FORMATION: Lower half of the Hell Creek Formation.

EXCAVATED: Bob Harmon, Joe Coolie, and a Museum of the Rockies field crew, 2000–2001.

REPOSITORY: Museum of the Rockies, Bozeman, MT.

SKELETAL REMAINS: The skeleton consists of a left prearticular and surangular, 20 dorsal ribs of varying completeness, 3 partial dorsal vertebrae, and a chevron (P. Larson, personal communication 2005).

COMPLETENESS: Twenty-six bones, or 9% of a skeleton by bone count.

ON DISPLAY? No.

COMMENTS: C-rex was named in honor of Celeste Horner, after the first letter in her name. Beginning in 2001, the Museum of the Rockies began naming (or referring to) each of their different *Tyrannosaurus rex* specimens after the first letter of the first name of its discoverer. Although most of these *T. rex* specimens are made up of only a bone or two, some (such as B-rex, C-rex, and G-rex) are much more complete.

Figure 1.23. UCRC PV1, as found in large, weathered blocks (A); left arm, hand scapula, coracoid, and furcula undergoing preparation (B). Photos: (A) Wendy Taylor; (B) Paul Sereno.

UCRC PV1

DISCOVERED: The year 1950 (or earlier), Zerbst family (Arlene Zerbst, personal communication 2005), but not identified as a *Tyrannosaurus rex* skeleton until 1997 by Craig Derstler (P. Larson, personal communication 2005).

LOCATION: A few kilometers south of the Zerbst ranch house on Schneider Creek, a branch of the Cheyenne River, Niobrara County, WY (Fig. 1.1, site 36).

FORMATION: Lance Formation.

EXCAVATED: Collected by a field crew led by Paul Sereno, University of Chicago, students, and Project Exploration 2001.

REPOSITORY: University of Chicago Research Collection, University of Chicago, Chicago, IL.

DESCRIBED: Furcula and pectoral girdle by Lipkin and Sereno (2004).

SKELETAL REMAINS: Paul Sereno (written communication) relates that "the skeleton preserves a complete, articulated and in-the-round torso including both pectoral girdles and forelimbs (including the furcula), gastral basket, ribcage, and the cervical-dorsal column. Only fragments of the hind-limbs, folded under the torso, remain. A body outline can be seen in cross-section, with the inside of the torso filled with finer-grained siltstone than the outside (medium-to-course sandstone)."

COMPLETENESS: Estimated 60 bones, or 20% of a skeleton by bone count.

ON DISPLAY? No.

COMMENTS: Originally published as UCPC V1. The body portion of an articulated *Tyrannosaurus rex* skeleton was known for many years before anyone identified that it was anything more than just another duckbill skeleton. It lay on a parcel of BLM land within the pasture of Leonard Zerbst, and often became the stop for geological field trips to view a dinosaur in the rough.

Figure 1.24. *UMNH 110000, left dentary on display at the New Mexico Museum of Natural History.*

According to Paul Sereno (written communication), "the specimen lay on its side in a large, extremely hard, sideritic, sandstone concretion. The earliest photo (in a 1952 rancher's gazette) shows a complete concretion set atop a footing of softer siltstone. The specimen was thought to be a duckbill as *Edmontosaurus* is common in the area. Since that time, the concretion broke into about 5 pieces, the largest containing the intact torso.

"There is no evidence that the skull was ever present, as traces should have been found in the flat terrain surrounding the concretion. A 20cm-diameter tree trunk and eroded hadrosaur bones are preserved lodged against the torso; the individual appears to have fallen suddenly into a channel. With partial preparation, it was possible to identify the specimen as *Tyrannosaurus rex*, an individual approximately 66% the size of 'Sue,' FMNH PR2081."

UMNH 110000

DISCOVERED: The year 2001 by Rose Difley (student) and Quintin Saharatian (technician) from the University of Utah (Sampson and Loewen 2005).

LOCATION: Near Price, Carbon County, UT, in the Manti La-Sal National Forest (Fig. 1.1, site 37).

FORMATION: North Horn Formation.

EXCAVATED: Collected by the Utah Museum of Natural History field crew, 2001.

REPOSITORY: Utah Museum of Natural History, Salt Lake City, UT.

DESCRIBED: Sampson and Loewen (2005).

SKELETAL REMAINS: The skull is very incomplete, consisting of a right postorbital and squamosal. The skeleton consists of 2 cervical 3 sacral and a series of 6 midcaudal vertebrae, along with 6 chevrons and a rib. Pelvic girdle elements include: the left pubis and ischium and the distal blade of the right ilium. There is a partial leg consisting of the left tibia, fibula, and the astragalus (Sampson and Loewen 2005).

COMPLETENESS: Twenty-six bones, or about 9% of a skeleton, by bone count.

ON DISPLAY? Yes, at the Utah Museum of Natural History, Salt Lake City, UT.

COMMENTS: This discovery extended the known geographic range of *Tyrannosaurus rex* westward into central Utah. This indicates that *T. rex* spanned "habitats from wet lowland coastal plain environments to cooler alluvial plain settings and semi-arid, upland intermontane basins" (Sampson and Loewen 2005). Because *Alamosaurus* has been found in the same formation, Sampson and Loewen (2005) suggested that perhaps *T. rex* may have exploited this potential food source.

TCM 2001.90.1
(Bucky; formerly BHI 4960)

DISCOVERED: The year 1998, Bucky Derflinger, rancher, amateur fossil collector.

LOCATION: Wade Derflinger Ranch, original Usta town site, Perkins County, SD (Fig. 1.1, site 38).

FORMATION: Hell Creek Formation, not far above the contact of the Fox Hills Formation.

EXCAVATED: Black Hills Institute field crew, summer of 2001 and 2002.

REPOSITORY: The Children's Museum, Indianapolis, IN.

DESCRIBED: Furcula by Larson and Rigby (2005) and Lipkin and Carpenter (this volume); partial description of excavation and elements by Glut (2006).

ACQUISITION: Purchased from the Black Hills Institute of Geological Research, 2001.

Figure 1.25. *Bucky*, TCM 2001.90.1, portion of the large Bucky excavation (A); skeleton as mounted (B).

SKELETAL REMAINS: There were no skull elements or major leg bones found with the skeleton. The skeleton consists of 8 cervical, 9 dorsal, 5 sacral, and 15 caudal vertebrae; 14 chevrons; 11 cervical ribs; 16 dorsal ribs; 24 gastralia; a complete yet pathological furcula; both scapulas, right coracoid, left ulna, 2 manus phalanges; both ilia, 1 ischia, 4 metatarsals, and 9 pes phalanges.

COMPLETENESS: A total of 101 bones, or 34% of a skeleton by bone count.

ON DISPLAY? Yes, the original Bucky skeleton is mounted, along with a cast of Stan, as if attacking a *Triceratops* skeleton (Kelsey).

COMMENTS: Bucky Derflinger discovered a pes phalange from this *Tyrannosaurus rex* while he was breaking in a young horse on his father's ranch about 8 miles east of the original Sue dig site. This could be considered the first discovery of a *T. rex* by a horse. The specimen was named Bucky in honor of its discoverer, but as fate would have it, this is a robust morphotype and therefore most likely female.

The carcass originally lay on the ground, decomposing, and became disarticulated. Soon after, it was transported by water and deposited in a low, shallow valley. The size of the excavation is enormous, about 150 feet by 30 feet, and with 12 to 30 feet of overburden. The skeleton was discovered along with numerous bones from an *Edmontosaurus* (with bite marks in the sacrum), several *Triceratops* bones, turtle elements, a diverse Late Cretaceous fauna of fish, reptiles, mammals, dinosaurs, and plants.

MOR 1128
(G-rex)

DISCOVERED: The year 2001, Greg Wilson, then a student at University of California, Berkeley.

LOCATION: Near the Fort Peck Lake, Garfield County, MT (Fig. 1.1, site 39).

FORMATION: Hell Creek Formation.

EXCAVATED: Nels Peterson and the Museum of the Rockies field crew, 2001.

REPOSITORY: Museum of the Rockies, Bozeman, MT.

SKELETAL REMAINS: The skeleton consists of an incomplete dentary, 7 robust ribs, 4 dorsal and 1 caudal vertebrae, 3 chevrons, a partial scapula, both ischia, both pubes, the left femur, and left tibia (P. Larson, personal communication 2005).

COMPLETENESS: Twenty-three bones, or 8% of a skeleton by bone count.

ON DISPLAY? A single tooth from the specimen is all that is currently on display (Bob Harmon, personal communication 2005).

MOR 1152
(F-rex)

DISCOVERED: The year 2001 by Frank Stewart.

LOCATION: Near the Fort Peck Lake, Garfield County, MT (Fig. 1.1, site 40).

FORMATION: Lower half of the Hell Creek Formation.
EXCAVATED: Museum of the Rockies field crew, 2001.
REPOSITORY: Museum of the Rockies, Bozeman, MT.
SKELETAL REMAINS: The skeleton consists of a leg, pelvis, posterior ribs, some posterior dorsal vertebrae (all heavily weathered), a metatarsal, 7 caudal vertebrae, and 4 chevrons.
COMPLETENESS: A total of 25(?) bones, or 8(?)% of a skeleton by bone count.
ON DISPLAY? No.
COMMENTS: The legs, pelvis, and ribs are heavily weathered and fragmented.

Otto

DISCOVERED: The year 2001 by Craig Pfister, a professional collector.
LOCATION: Near Ekalaka, Carter County, MT (Fig. 1.1, site 41).
FORMATION: Hell Creek Formation.
EXCAVATED: The years 2001 and 2002 by Craig Pfister.
REPOSITORY: Currently housed at Great Plains Paleontology, Madison, WI.
SKELETAL REMAINS: According to Craig Pfister, the skeleton consists of both femora, both tibia, 2 metatarsals, 1 fibula, multiple cervical and dorsal ribs, and multiple caudal vertebrae.
COMPLETENESS: Thirty-two bones, or 10% of a skeleton by bone count.
ON DISPLAY? No.
COMMENTS: The site was a point bar deposit and measured 40 m by 32 m and 15 m deep (Craig Pfister, written communication 2005).

MOR/USNM
(*N-rex*)

DISCOVERED: The year 2001, Nathan Myhrvold, computer businessman, enthusiastic amateur, Hell Creek Project underwriter.
LOCATION: South of Fort Peck Lake, Charles M. Russell National Wildlife Refuge, Garfield County, MT (Fig. 1.1, site 42).
FORMATION: Hell Creek Formation.
EXCAVATED: Michael Brett-Surman and a Smithsonian Institution field crew, 2002, 2003.
REPOSITORY: According to Michael Brett-Surman (personal communication 2005), the specimen is currently still held by the Museum of the Rockies, but it is in the process of being transferred to the National Museum of Natural History, Smithsonian Institution, Washington, DC.
SKELETAL REMAINS: The skeleton consists of an incomplete dentary, an angular, 1 cervical vertebrae, 2 cervical ribs, 2 dorsal spines, 2 dorsal ribs, 3 caudal vertebrae, 3 chevrons, an ilium (weathered), ischium, pubis, right leg, and articulated foot (nearly complete).
COMPLETENESS: About 40 bones, or 13% of a skeleton by bone count.

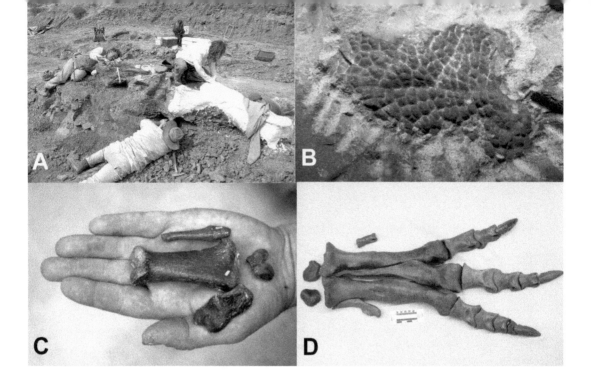

Figure 1.26. *Wyrex*, BHI 6230, excavation (A); skin impression (B); left carpals and metacarpals (C); right pes (D).

ON DISPLAY? No.

COMMENTS: This skeleton was named for first initial of the discoverer's name. Nathan Myhrvold was working as a volunteer with the Museum of the Rockies in 2001 when he found the skeleton. The Smithsonian Institute, which has only a cast of Stan on display, has been trying to obtain an original skeleton and is working on acquiring this specimen from the Museum of the Rockies; thus, at this time, it has no catalog number.

BHI 6230
(Wyrex)

DISCOVERED: The year 2002, Dan Wells, policeman and amateur collector, and Don Wyrick, landowner, rancher, and amateur collector.

LOCATION: Don and Allison Wyrick Ranch, north of Baker, Fallon County, MT (Fig. 1.1, site 43).

FORMATION: Hell Creek Formation.

EXCAVATED: Dan Wells 2002 and 2003, Black Hills Institute field crew 2004.

REPOSITORY: Currently housed at the Houston Museum of Nature and Science, Houston, TX.

DESCRIBED: Partial description by Larson and Donnan (2004).

SKELETAL REMAINS: The skull consists of a partial braincase, the right squamosal, postorbital, jugal, surangular, articular, prearticular, and angular. The postcranial skeleton consists of 2 nearly complete legs and feet (lacking to date only the left tibia, astragalus, calcaneum, 1 left tarsal, both metatarsal I, and 5 pes phalanges from the right foot). There are 22 vertebrae (5 dorsal, 1 cervical, and 11 caudal), 20 ribs (5 cervical and 15 dorsal), both ischia, right pubis, and right ilium. It also has a left scapula, coracoid, humerus, ulna, carpals, and all 3 metacarpals. Seventeen gastralia have been found with the skeleton to date.

COMPLETENESS: A total of 114 bones (to date), or 38% of a skeleton by bone count.

ON DISPLAY? No, but preparation of the skeleton may be viewed at the Black Hills Institute of Geological Research, in Hill City, SD.

COMMENTS: Intrigued by the find, Wells took a bone to the Black Hills Institute for identification. He returned to the site the next summer and began to excavate in earnest. Don Wyrick realized the importance of the find when legs, foot bones, and vertebrae started appearing. He called a halt to the digging, contacted the Black Hills Institute, and arranged to have them finish the excavation. The name "Wyrex" is a combination of the first 2 letters of Don's last name and "rex."

In May 2004, Black Hills Institute began the first live online *Tyrannosaurus rex* excavation. With daily reports, photos, and video segments of the days digging, this dinosaur excavation extended far beyond the quarry and into schools, homes, and businesses for the next 3 weeks. Thousands of people a day went to the Web site to see what new bones or discoveries had been made.

To date, 22 turtles (*Plesiobaena antiqua*) have been unearthed at this site, from complete to disarticulated and fragmentary shells. Wyrex was buried on the edge of an ancient pond or lake, and these turtles had perhaps been feeding on the carcass. It is unknown at this point what caused the death of either the *T. rex* or the turtles.

Some of the exciting discoveries with Wyrex are carpals along with 3 metacarpals from the left hand (see Lipkin and Carpenter this volume), and 2 nearly complete feet. During preparation, several patches of skin (Fig. 1.26B) were found with the skeleton. Most of the skin patches (more than a dozen) were found on the bottom side of the articulated tail. The discovery of skin with Wyrex is a first for *T. rex*. Plans are currently underway for further excavation at the site in hope of finding more of the skeleton. The preparation of the skeleton may be viewed by the public at Black Hills Institute of Geological Research in Hill City, SD. Ownership of the specimen is currently being transferred the Houston Museum of Nature and Science, where it is scheduled to be on display by 2009.

LACM 7509/150167
(Thomas)

DISCOVERED: The year 2003, Bob Curry, schoolteacher and amateur fossil collector.

LOCATION: BLM land near Ekalaka, Carter County, MT (Fig. 1.1, site 44).

FORMATION: Hell Creek Formation.

EXCAVATED: An international crew from the Natural History Museum of Los Angeles County, 2003–2005.

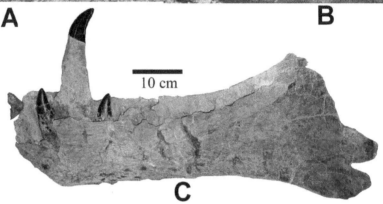

Figure 1.27. Thomas, LACM 7509/150167, during excavation (A); right femur during excavation (B); right dentary (C). Photos: (A) Doug Goodreau; (B) Ursula Goelich; (C) Gary Takeuchi. All photos courtesy Natural History Museum of Los Angeles County.

REPOSITORY: The Natural History Museum of Los Angeles County, Los Angeles, CA.

SKELETAL REMAINS: From Luis Chiappe (written communication 2005): "The skull consists of a complete braincase, postorbital, both jugals, squamosal, quadratojugal, quadrate, lacrimal, both frontals, both maxillae, ectopterygoid, both dentaries, and at least one set of articulated postdentary bones (there are other skull bones but in jackets but they cannot be identified at this time). There are also 30–35 loose teeth and a few more in situ. The skeleton consists of complete legs and feet (missing 1 metatarsal and some phalanges, a few dorsals, the sacrum, about 20 caudals, along with many ribs and gastralia. The pelvis includes both ilia and ischia. Both scapula and coracoids are preserved although no forelimb elements are known at the moment."

COMPLETENESS: A total of more than 110 bones (to date), or more than 37% of a skeleton by bone count.

ON DISPLAY? No.

COMMENTS: The dinosaur was named Thomas by the discoverer, Bob Curry, in honor of his brother. Frankie Jackson of Montana State University informed Luis Chiappe (director of the Dinosaur Institute at the LACM) of the discovery. The excavation took 3 seasons, during which they collected about 130 blocks, some containing several elements. Much of the specimen remains unprepared, so the number of recognized elements is likely to increase (Luis Chiappe, written communication 2005).

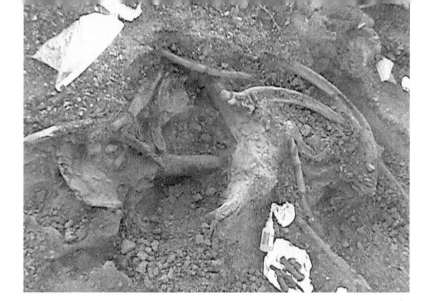

Figure 1.28. *Ivan, during excavation showing scapula-coracoid, ribs, and other bones. Photo courtesy Gary Olson.*

Wayne

DISCOVERED: The year 2004, Gary Olson, professional collector.
LOCATION: South of Rhame, southwest Bowman County, SD (Fig. 1.1, site 45).
FORMATION: Hell Creek Formation.
EXCAVATED: Gary Olson and Alan Komrosky, 2004–2005.
REPOSITORY: None; currently in possession of the collector, Fargo, ND.
SKELETAL REMAINS: This incomplete specimen consists of 19 caudal and 1 dorsal vertebrae, 2 chevrons, several rib (and/or gastralia) segments, plus some other unidentified parts (Gary Olson, personal communication).
COMPLETENESS: Twenty-four bones, or 8% of a skeleton by bone count.
ON DISPLAY? No.
COMMENTS: A disarticulated, weathered skeleton.

Ivan

DISCOVERED: The year 2005, Gary Olson, professional collector.
LOCATION: North of Buffalo, northwest Harding County, SD (Fig. 1.1, site 46).
FORMATION: Hell Creek Formation.
EXCAVATED: Gary Olson and Alan Komrosky, 2005.
REPOSITORY: None; currently in possession of the collector, Fargo, ND.
SKELETAL REMAINS: Gary Olson (personal communication 2005) relates that the skeleton consists of a nearly complete pelvis, with paired ilia (top half eroded), paired ischia, and paired pubes articulated to the sacrum. Ivan has one leg, consisting of a femur, tibia, fibula, and astragalus. The feet contain 2 metatarsals and 6 pes phalanges. The body has a fused scapula coracoid, an estimated 15 cervical and dorsal vertebrae, and approximately 25 cervical and dorsal ribs (this rib count may also include some gastralia), approximately 25 caudal vertebrae, and an estimated 30 chevrons.

COMPLETENESS: A total of 116 bones, or 39% of a skeleton by bone count (Gary Olson inventory).
ON DISPLAY? No.
COMMENTS: Gary Olson was hunting for fossils when he happened upon several bones of a large theropod weathering out of the hillside. Exposed were the femur, tibia, metatarsals, pes phalanges, and large number of rib and bone scraps. He planned to dub it Ivan the Terrible, but unfortunately, the specimen was not found with a skull, so he called it just Ivan. Olson explained that he dug all around the skeleton, but it appears that if Ivan was originally buried with its skull, it weathered out long ago. The preservation is excellent.

Additional Important Specimens

The holotype of *Manospondylus gigus* Cope, 1892, consisted of 2 large theropod cervical vertebrae (only 1 vertebra is known to exist today). It was later discussed and redefined as *Tyrannosaurus rex* by Osborn in 1916. In 1905, Barnum Brown collected another, very partial, *Tyrannosaurus rex* specimen from Hell Creek, MT. According to Osborn (1906), the specimen (AMNH 5881) consisted of the left femur, both tibiae, left fibula, and 4 metatarsals. Brown in 1908 collected a braincase of a *T. rex* in MT (AMNH 5029 = CM 9379); it also has a splenial, articular, and a prearticular. The American Museum has another individual *T. rex* braincase (AMNH 5117), collected by Sternberg and figured by Osborn (1912).

The Museum of Paleontology, University of California, Berkeley, has several different *Tyrannosaurus rex* specimens, one a maxilla (UCMP 118742), the other a maxilla with dentaries (UCMP 131583), both from Montana. The Science Museum of Minnesota has the anterior portion of a braincase from South Dakota (MMS 51-2004) (Molnar 1991). The Museum of the Rockies collected a portion of a braincase in MOR 1131 (also referred to as J-rex). There is also a large partial theropod specimen from the Tornillo Formation of West Texas that was described as *T. rex* (Lawson 1976), but Carpenter (1990) thought that it was morphologically too different to include it in the species (although Carr and Williamson 2004 believe it is).

Williamson and Carr (2005) reported on an additional, yet fragmentary, *Tyrannosaurus rex* specimen (with dentary) from the Naashoibito Member of the Kirtland Formation in New Mexico.

There are reports of 2 additional *T. rex* skeletons discovered during the summer of 2005 from the Hell Creek Formation in Harding County. Although I have seen a few bones from these specimens, it is still premature to provide any details on their completeness or significance.

There are numerous less complete *Tyrannosaurus rex* specimens (10 bones or fewer) in various collections across North America and Europe (e.g., Natural History Museum, London, England, UK) that have not been included in this summary. It certainly appears probable that more *T. rex* discoveries will be made in the near future. With so many new specimens to study, there is still the potential to learn much more about the most famous of all the dinosaurs, *T. rex*.

Specimen	No. of Bones	Percentage of Skeleton
Sue FMNH PR2081	219	73%
Stan BHI 3033	190	63%
MOR 555	146	49%
AMNH 5027	143	48%
Ollie*	124	41%
Scotty RSM 2523.8*	120+	40%+
Samson (Z-rex)*	121	40%
Pecks rex MOR 980*	120	40%
Ivan*	116	39%
Wyrex BHI 6230	114	38%
MOR 1125	111	37%
Thomas LACM 7509/10167*	110+	37%+
Bucky TCM 2001.90.1	101	34%
Black Beauty RTMP.81.6.1	85	28%
SDSM 12047	82	27%
Duffy BHI 4100	79	26%
LACM 23844	74	25%
Tinker, SD, collected by Eatman and Ferrel*	73	24%
UCRC V1*	60	20%
MOR 009	58	19%

Table 1.2. *Twenty Most Complete* **Tyrannosaurus Rex** *Skeletons*

Note.—See text for details.

* The bone count was provided by others, or the number is estimated because the preparation is not yet complete. For some specimens, the actual bone count may go up (with further preparation) or down (if gastralia were identified as ribs).

From all of the information that could be learned from contacting people personally, visiting museums, reading publications, and researching on the Internet, it appears that there are currently 46 known *Tyrannosaurus rex* specimens with 10 bones or more (Table 1.1). As can be seen from Figure 1.1 and Table 1.1, most *Tyrannosaurus rex* skeletons were collected just south of Fort Peck Lake, near the type of Hell Creek Formation, with a second large conglomeration near the border of Montana, North Dakota, and South Dakota. The exposures of the Hell Creek Formation are extensive and well exposed in these regions, which most likely is the reason for the discovery of so many specimens from these areas. Although fewer specimens are known from other regions, there is every reason to believe that additional specimens will be found. In summary, then, the range of *T. rex* is therefore known to include nearly the entire Western Interior.

There is always much discussion as to which is the most complete *Tyrannosaurus rex* skeleton, or where each skeleton lies according to the amount of completeness. For that reason, the 20 most complete *T. rex* skeletons are summarized in Table 1.2.

Acknowledgments

Thanks to Peter Larson and Robert Farrar for providing me with quite a bit of this information and also for the much-needed editorial help during the drafting of this chapter. Thanks to Kenneth Carpenter and Peter Larson for their patience, editing, and the organization of this publication. Thanks to the Black Hills Museum of Natural History for organizing the 100 Years of *Tyrannosaurus rex* Symposium in June 2005. Thanks to John Babiarz, Japh Boyce, Kenneth Carpenter, Luis Chiappe, Phil Currie, Kraig Derstler, Bob Harmon, Rick Hebdon, Barry James, Christine Lipkin, Nate Murphy, Gary Olson, Scott Sampson, Richard Slaughter, Craig Sundell, Tom Williamson, and the Black Hills Institute for providing me with much-needed information on the specimens. Larry Shaffer helped by cleaning up and organizing the many photos of *T. rex*. Thanks to all of the different photographers, who are credited in the figure captions. And finally, a big, special thank you to each and every one of the landowners (including the U.S. and Canadian governments), who allowed access to the discoverers and the collectors of these wonderful and fascinating creatures.

References Cited

Bakker, R. T. 1986. *The Dinosaur Heresies: New Theories Unlocking the Mystery of the Dinosaurs and Their Extinction*. Zebra Books, New York.

Bakker, R. T., Williams, M., and Currie, P. 1988. *Nanotyrannus*, a new genus of pygmy tyrannosaur, from the latest Cretaceous of Montana. *Hunteria* 1(5):26.

Bjork, P. R. 1982. On the occurrence of *Tyrannosaurus rex* from northwestern South Dakota (abstract). *Proceedings of the South Dakota Academy of Science* 61:161–162.

Brochu, C. A. 2003. Osteology of *Tyrannosaurus rex*: insights from a nearly complete skeleton and high-resolution computed tomographic analysis of the skull. *Journal of Vertebrate Paleontology* 22, Memoir 7, Supplement to 4.

Carpenter, K. 1990. Variation in *Tyrannosaurus rex*. P. 141–145 in Carpenter, K., and Currie, P. J. (eds.). *Dinosaur Systematics: Approaches and Perspectives*. Cambridge University Press, Cambridge.

———. 2004. Redescription of *Ankylosaurus magniventris* Brown 1909 (Ankylosauridae) from the Upper Cretaceous of the Western Interior of North America. *Canadian Journal of Earth Science* 41: 961–986.

Carpenter, K., and Smith, M. 2001. Forelimb osteology and biomechanics of *Tyrannosaurus rex*. P. 90–116 in Tanke, D. H., and Carpenter, K. (eds.). *Mesozoic Vertebrate Life*. Indiana University Press, Bloomington.

Carpenter, K., and Young, D. B. 2002. Late Cretaceous dinosaurs from the Denver Basin. *Colorado, Rocky Mountain Geology* 37(2): 237–254.

Carr, T. D. 1999. Craniofacial ontogeny in Tyrannosauridae (Dinosauria, Coelurosauria). *Journal of Vertebrate Paleontology* 19(3): 497–520.

Carr, T. D., and Williamson, T. E. 2000. A review of Tyrannosauridae (Dinosauria: Coelurosauria) from New Mexico. P. 113–145 in Lucas, S. G., and Heckert, A. B. (eds). Dinosaurs of New Mexico. *New Mexico Museum of Natural History and Science Bulletin* 17.

Carr, T. D., and Williamson, T. E. 2004. Diversity of Late Maastrichtian Tyrannosauridae (Dinosauria: Theropoda) from western North America. *Zoological Journal of the Linnean Society* 142: 479–523.

Cope, E. D. 1892. Fourth note on the dinosauria of the Laramie. *American Naturalist* 26: 756–758.

Counter, D. 1996. *T-rex: The Real World* (video). Off Line Video, Butte, MT.

Currie, P. J. 1993. Black Beauty. *Dino Frontline* 4: 22–36.

———. 2003. Cranial anatomy of tyrannosaurid dinosaurs from the Late Cretaceous of Alberta, Canada. *Acta Paleaeontoligica Polonica* 48(2): 191–226.
Davies, M. J. 1997. The curse of *T. rex*. *Nova* (video). WGBH Boston Video, South Burlington, VT.
Derstler, K., and Myers, J. 2005. Preliminary account of the tyrannosaurid "Pete" from the Lance Formation of Wyoming. P. 20 in *100 Years of Tyrannosaurus rex: A Symposium, Abstracts*. Black Hills Museum of Natural History, Hill City, SD.
Dingus, L. 2004. *Hell Creek, Montana: America's Key to the Prehistoric Past*. St. Martin's Press, New York.
Donnan, K., and Counter, D. 1999. *The Rex-Files: STAN* (video). Counter Productions & Black Hills Institute of Geological Research, Hill City, SD.
———. 2000. *The Rex-Files: SUE* (video). Counter Productions & Black Hills Institute of Geological Research, Hill City, SD.
Erickson, G. M., Lappin, A. K., and Larson, P. L. 2005. Androgynous *rex*—the utility of chevrons for determining the sex of crocodilian and non-avian dinosaurs. *Zoology* 108: 277–286.
Fiffer, S. 2000. *Tyrannosaurus Sue*. W. H. Freeman, New York.
Gillette, D. D., Wolberg, D. L., and Hunt, A. P. 1986. *Tyrannosaurus rex* from the McRae Formation (Lancian, Upper Cretaceous), Elephant Butte Reservoir, Sierra County, New Mexico. P. 235–238 in Clemons, R. E., King, W. E., Mack, G. H., and Zidek, J. (eds.). *New Mexico Geological Society Guidebook, 37th Field Conference, Truth or Consequences Region*. New Mexico Geological Society, Socorro, NM.
Glut, D. F. 1997. *Dinosaurs: The Encyclopedia*. McFarland, Jefferson, NC.
———. 2000. *Dinosaurs: The Encyclopedia, Supplement 1*. McFarland, Jefferson, NC.
———. 2002. *Dinosaurs: The Encyclopedia, Supplement 2*. McFarland, Jefferson, NC.
———. 2003. *Dinosaurs: The Encyclopedia, Supplement 3*. McFarland, Jefferson, NC.
———. 2006. *Dinosaurs: The Encyclopedia, Supplement 4*. McFarland, Jefferson, NC.
Holtz, T. R., Jr. 2004. Tyrannosauroidea. P. 111–136 in Weishampel, D. B., Dodson, P., and Osmólska, H. *The Dinosauria*. 2nd ed. University of California Press, Berkeley.
Horner, J. R., and Lessem, D. 1993. *The Complete T. rex*. Simon & Schuster, New York.
Hurum, J. H., and Sabath, K. 2003. Giant theropod dinosaurs from Asia and North America: skulls of *Tarbosaurus bataar* and *Tyrannosaurus rex* compared. *Acta Palaeontologica Polonica* 48(2): 161–190.
Johnson, K. R. 1996. Description of seven common fossil leaf species from the Hell Creek Formation (Upper Cretaceous: Upper Maastrichtian), North Dakota, South Dakota, and Montana. *Denver Museum of Natural History Proceedings*, Series 3(12).
Larson, P. L. 1994. *Tyrannosaurus* sex. P. 139–155 in Rosenberg, G., and Wolberg, D. (eds.). *Dino Fest Proceedings*. Paleontological Society Special Publication 7.
———. 1997. The king's new clothes: a fresh look at *Tyrannosaurus rex*. P. 65–71 in Wolberg, D. L., Stump, E., and Rosenberg, G. D. *Dinofest International Proceedings*. Academy of Natural Sciences, Philadelphia.
———. 2000. Cranial morphology, mechanics, kinesis, and variation in *Tyrannosaurus rex*. In *The Rex Files: Scientific Papers and Popular Articles and Miscellaneous Information on Tyrannosaurus rex*. Black Hills Institute of Geological Research, Hill City, SD.
Larson, P. L., and Donnan, K. 2002. *Rex Appeal, the Amazing Story of Sue, the*

Dinosaur that Changed Science, the Law and My Life. Invisible Cities Press, Montpelier, VT.

———. 2004. *Bones Rock.* Invisible Cities Press, Montpelier, VT.

Larson, P. L., and Rigby, K. 2005. Furcula of *Tyrannosaurus rex.* P. 247–255 in K. Carpenter (ed.). *The Carnivorous Dinosaurs.* Indiana University Press, Bloomington.

Lawson, D. 1976. *Tyrannosaurus* and *Torosaurus*, Maestrichtian dinosaurs from Trans-Pecos, Texas. *Journal of Paleontology* 50: 158–164.

Lipkin, C., and Sereno, P. C. 2004. The furcula of *Tyrannosaurus rex. Journal of Vertebrate Paleontology* 24(Suppl. to 3): 83A–84A.

Maleev, E. A. 1974. Gigantic carnosaurs of the family Tyrannosauridae. *Results of the Soviet-Mongolian Paleontological Expedition* 1: 132–191.

McIntosh, J. S. 1981. Annotated catalogue of the dinosaurs (Reptilia, Archosauria) in the collections of Carnegie Museum of Natural History. *Carnegie Museum of Natural History Bulletin* 18: 1–67.

Molnar, R. E. 1991. The cranial morphology of *Tyrannosaurus rex. Palaeontographica* 217: 137–176.

Newman, B. H. 1970, Stance and gait in the flesh-eating *Tyrannosaurus. Biological Journal of the Linnean Society* 2:119–123.

Olshevsky, G., Ford, T. L., and Yamamoto, S. 1995. The origin and evolution of the tyrannosaurids, part 1. *Kyoryugaku Saizensen* [*Dino Frontline*] 9: 92–199.

Osborn, H. F. 1905. *Tyrannosaurus* and other Cretaceous carnivorous dinosaurs. *Bulletin of the American Museum of Natural History* 21: 259–265.

———. 1906. *Tyrannosaurus*, Upper Cretaceous carnivorous dinosaur (second communication). *Bulletin of the American Museum of Natural History* 22: 281–296.

———. 1912. Crania of *Tyrannosaurus* and *Allosaurus* (*Tyrannosaurus* contribution No. 3). *Memoirs of the American Museum of Natural History* 1: 1–30.

———. 1916. Skeletal adaptations of *Ornitholestes, Struthiomimus, Tyrannosaurus. Bulletin of the American Museum of Natural History* 33: 733–771.

Paul, G. S. 1988. *Predatory Dinosaurs of the World: A Complete Illustrated Guide.* Simon and Schuster, New York.

Russell, D. A. 1970. Tyrannosaurs from the Late Cretaceous of Canada. *National Museum of Natural Sciences, Publications in Paleontology* 1: 1–34.

Sampson, S. D., and Loewen, M. A. 2005. *Tyrannosaurus rex* from the Upper Cretaceous (Maastrichtian) North Horn Formation of Utah: biogeographic and paleoecological implications. *Journal of Vertebrate Paleontology* 25(2): 469–472.

Schweitzer, M. H., Wittmeyer, J. L., and Horner, J. R. 2004. A novel dinosaurian tissue exhibiting unusual preservation. *Journal of Vertebrate Paleontology* 24(Suppl. to 3): 111A.

———. 2005. One pretty amazing *Tyrannosaurus rex*: a presentation celebrating 100 years of *Tyrannosaurus rex.* P. 36 in *100 Years of Tyrannosaurus rex: A Symposium, Abstracts.* Black Hills Museum of Natural History, Hill City, SD.

Smith-Hill, P. 1983. Haystack Butte surrenders terrible lizard: South Dakota ranchers dig dinosaurs. *American West* (March/April): 23–29.

Williamson, T. E., and Carr, T. D. 2005. Latest Cretaceous tyrannosaurs from the San Juan Basin, New Mexico. P. 38 in *100 Years of Tyrannosaurus rex: A Symposium, Abstracts.* Black Hills Museum of Natural History, Hill City, SD.

Appendix. The Skeletal Elements of *Tyrannosaurus rex*

The number of bones in an adult *Tyrannosaurus rex* skeleton is estimated at 300. Of these, 55 are skull bones, which include 41 cranial and 14 mandible elements. The cranial elements consist of paired premaxillae, maxillae, nasals (fused), lacrimals, jugals, postorbitals, squamosals, quadrates, quadratojugals, palatines, pterygoids, ectopterygoids, epipterygoids, the unpaired vomer, and the braincase. The braincase consists of the unpaired basioccipital, supraoccipital, basisphenoid, and parasphenoid, along with the paired parietals (fused), exoccipital-opisthotics, prootics, laterosphenoids, and frontals. The lower jaws consist of paired dentaries, coronoids (also referred to as "supradentaries" by some other authors), splenials, angulars, surangulars, prearticulars, and articulars. Hyoids remain undescribed in tyrannosaurids, although there were probably at least 2 in *Tyrannosaurus rex*. Teeth are not counted because they are shed.

The axial skeleton (minus the skull) contains 114 bones. It consists of 10 cervical vertebrae, 2 proatlas, 18 cervical ribs, 13 dorsal vertebrae, 22 dorsal ribs (the 12th and 13th dorsal vertebrae do not have ribs), 5 sacral vertebrae, 44 caudal vertebrae (estimated, the actual caudal count for *Tyrannosaurus rex* is yet unknown), and 40 to 42 chevrons (estimated). Osborn (1916) estimated that *T. rex* had 56 caudals, Maleev (1974) estimated 40 and 45 in the Asian tyrannosaurid, *Tarbosaurus*, Holtz (2004) believed that tyrannosaurs had from 35 and 44 caudals, and Paul (1988) estimated that there are 39 caudals in *T. rex*. The most complete tail, FMNH PR2081 (Sue), has 36 caudal vertebrae and was restored with 47 caudal vertebrae (Brochu 2003). The number of *T. rex* chevrons is also unknown because this number ultimately relies on the caudal count. Chevrons probably begin with the second caudal vertebra, and they may or may not have extended to the last caudal vertebra. The placement of the first chevron varies in living crocodilians and nonavian dinosaurs, and the chevron count varies as well (Erickson et al. 2005). I consider gastralia as dermal elements and have not included them in this bone count. FMNH PR2081 has a fairly complete set of gastralia of 13 pairs (Brochu 2003). Both TCM 2001.90.1 (Bucky) and MOR 980 (Peck's Rex) also have fairly complete gastralia baskets. Some *T. rex* gastralia are as large as dorsal ribs and are often confused or misidentified as dorsal ribs.

The appendicular skeleton contains 89 bones. The forelimbs have 31 bones consisting of the furcula, 4 shoulder girdle (paired scapulae and paired coracoids), 6 forelimb (paired humeri, paired ulnae, and paired radii), 4 wrist (2 paired carpals), 16 manus bones (6 metacarpals and 10 manus phalanges—i.e., 5 in each hand). On the basis of MOR 980 (Peck's Rex) and BHI 6230 (WYREX), *Tyrannosaurus rex* had 3 metacarpals in each hand, as in *Daspletosaurus* (Russell 1970), yet only 2 functional fingers on each hand (see Lipkin and Carpenter this volume). The hind limbs consist of 58 bones, which include the following: 6 pelvic (paired ilia, pubes, and ischia), 6 leg (paired femora, tibiae, fibulae), 8 ankle (paired calcanea, astragali, and 2-paired tarsals), and 38 bones in the feet (5 pairs of metatarsals and 14 pairs of pes phalanges). The fifth metatarsal was apparently nonfunctional and had no phalanges.

Figure 2.1. *Tyrannosaurus rex* skeletal drawing (by W. D. Matthew) from Osborn (1905), based on a skeleton not fully collected or prepared from Montana.

Figure 2.2. *Tyrannosaurus rex* skeletal drawing (by L. M. Sterling) from Osborn (1906), based on skeletal remains from Wyoming and Montana (AMNH 5886, AMNH 973, and AMNH 5881).

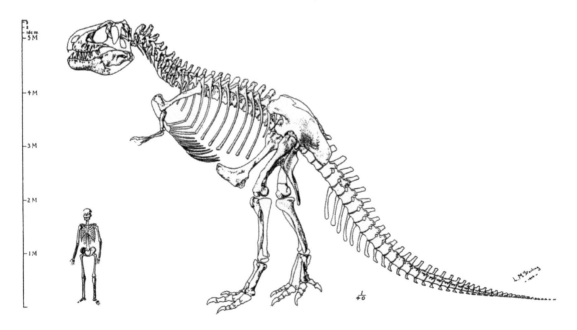

WYOMING'S *DYNAMOSAURUS IMPERIOSUS* AND OTHER EARLY DISCOVERIES OF *TYRANNOSAURUS REX* IN THE ROCKY MOUNTAIN WEST

2

Brent H. Breithaupt, Elizabeth H. Southwell, and Neffra A. Matthews

Introduction

The first remains of tyrannosaurids in North America were discovered in 1855, when the famous western explorer, Ferdinand V. Hayden, found some large teeth in the Cretaceous units of Montana. The following year, Philadelphia paleontologist Joseph Leidy (1856) described these teeth as those of the carnivorous dinosaur *Deinodon horridus* ("horrifying terrible tooth"). Currently, most of these teeth represent the Late Cretaceous theropod *Albertosaurus*.

The discovery of the best-known representative of the tyrannosaurids, *Tyrannosaurus rex*, dates back to 1874, when Colorado schoolteacher Arthur Lakes found a "Fossil Saurian Tooth" (YPM 4192) in the Late Cretaceous Denver Formation near Golden, CO (Carpenter and Young 2002, p. 239). This specimen was sent to Yale paleontologist Othniel C. Marsh, and although it was not described by him, it still resides in the collections at the Yale Peabody Museum in New Haven, CT. From this same site in the Denver Basin, mention is made by geologist George L. Cannon of a large theropod jaw in 1888 (Cannon 1888), but further information and the specimen are unavailable. Over the years, other specimens (later to be identified as *Tyrannosaurus rex*) would be made available to Marsh by his collector, John B. Hatcher, from the uppermost Cretaceous of Wyoming. These and subsequent discoveries are presented below.

Dynamosaurus and other Tyrannosaurus Specimens

In 1890, Hatcher found a partial right metatarsal IV (USNM 2110) in the Lance Formation from Lance Creek, Converse County (now Niobrara County), WY (Marsh 1890). The following year, Hatcher continued to collect in the latest Cretaceous units of eastern Wyoming. Along Alkali Creek, he found a left femur, tibia, and partial fibula (USNM 8064) at one locality, and (with the help of A. E. Sullins) a right ilium (USNM 6183) at another site along this drainage (Marsh 1892). Marsh (1896) identified all of these Wyoming specimens as representing a large form of *Ornithomimus* (*O. grandis*). This "grand bird-mimic" was "one of the most destructive enemies of the herbivorous Ceratopisidea," according to Marsh (1896, p. 206). At the same time that Marsh's collectors were exploring the Rocky

Mountain West, Marsh's archrival, Edward D. Cope of Philadelphia, was also collecting in similar areas. While in South Dakota in 1892, Cope discovered in the Upper Cretaceous Ceratops Beds (see Marsh 1896; Hatcher 1893; Hatcher et al. 1907), two large, weathered, vertebral fragments (AMNH 3982) that he named *Manospondylus gigas*, "giant porous vertebra" (Cope 1892). Although he believed these were from one of the large agathaumid ceratopsians that he had studied previously (Breithaupt 1999), the single specimen currently residing in at the American Museum of Natural History in New York, a dorsal vertebral centrum (AMNH 3982) is from a *Tyrannosaurus rex* (see N. L. Larson this volume).

The lure of the Ceratops Beds of the West would lead to some of the most important discoveries of *Tyrannosaurus rex* in the early 1900s. In 1900, paleontologist Henry F. Osborn sent famed dinosaur hunter Barnum Brown to the Cretaceous Ceratops Beds to find *Triceratops* specimens for display at the American Museum of Natural History in New York. Brown, H. M. Smith, and their cook, Armstrong, explored the similar areas of eastern Converse County (now Niobrara County) in eastern Wyoming, as had Hatcher previously. After getting off the train at Edgemont, SD, they traveled up the Cheyenne River to the junction with Alkali Creek. In the Cretaceous outcrops of this area, they were successful in finding some fragmentary *Triceratops* remains. Continuing up a small tributary of Seven Mile Creek, Brown and crew found in the latest Cretaceous deposits of the area leaf impressions, a fossil turtle, and a *Triceratops* pubis, as well as a disassociated partial skeleton (approximately 13% complete) of a large "*Ceratosaurus*-like" carnivorous dinosaur, according to Brown (see Osborn 1905).

This specimen (AMNH Field 12) consisted of the lower jaws and teeth, various cervical and dorsal vertebrae, ribs, and portions of the hips and limbs (formerly AMNH 5866, now BMNH R7995; see N. L. Larson this volume), as well as numerous dermal plates (BMNH R8001; see Carpenter 2004). Mixed in with this skeleton were the teeth and jaw of a hadrosaur, the frill of a ceratopsian, and the teeth of an ankylosaur, as well as the scales of fish and other undetermined bones, "all evidence of the animals last meal," according to Brown (see Brown and crew opened up a 4 by 12 m (14 by 40 foot) quarry in the soft, fined-grained claystones and siltstones in the Lance Formation, approximately 4 km (2.5 miles) north of Cheyenne River (not 6 miles in Weston County, as noted by Osborn 1905; see Breithaupt et al. 2006).

Osborn (1905) named this partial skeleton *Dynamosaurus imperiosus* ("powerful imperial lizard"), in particular because of the osteoderms (now known to belong to an ankylosaur—see Carpenter 2004). Osborn (1905) also named *Tyrannosaurus rex* ("king of the tyrant lizards") for a partial "*Deinodon*-like" skeleton (formerly AMNH 973, now CM 9380) found in the uppermost Cretaceous Hell Creek Formation of Montana by Brown in 1902 (see N. L. Larson this volume). Osborn was spurred to publish his 1905 paper on Cretaceous carnivorous dinosaurs before all of his material had been fully collected and prepared (11%) from the hard concretionary sandstone matrix because the Carnegie Museum was also preparing a partial (9%) skeleton of a large theropod at the same time). This specimen (CM 1400), which included parts of the skull and lower jaws, various vertebrae, ribs, and hip and limb bones, as well as various ornithischian fragments, was also found in the Lance Formation of Converse County (now Niobrara County, WY) along Schneider Creek by former AMNH employee Olaf A. Peterson in 1902 (McIntosh 1981; see N. L. Larson this volume). Because the Carnegie specimen was thought to include a skull, Osborn was concerned that Peterson would publish on this large carnivorous dinosaur before him (Breithaupt et al. 2006).

Once the Wyoming and Montana material had been prepared (by Richard S. Lull and by Paul Miller and Peter Kaisen, respectively), Osborn (1906) synonymized *Dynamosaurus imperiosus* with *Tyrannosaurus rex*, utilizing both specimens in his descriptions and figures (Fig. 2.2). However, even as late as 1916, Osborn still contended that the dermal scutes of the "*Dynamosaurus imperiosus*" specimen were unlike those of any ornithischian, and he believed that they may have extended down the back and along the sides of *T. rex*, "the most superb carnivorous mechanism among the terrestrial Vertebrata, in which raptorial destructive power and speed are combined" (Osborn 1916, p. 762). Interestingly, close examination of these osteoderms shows a series of depressions that resemble bite marks from a large theropod. In fact, these anklyosaurian scutes may have represented the last meal of the *D. imperiosus* specimen.

Brown continued to work in Montana, and in 1906, he uncovered an even more complete skeleton of *T. rex* in Montana (Osborn 1912), including the first complete skull and most (45%) of the skeleton (AMNH 5027). After the 1908 field season, the American Museum of Natural History had 2 fairly complete skeletons of *Tyrannosaurus rex*, which they planned to mount in a dramatic, interacting pose (see Osborn 1913). However, because of the high cost, only the specimen (AMNH 5027) found in 1906 was exhibited (see Osborn 1916).

By 1912, the museum had obtained a total of 8 specimens of *Tyrannosaurus rex* from Wyoming and Montana. However, many of the best specimens were later sent to other museums. The type specimen (AMNH 973) of *Tyrannosaurus rex* was sold to Pittsburgh in 1941 (CM 9380) and in 2008 was remounted for display. The type skeleton of *Dynamosaurus imperiosus* (AMNH 5866), along with parts of AMNH 973, 5027, and 5881, were sold to the British Museum (Natural History) (BMNH R799) in 1960. The *D. imperiosus* material was used in an interesting half-mount display of this dinosaur, where it exhibited one of the first state-of-the-art poses (i.e., a shorter tail carried in the air) by Barney Newman (1970). In addition, *D. imperiosus* was one of the first *T. rex* specimens to undergo bone histological study (Reid 1984). After being on display for a number of years, most of the composite half-mount skeleton is now in the research collections of the Natural History Museum, although one dentary remains on display.

In 1996, a partial (12%) skeleton was found in Weston County, WY. Although it was touted as being more of the "*Dynamosaurus imperiosus*" specimen (Bonhams and Butterfields 2004) found by Brown, it was actually found many miles north in a different county of eastern Wyoming (see N. L. Larson this volume).

Building on the foundational research done in the late 1800s and early 1900s, continued work in the uppermost Cretaceous formations of the Western Interior has produced many important additional specimens of *Tyrannosaurus rex*. Currently known from dozens of skeletons from the Western Interior (summarized by N. L. Larson this volume), *Tyrannosaurus rex* was clearly one of the most impressive creatures ever to walk our planet. After over a century of research, discoveries during the last 15 years provide interesting insights about this dinosaur, and illustrate how science progresses and how our ideas on prehistoric beasts change through time as new information, technology, and interpretations become available (e.g., Schweitzer et al. 2005; Paul 2000; Hutchinson and Garcia 2002; Horner and Lessem 1993; Farlow at al. 1995; Carpenter 1990; Brochu 2003). Although much has been written about this well-known dinosaur, many questions still remain. Although *Tyrannosaurus rex* itself is extinct, ideas regarding its life and times continue to evolve.

References Cited

Bonhams and Butterfields. 2004. *Catalogue of Natural History Auction*. San Francisco.

Breithaupt, B. H. 1999. The first discoveries of dinosaurs in the American West. P. 59–65 in Gillette, D. D. (ed.). *Vertebrate Paleontology in Utah*. Utah Geological Survey Miscellaneous Publication 99(1).

Breithaupt, B. H., Southwell, E. H., and Matthews, N. A. 2006. *Dynamosaurus imperiosus* and the earliest discoveries of *Tyrannosaurus rex* in Wyoming and the West. P. 257–258, Lucas, S. G., and Sullivan, R. (eds.). *Late Cretaceous Vertebrates from the Western Interior*. New Mexico Museum of Natural History and Science 35.

Brochu, C. A. 2003. Osteology of *Tyrannosaurus rex*: insights from a nearly complete skeleton and high-resolution computed tomographic analysis of the skull. *Journal of Vertebrate Paleontology*, Memoir 7.

Cannon, G. L. 1888. On the Tertiary Dinosauria found in Denver beds. *Colorado Scientific Society Proceedings* 3: 140–147.

Carpenter, K. 1990. Variation in *Tyrannosaurus rex*. P. 141–145 in Carpenter, K., and Currie, P. J. (eds.). *Dinosaur Systematics: Perspectives and Approaches*. Cambridge University Press, Cambridge.

———. 2004. Redescription of *Ankylosaurus magniventris* Brown 1908 (Ankylosauridae) from the Upper Cretaceous of the Western Interior of North America. *Canadian Journal of Earth Sciences* 41: 961–986.

Carpenter, K., and Young, D. B. 2002. Late Cretaceous dinosaurs from the Denver Basin, Colorado. *Rocky Mountain Geology* 37(2): 237–254.

Cope, E. D. 1892. Fourth note on the Dinosauria of the Laramie. *American Naturalist* 26: 756–758.

Farlow, J. O., Smith, M. B., and Robinson, J. M. 1995. Body mass, bone strength indication, and cursorial potential of *Tyrannosaurus rex*. *Journal of Vertebrate Paleontology* 15(4): 713–725.

Hatcher, J. B. 1893. The Ceratops Beds of Converse County, Wyoming. *American Journal of Science* (ser. 3) 145: 135–144.

Hatcher, J. B., Marsh, O. C., and Lull, R. S. 1907. *The Ceratopsia*. U.S. Geological Survey Monographs 49.

Horner, J. R., and Lessem, D. 1993. *The Complete T. rex*. Simon & Schuster, New York.

Hutchinson, J. R., and Garcia, M. 2002. *Tyrannosaurus* was not a fast runner. *Nature* 415: 1018–1021.

Leidy, J. 1856. Notice of the remains of extinct reptiles and fishes, discovered by Dr. F. V. Hayden in the badlands of the Judith River, Nebraska Territory. *Academy of Natural Sciences, Proceedings* 1856: 72–73.

Marsh, O. C. 1890. Description of new dinosaurian reptiles. *American Journal of Science* 39: 83–86.

———. 1892. Notice of new reptiles from the Laramie Formation. *American Journal of Science* 143: 449–453.

———. 1896. The dinosaurs of North America. *Annual Report, United States Geological Survey* 16: 143–244.

McIntosh, J. S. 1981. Annotated catalogue of the dinosaurs (Reptilia, Archosauria) in the collections of Carnegie Museum of Natural History. *Carnegie Museum of Natural History Bulletin* 18: 67.

Newman, B. H. 1970. Stance and gait in the flesh-eating *Tyrannosaurus*. *Biological Journal of the Linnean Society* 2: 119–123.

Osborn, H. F. 1905. *Tyrannosaurus* and other Cretaceous carnivorous dinosaurs. *Bulletin of the American Museum of Natural History* 21: 259–265

———. 1906. *Tyrannosaurus*, Upper Cretaceous carnivorous dinosaur. *Bulletin of the American Museum of Natural History* 22: 281–296.

———. 1912. Crania of *Tyrannosaurus* and *Allosaurus*. *Memoirs of the American Museum of Natural History* (n.s.) 1:1–30.

———. 1913. *Tyrannosaurus*, restoration and model of the skeleton. *Bulletin of the American Museum of Natural History* 32: 91–92.

———. 1916. Skeletal adaptations of *Ornitholestes, Struthiomimus, Tyrannosaurus*. *Bulletin of the American Museum of Natural History* 35:733–771.

Paul, G. 2000. Limb design, function and running performance in ostrich-mimics and tyrannosaurs. P. 257–270 in Pérez-Moreno, B. P., Holtz, T. J., Sanz, J. L., and Moratalla, J. (eds.). *Aspects of Theropod Paleobiology. Gaia: Revista de Geociencias, Museu Nacional de Historia Natural, Lisbon*, 15.

Reid, R. E. H. 1984. The histology of dinosaurian bone, and its possible bearing on dinosaurian physiology. *Zoological Society of London Symposia* 52: 629–663.

Schweitzer, M. H., Wittmeyer, J. L., Horner, J. R., and Toporski, J. K. 2005. Soft-tissue vessels and cellular preservation in *Tyrannosaurus rex*. *Science* 307: 1952–1955.

HOW OLD IS *T. REX*? CHALLENGES WITH THE DATING OF TERRESTRIAL STRATA DEPOSITED DURING THE MAASTRICHTIAN STAGE OF THE CRETACEOUS PERIOD

3

Kirk Johnson

Intense interest in *Tyrannosaurus rex* and other dinosaurs of the Hell Creek, Laramie, Lance, Scollard, Willow Creek, Frenchman, Denver, and Ferris formations (the *Triceratops* fauna) raises the obvious question of the age and duration of these formations. At the present time, this is a surprisingly difficult question to answer. It is useful to divide the question in half: Over what time span did these dinosaur-bearing formations form? And what are the ages of specific dinosaur fossils?

Because several of these formations span the Cretaceous-Paleogene (K-T) boundary, and because no in situ nonavian dinosaurs are known from above the K-T boundary, a definition of the K-T boundary serves as a usable minimum age for the *Triceratops* fauna. The K-T boundary is based on a stratotype in Tunisia and is radiometrically dated at 65.5 ± 0.3 Ma. The K-T boundary can be recognized in terrestrial rocks of the Great Plains and Rocky Mountains by a combination of criteria, including disappearance of the palynological *Wodehouseia spinata* Assemblage Zone, presence of iridium and shocked mineral anomalies, and reversed magnetic polarity (elaborated in Hartman 2002; Hicks et al. 2002; Nichols and Johnson 2002; Johnson et al. 2002). All of these methods require laboratory analysis but can resolve the stratigraphic position of the boundary to a few millimeters.

Age resolution within Maastrichtian rocks is much less precise and involves the use of magnetostratigraphy, radiometric dating, pollen/spore, mammal, and ammonite biostratigraphy. The Maastrichtian Stage (final stage of the Cretaceous) begins at 70.6 ± 0.6 Ma, on the basis of correlation of its stratotype in a quarry in southwest France to the marine strontium isotope curve, and ends at the K-T boundary. Parts of five geomagnetic polarity subchrons occur in the Maastrichtian (part of C29R, C30N, C30R, C31N, and part of C31R). Because C30R is brief (~10 ky), the 5.1 Ma of the Maastrichtian is essentially represented by a normal interval bounded by 2 reversals. Great Plains Maastrichtian stratigraphy is complicated by the Bearpaw marine transgression, which results in a section that is marine at its base and terrestrial at the top and to the west. Because magnetostratigraphy and radiometric dating can be applied to both marine and terrestrial strata, correlation should be possible. However, only four $Ar^{40/39}$ radiometric

Dating the *T. rex* Beds

dates are known from formations that contain the *Triceratops* fauna: 66.56 ± 0.1 Ma from the Kneehills Tuff just below the Scollard Formation in Alberta (45 m below the K-T); 66.8 ± 1.2 Ma from the upper Meeteetse Formation (below the Lance Formation and 200 m below the K-T) in Elk Basin, WY; and 65.73 ± 0.13 and 65.96 ± 0.21 Ma from the Denver Formation (5 and 20 m below the K-T boundary, respectively). Meanwhile, the youngest dated ammonite zones are *Baculites grandis* at 70.15 ± 0.65 Ma and *B. clinolobatus* at 69.57 ± 0.37 Ma, both from Red Bird, WY. Thus, there is a gap of at least 2.77 Ma in the middle Maastrichtian where no radiometric dates are known.

This gap coincides with the mid-Maastrichtian polarity normal (C30N/C31N), so magnetostratigraphy does not help resolve the issue. There are only two palynostratigraphic zones in the Maastrichtian, one in the very early part (*Aquilapollenites striatus* IZ) and one in the very late part (*Wodehouseia spinata* AZ). These palynological zones are only know in superposition in two cores in the Denver Basin. Mammals of the Lancian North American Land Mammal age co-occur with the *Triceratops* fauna, so they share its poor time resolution rather than resolving it. A short-term solution has been to use the duration (from marine cyclostratigraphy) and thickness of polarity subchron C29R to project an age model downsection from the K-T boundary to date Maastrichtian fossils and formations. This method yields an age range of 66.86 to 68.0 Ma for the base of the Hell Creek Formation in North Dakota and Montana, and 68 to 69 Ma for the *Triceratops*-bearing Laramie Formation in the Denver Basin. This type of projection is prone to inaccuracy because the K-T boundary and the top of C29R are separated in several sections by a thickness that is ~10% of the thickness of the Hell Creek Formation. Thus, small errors could be greatly magnified when using two closely spaced points to extrapolate the distance to a much more distant point.

Clearly, more radiometric dates near the base of the *Triceratops* fauna are needed to refine age models for Maastrichtian terrestrial fossils. The search for volcanic tuffs and bentonites near the base of the Hell Creek and in the top of the underlying Fox Hills Sandstone is a priority for future research.

References Cited

Hartman, J. H. 2002. Hell Creek Formation and the early picking of the Cretaceous-Tertiary boundary in the Williston Basin. P. 1–7 in Hartman, J. H., Johnson, K. R., and Nichols, D. J. (eds.). *The Hell Creek Formation and the Cretaceous-Tertiary Boundary in the Northern Great Plains.* Geological Society of America Special Paper 361.

Hicks, J. F., Johnson, K. R., Obradovich, J. D., Tauxe, L., and Clark, D. 2002. Magnetostratigraphy and geochronology of the Hell Creek and basal Fort Union Formations of southwestern North Dakota and a recalibration of the age of the Cretaceous-Tertiary boundary. P. 35–55 in Hartman, J. H., Johnson, K. R., and Nichols, D. J. (eds.). *The Hell Creek Formation and the Cretaceous-Tertiary Boundary in the Northern Great Plains.* Geological Society of America Special Paper 361.

Johnson, K. R., Nichols, D. J., and Hartman, J. H. 2002. Hell Creek Formation: A 2001 synthesis. P. 503–510 in Hartman, J. H., Johnson, K. R., and Nichols, D. J. (eds.). *The Hell Creek Formation and the Cretaceous-Tertiary Boundary in the Northern Great Plains.* Geological Society of America Special Paper 361.

Nichols, D. J. and Johnson, K. R. 2002. Palynology and microstratigraphy of Cretaceous-Tertiary boundary sections in southwestern North Dakota. P. 95–143 in Hartman, J. H., Johnson, K. R., and Nichols, D. J. (eds.). *The Hell Creek Formation and the Cretaceous-Tertiary Boundary in the Northern Great Plains.* Geological Society of America Special Paper 361.

Figure 4.1. *Discovery site for LDP 977-2.*

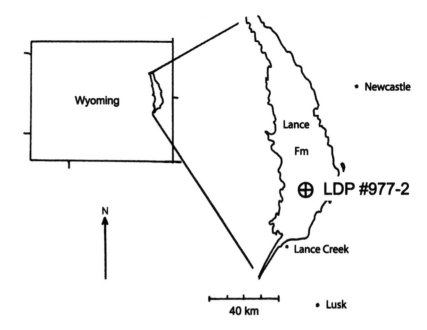

Figure 4.2. *Stratigraphic occurrence of LDP 977-2.*

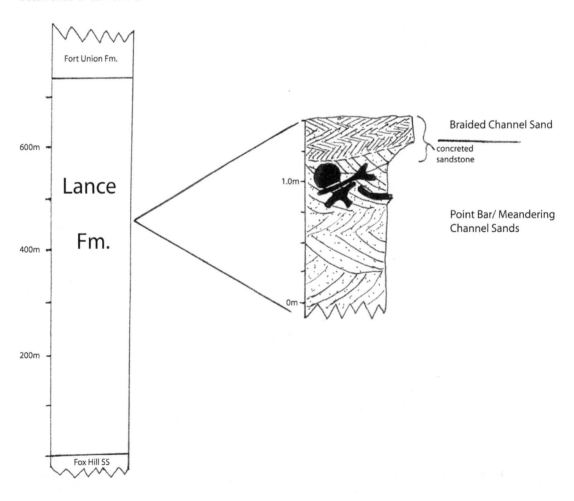

PRELIMINARY ACCOUNT OF THE TYRANNOSAURID PETE FROM THE LANCE FORMATION OF WYOMING

4

Kraig Derstler and John M. Myers

Introduction

During the 1997 expedition of the Lance Dinosaur Project, Rob Patchus discovered a tyrannosaurid skeleton in the Lance Formation of eastern Wyoming. Nicknamed "Pete" to honor both Peter Larson and Rob's late grandfather, the skeleton was excavated in July and August 1997. The next summer, Peter Larson and several volunteers helped K.D. bulldoze the site in an unsuccessful effort to locate additional bones. Presently, the fossil, LDP 977-2, is housed at the University of New Orleans, with some portions on loan to Kansas State University, Manhattan, where J.M. is preparing and studying it as part of his graduate research program.

Occurrence

The specimen was discovered in a small butte on the Swanson Ranch in northern Niobrara County, WY (Fig. 4.1). The butte was removed in the course of the excavation but was a prominent feature of the local terrane; it lay within 200 m of an outlook where contemporary ranchers scan the region for range fires. The precise location is available from K.D.

The type for the Lance Formation in northern Niobrara County and southern Weston County is roughly 720 m (2360 feet) thick (Clemens 1960). The formation dips nearly 1° west. Unfortunately, with such a small angle of dip, it is not possible to directly measure the thickness of the Lance (and we are not aware of any complete cores or useful well logs for this purpose). As a result, it is necessary to use geometric constructions of the stratigraphic column. Stratigraphic positions must then be expressed as percentages above the base or below the top of the formation. When this technique is used, and if we assume 720 m thickness, LDP 977-2 occurs 460 m (63%) above the base of the Lance, or 460 m below the top of the formation (Fig. 4.2).

At the site, approximately 3 m of Lance are exposed. All but the upper 30 cm consist of light brown, friable, medium-grained sandstone, with 30–80 cm crossbed sets and scattered pieces of carbonized wood. LDP 977-2 occurred in the upper 50 cm of this sand. The outcrop was capped by a 30-cm-thick layer of dark brown, well-cemented, medium- to fine-grained sandstone that had 5–10-cm crossbeds arranged as climbing ripples. This sandstone body was elongate north-south and, despite the weathering, thinned noticeably to the east and west. The contact between the underlying light-colored sandstone and the dark one above is obscured by the fact that the dark sand was concreted and the zone of heavy cement extended down into the fossiliferous sandstone. The tops of many of the bones were cemented within this concreted zone.

The bone-bearing sandstone is interpreted as point bar and meandering channel deposits. The hard cap rock is interpreted as a surviving segment of a braided stream deposit formed during the dry season on the much larger meandering channel sands. This is not to say that the rocks necessarily represent a single wet-dry seasonal couplet. One of us (K.D.) has observed similar concreted, braided-stream segments throughout the thicker meandering channel sands of the Lance and contemporaneous Hell Creek deposits of Wyoming, the Dakotas, and Montana. When they weather out, they are frequently misidentified as petrified logs by non-geologists. Occasionally, a bit of the original anastomosing pattern of a braided stream is preserved.

The outcrop contained too little information to determine whether the bones specifically accumulated on a point bar or came to rest within the channel of the meandering stream.

Description

The excavation covered roughly 150 m^2 (Fig. 4.3). The entire set of in situ bones came from the north-central 10 m^2. A huge debris field extended from the exposed edge of the outcrop and continued for at least 100 m into a deep canyon east of the outcrop. Tens of thousands of bones scraps were recovered, but only a small percentage was osteologically identifiable. The bones are slightly permineralized with calcite and traces of pyrite, which is typical of bones from the Lance.

As shown in Figure 4.3, the fossil includes 2 short segments of semi-articulated cervical and dorsal vertebrae containing 5 vertebrae each. An 11th isolated vertebra was also recovered. The outcrop also produced at least 1 cervical rib, 10 dorsal ribs, and 4 gastralia. In the field, crew mem-

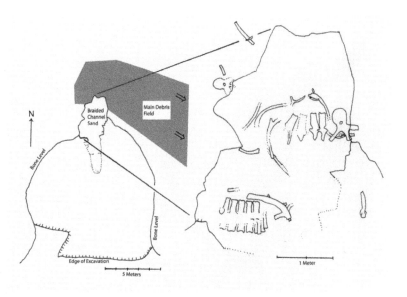

Figure 4.3. *Bone map of LDP 977-2. Only the upper portion of the debris field is shown.*

Figure 4.4. Distribution of bones found with Pete (LDP 977-2). Drawing modified from Derstler (1994), scaled to an animal with a 31-foot length; 180-cm-tall human for scale.

bers identified fragments of numerous dorsal ribs and gastralia, at least 2 proximal caudal vertebrae, 1 distal caudal vertebra, dorsal/cervical vertebrae, the shaft of a hind limb long bone, and several heavy pieces of the pectoral girdles. No skull elements or teeth were identified. The distribution of identified bones is shown in Figure 4.4. Interestingly, all of the ribs appear to have their ventral ends broken before final burial.

The dorsoventral height of the last cervical is 72 cm, whereas the anteroposterior thickness of the second dorsal centrum is 11 cm. These measurements are three-fourths the size of those corresponding to FMNH PR2081 (Sue). Thus, Pete was probably an animal 9.4 m (31 feet) long.

Until LDP 977-3 is prepared, the fossil will be difficult to identify beyond noting that it is a large tyrannosaurid. The bones differ in no significant way from those of definitive specimens of *Tyrannosaurus rex* (e.g., CM 9380, MOR 555, and MOR 980, BHI 3033). As such, the specimen is tentatively identified as *T. rex*.

Historical Consideration

LDP 977-3 was found in a region of the Lance Creek area known to have been examined by several early paleontology expeditions. Hatcher collected in the region for O. C. Marsh in the late 1800s and for the Carnegie Museum in the early 1900s. Although he did not keep notes on the areas he prospected during the initial explorations of the Lance dinosaur fields, it is possible to reconstruct this information for each year (1888–1892) by noting the year of collection for each of the excavation sites he placed on his hand-drawn map (Hatcher et al. 1907). Although these maps cannot be completely reconciled with more recent maps, it is possible to approximately locate LDP 977-2 on Hatcher's map (Fig. 4.5). From this, Hatcher

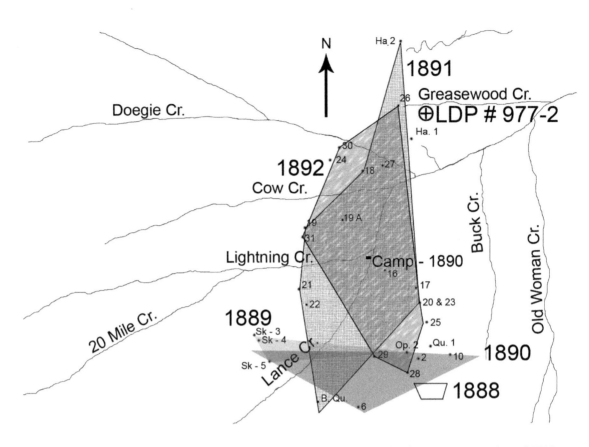

Figure 4.5. Hatcher's Lance fieldwork for Yale Peabody Museum, reconstructed from museum records by year. Base map modified from an unpublished hand-drawn map in Hatcher's handwriting found in Yale Peabody Museum files. Shaded polygons enclose sites visited by Hatcher and his crew during each of the years 1888 through 1892.

nearly reached LDP 977-2 during each of his last 2 seasons, 1891 and 1892, but passed slightly west.

In 1895, Samuel Williston lead an expedition into the Lance. He briefly camped at Hatcher's 1890 campsite and collecting 2 miles further north (Clemens 1963). Williston reportedly traveled much farther north while prospecting, but it would be an exaggeration to claim that Williston barely missed LDP 977-2. Hatcher returned to the Lance in 1900 to collect for the Carnegie Museum of Natural History (Clemens 1963). He limited his work to Marcus Draw, as shown in Figure 4.6. This time, he barely missed LDP 977-2 from the southwest.

In 1900, Barnum Brown, who was on the Williston trip, also ventured into the Lance Formation, collecting for the American Museum of Natural History (Clemens 1963). He prospected along Alkali and Seven Miles Creeks. Although there is no evidence that he got closer than 15 km (9 miles) to the northeast of LDP 977-2, it is worth noting that Brown did collect the type specimen for *Dynamosaurus imperiosus* along Seven Miles Creek at this time (see N. L. Larson this volume).

By piecing together information from museum labels and sketchy published accounts (e.g., Sternberg 1990, 1991), it is possible to reconstruct the area prospected by the Sternberg family during the 3 years that they worked in the Lance, 1908 through 1910. This information is presented in Figure 4.7. They worked further north and west. They too almost ventured into the vicinity of LDP 977-2.

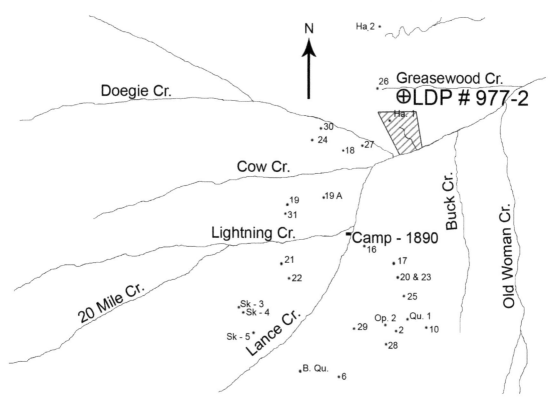

Figure 4.6. *Hatcher's Lance fieldwork for Carnegie Museum of Natural History in 1900. Map as in Figure 4.5.*

Figure 4.7. *Sternberg family fieldwork, combining the years 1908 through 1910. Base map modified from BLM Lance Creek Surface Management Map, 1:100,000 scale, 1991 edition.*

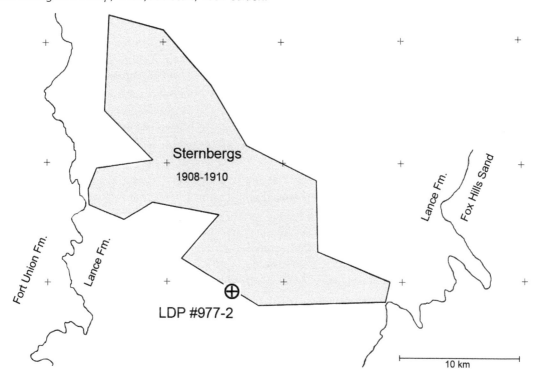

Figure 4.8. *Photograph of LDP 977-1 outcrop and the western skyline before excavation in July 1997.*

Although other paleontologists passed near LDP 977-2 between 1910 and 1997, we are only aware of one other near miss. In the late 1940s, a field party from the University of Wyoming camped roughly half a mile to the north-northeast while they excavated an adult *Edmontosaurus annectens* on another nearby ranch (Leonard Zerbst, personal communication to K.D. 1995). There is no record of any prospecting by this crew.

Naturally there are various sources of uncertainty in the places reached by each expedition. Within these bounds, it appears that Hatcher, Williston, Brown, and the Sternbergs approached close to the location of LDP 977-2 without actually reaching it. It seems likely that any fossil prospector in the area would have searched the precise spot where LDP 977-2 lay. As mentioned above, the discovery butte was a prominent part of the local landscape (Fig. 4.8) and it lay within a few hundred meters of the best lookout in the area. It is an easy spot to reach, so the site would have been irresistible to anyone searching for dinosaurs.

We do not have any direct information on the amount of time that LDP 977-2 was exposed to erosion. However, the size of the debris field suggests that the skeleton originally contained many intact bones and that it weathered for a considerable period of time. Perhaps more importantly, comparison between photographs of nearby Lance outcrops taken by Levi Sternberg in 1908 (Sternberg 1990) and the modern terrane show that hard sandstone caprocks and the softer sands immediately underlying have changed little in 95 years. This strongly suggests that the similar caprock and softer sands associated with LDP 977-2 have existed for well over a century. In short, the skeleton was almost certainly exposed when Hatcher, Brown, and the Sternbergs explored the Lance.

We find it fascinating to imagine how the history of *Tyrannosaurus rex* research might have differed if Hatcher had provided Marsh (or his successors) with a good skeleton in the early 1890s, or if he had done the same for Carnegie's museum in 1900 (years before they purchased one from AMNH), or if the Sternbergs had been able to sell another skeleton to Osborn in 1908.

Acknowledgments

We thank Rob Patchus, who discovered the fossil, and the other members of the 1997 Lance Dinosaur Expedition: Ron Francis, Marcus Eriksen, and Joey Masters. We also thank Stanley and Gloria Swanson, Roy Rassbach, and the town of Newcastle, WY. We gratefully acknowledge the assistance of Peter Larson and his volunteers, who helped bulldoze the overburden at the site in 1998. Finally, Larry and Joyce Tuss, Winifred, MT, graciously hosted us while we prepared this chapter in the aftermath of Hurricane Katrina.

References Cited

Clemens, W. A. 1960. Stratigraphy of the type Lance Formation. *Report of the XXI International Geological Congress (Norden), Part V (Proceedings of Section 3—The Cretaceous-Tertiary Boundary)*: 7–13.

———. 1963. Fossil mammals of the type Lance Formation, Wyoming. Part I. Introduction and multituberculata. *University of California Publications in Geological Sciences* 48: 1–105.

Derstler, K. 1994. Dinosaurs of the Lance Formation in eastern Wyoming. P. 127–146 in Nelson, G. E. (ed.). *Forty-fourth Annual Field Conference Guidebook*. Wyoming Geological Association, Casper.

Hatcher, J. B., Marsh, O. C., and Lull, R. S. 1907. *The Ceratopsia*. United States Geological Survey Monograph 49.

Sternberg, C. H. 1990. *The Life of a Fossil Hunter*. 1908. Indiana University Press, Indianapolis.

———. 1991. *Hunting Dinosaurs in the Badlands of the Red Deer River, Alberta, Canada*. 1932. NeWest Press, Edmonton, Alberta.

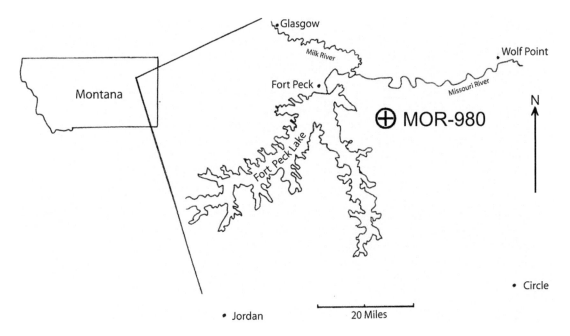

Figure 5.1. *Location map for MOR 980 (Peck's Rex).*

Figure 5.2. *Stratigraphic section. (Left) Measured section (information from Rigby et al. 2001). (Right) Measured section within the MOR 980 quarry.*

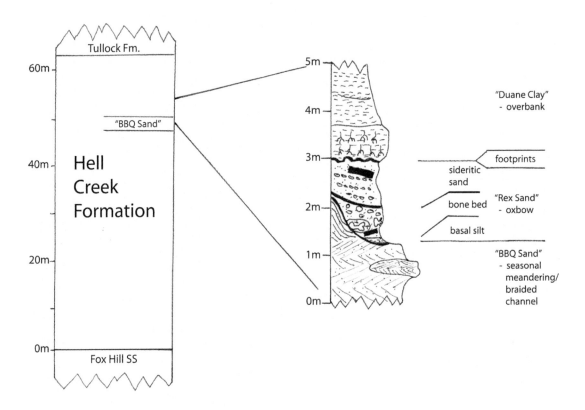

TAPHONOMY OF THE *TYRANNOSAURUS REX* PECK'S REX FROM THE HELL CREEK FORMATION OF MONTANA

5

Kraig Derstler and John M. Myers

Introduction

During summer 1997, while excavating a hadrosaur skeleton from the Hell Creek Formation in McCone County, MT, a field crew led by Keith Rigby discovered the disarticulated remains of a large tyrannosaurid. Occurring near the eastern shore of Lake Fort Peck, this theropod quickly became known as Peck's Rex. (The name is trademarked by Fort Peck Paleontology Inc., Fort Peck, MT.) That autumn, a local resident looted part of the skeleton. Within a few weeks, federal agents recovered the looted material and conducted a salvage excavation at the site. Resultant hostilities hampered subsequent piecemeal excavation by Rigby and associates. It was not until 2004 that we, aided by a field crew, essentially completed the excavation.

The skeleton is cataloged by the Museum of the Rockies in Bozeman, MT, as MOR 980, and it is on loan to Fort Peck Paleontology Inc. (FPPI), Fort Peck, MT. Major portions of the skeleton have been prepared by staff and volunteers of FPPI. Damaged portions were restored and the missing elements reconstructed by FPPI in 2003–2004. Substantial portions of the fossil remain unprepared, including vertebrae embedded within concretions and most of the bones collected in 2004. Nevertheless, as of June 2004, casts of Peck's Rex had been mounted for exhibit in Fort Peck, MT, and Baltimore, MD.

The humeri were later prepared and studied by us at the University of New Orleans. Other portions of the skeleton are being studied in different laboratories, and some observations are presented elsewhere in this volume. However, the purpose of this chapter is to consider the events between death and final burial for MOR 980.

Occurrence

The quarry for MOR 980 is located on federal land administered by the U.S. Army Corp of Engineers, approximately 32 km (20 miles) southeast of the Fort Peck Dam in McCone County, MT (Fig. 5.1). It lay approximately 1 mile west of State Route 24. More specific locality information is available from us and the staff at FPPI.

The skeleton was collected from the upper part of the upper Maastrichtian Hell Creek Formation, 6 m below the top (Fig. 5.2, left). The quarry lay immediately above the informally named BBQ Sand (Rigby et al. 2001). Details of the stratigraphy, including a composite measured section and a photograph of the locality, can be found in this reference.

In the measured section presented here (Fig. 5.2 right), Peck's Rex lay near the base of the local Rex Sand, a 0.1–1.8-m-thick sand and silt unit that is particularly rich in sideritic nodules. In turn, it is overlain by the Duane Clay. The base of the Rex Sand is erosional, whereas the upper contact is conformable but heavily bioturbated by large (ceratopsian?) footprints.

Preservation

In general, the bones are permineralized with calcite and minor amounts of pyrite. A few bones are permineralized with siderite instead of calcite. Between one-third and one-half of the bones were embedded in large (0.3–2.0 m) silty limestone concretions.

MOR 980 includes well over 85% of the skeleton, mostly disarticulated. However, a string of 11 distal caudal vertebrae lay in articulation near the eastern edge of the excavation. All of the teeth are rooted, indicating that they fell out of the dentaries, maxillae, and premaxillae as the carcass disintegrated. The fossil includes a number of seldom-seen elements—for example, a third metacarpal, the furcula, and both humeri (see Lipkin and Carpenter this volume). On the other hand, the right leg is almost entirely missing. Virtually all metatarsals are also missing from the collection, although two more were observed in situ (N. Murphy, personal communications to K.D. 2003 and 2004).

None of the bones are distorted by compaction, although many show signs of postmortem rotting and preburial breakage. Pieces of individual bones are scattered, and the breaks are not crisp. These observations indicate that at least some of the bones were softened, presumably by preburial rotting, then fell to the lake bottom as individual pieces were shed from the carcass. Modest movement of the carcass within the lake, rather than water flow, may explain the scattering.

Excavation Maps

The quarry and bone maps generated during 2004 by K.D. are shown in Figures 5.3 and 5.4. Individual bones are keyed to inventory lists on files with K.D. and FPPI. As excavation proceeded, it became apparent that the Rex Sand was deposited on a high-relief surface. Several dozen elevations were measured to the nearest centimeter on this surface using an arbitrary base elevation. A contour map of this surface is shown as Figure 5.5 using 10-cm contour intervals. Although information is not available for excavations before 2004, it is apparent that MOR 980 was concentrated in the deepest part of the oxbow.

Bone maps for pre-2004 excavations are sketchy at best. For example, we are not aware of any maps that were produced during the 1997 looting or the follow-up salvage excavation. In 2002, Bill Wagner reconstructed a bone map for all of the areas excavated up to that time using old photographs, sketch maps, preparation maps for several large bone-bearing concretions, and personal recollections. He provided K.D. with this map,

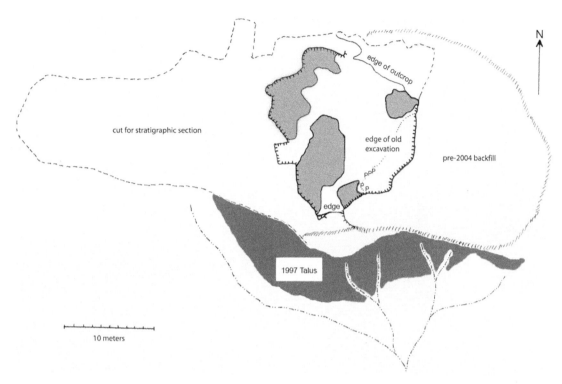

Figure 5.3. *The 2004 quarry map. Lighter gray areas are portions of the exposed Rex Sand that were left in place. P indicates old plaster on the side wall of previous excavation. Edge of outcrop and edge of old excavation are mapped only where observed. The 1997 talus was identified by Duane Sibley and others present during that year's excavation. The area includes debris produced by all 3 excavations that year.*

Figure 5.4. *Bone map for 2004 excavation of MOR 980. Unexcavated portions of the Rex Sand are less than 5 cm thick and judged unlikely to contain additional bones. C indicates bone-bearing limestone concretions; +, survey grid points.*

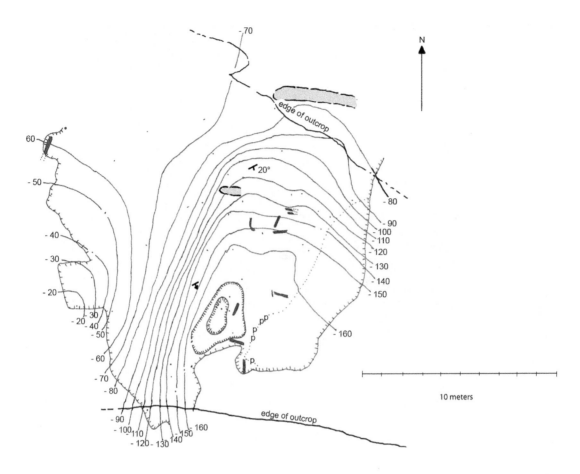

Figure 5.5. Contour map of the base of the Rex Sand. Small dots represent elevation stations. Contour interval is 10 cm. Carbonized logs within the Rex Sand are shown in solid black. Braided channel sands within the underlying BBQ Sand are shown as gray areas. Orientation and flow direction of asymmetrical ripples in a basal layer of the Rex Sand are shown by bearing symbol attached to a directional arrow. Other information taken from Figures 5.3 and 5.4.

copies of his original data, and another field map produced during the 2003 season. We combined all of this information onto a single, pre-2004 bone map, but unfortunately, we could not correlate this with the 2004 map. This correlation was made possible by uncovering about 9 m of the old excavation to use as a datum for linking the old and new maps. Plaster splatters on the exhumed excavations also helped align the two. The resulting composite map is shown in Figure 5.6.

Fossils from the Rex Sand

The only prominent vertebrate fossils in the Rex Sand are Peck's Rex and a partially articulated hadrosaur. The latter apparently lay several meters southwest of MOR 980 on a higher part of the same depositional surface. We are not aware, however, of any maps or other written records for the condition of this hadrosaur or its precise location. A few bones and tendons from the hadrosaur washed down into the deeper area and were mixed in with Peck's Rex. Understandably, none of the *Tyrannosaurus* bones washed upward to commingle with the hadrosaur.

Only rare scraps of other animals were recovered from the Rex Sand. They include a few internal molds of nondescript snails, several shed *Leidosuchus* teeth, a large shed tooth from a large dromaeosaurid, some gar and other fish debris, one baenid turtle fragment, and rounded fragments of

Triceratops bones and teeth. Unusually, the Rex Sand contained no soft-shell turtle remains. Although animal fossils are sparse, plant remains are abundant and often well preserved. They include carbonized wood, seeds, amber droplets, and foliage representing a diverse assemblage of dicots, palms, and conifers. The Rex Sand contains no obvious trace fossils other than the dinosaur tracks at the contact with the overlying Duane Clay (see below).

The Rex Sand is interpreted as an oxbow lake filling. It occupies a channel that was cut into the underlying BBQ Sand, and it grades upward into overbank silty claystone (dubbed the Duane Clay). The upper surface of the Rex Sand is planar and subhorizonal. The thickness of the Rex Sand corresponds to the depth of the underlying depositional surface. Internally, the Rex Sand can be divided into 3 subunits. The lowest is a several-centimeter-thick series of thin-bedded, yellow-green, silty, fine-grained sandstones. Each layer contains asymmetrical ripples that indicate current flowing toward the south-southeast. The second unit contains the bones of Peck's Rex. It is a 5–70-cm-thick, orange-brown, silty sand, with small shale chips, many pieces of carbonized wood, other plant debris, small limestone nodules, and occasional pyritic or sideritic concretions. Some

Depositional Model

Figure 5.6. *Composite bone map. Pre-2004 maps and information compiled by Bill Wagner. Dashed figures on old map are large, bone-bearing concretions. Note the articulated string of caudal vertebrae near the eastern edge of the map.*

of the sideritic concretions have a limestone core. The unit is massive but poorly cemented. Occasional imbricated pebbles indicate current flow toward the south-southeast. However, the lack of obvious bone orientation suggests that the currents were weak. The uppermost subunit is a 60–100-cm-thick, medium-bedded orange sand, with multiple layers of sideritic concretions and occasional large pieces of carbonized wood. It contains few other fossils and no internal sedimentary structures. In general, it is better cemented than the bone-bearing layer.

The underlying BBQ Sand is a meandering channel-and-point-bar unit deposited by a sizable meandering stream. Reflecting wet-dry seasonality in the region, the BBQ Sand contains a few braided channel sands deposited during the dry season. These braided channels, their internal crossbeds, and crossbeds within the meandering portions of the BBQ Sand all indicate water flowing to the east (Fig. 5.5).

In contrast, the Rex Sand has consistent indicators showing that the oxbow filled with sediment (and water) moving to the south-southeast. Most reasonably, this infilling is not related to the deposition of the underlying BBQ Sand. Instead, it may be tied to the same stream that cut and then abandoned the oxbow channel.

Despite the presence of flow indicators within the Rex Sand, several pieces of evidence indicate that the oxbow was generally stagnant. The lack of aquatic fauna, particularly soft-shell turtles, suggests hostile aquatic conditions. Preservation of plant material (particularly abundant, diverse, well-preserved foliage in the base of the bone-bearing part of the Rex Sand) indicates that anoxia prevented the leaves from decaying. Finally, the siderite found throughout the Rex Sand suggests that the bottom of the lake was chemically reducing, again consistent with the anoxia and stagnant water hypotheses. Presumably, oxygenated water and sediment reached the oxbow episodically, during floods. Between floods, the oxbow waters were inhospitable to life. Peck's Rex entered the fossil record through this setting.

Taphonomy

There is presently no meaningful information on the cause of death for MOR 980 or how it entered the Rex Sand oxbow. However, it seems reasonable to suggest that it entered as a relatively intact carcass. The high degree of skeletal completeness, the articulated caudals, and the presence of the bones that usually disappear early in postmortem history (manus, distal caudals, gastralia) all support this hypothesis. The prepared portions of the skeleton do not show any solid evidence of postmortem scavenging marks. This is consistent with the nearly complete absence of shed teeth of potential scavengers in the Rex Sand.

The bones are concentrated in the deepest portion of the oxbow where water was at least 1.5 m deep (based on the thickness of sediments). The lack of bones on the oxbow shelf to the west indicates that this area was too shallow for the carcass. Therefore, the skeleton arrived as a bloated carcass in the deepest part of the oxbow and was unable to move into shallow areas, or it was actually grounded there. In either case, the bones dropped from the carcass to the lake bottom as they rotted free.

Many of the bones are broken, with the pieces found meters apart. Such broken edges are never sharp, and they usually have matrix injected into the trabeculae. Such bones had to be soften before breakage so that they simply fell apart. In short, many of the bones were rotted before they fell from the carcass. Consistent with this hypothesis are hundreds of theropod bone shards scattered throughout the middle portion of the Rex Sand. These represent bones that rotted to the point where they simply disintegrated. The missing 15% of the skeleton cannot be explained as loss due to modern weathering because none of the skeleton was exposed at the time of discovery. Rather, the specimen was accidentally discovered in the course of excavation of a hadrosaur skeleton.

There are few alternatives to explain the missing portions. Perhaps a few smaller bones remain in the thin, unexcavated areas of the Rex Sand. Possibly a few are hidden within the unprepared concretions, or a few smaller bones may have rotted completely before burial. Finally, some of the bones could have been destroyed or otherwise lost during looting and the salvage excavation. Whatever the explanation, we are surprised that MOR 980 is not more complete because there seems to be no obvious way to dispose of the missing bones.

Acknowledgments

We thank the staff and volunteers at FPPI and their counterparts at the U.S. Army Corps of Engineers, particularly Duane Sibley, Lou Trembley, and Bill Wagner. We also acknowledge the exceptional field assistance of Douglas and Arch Van Belle, Dory Turnipseed, Teresa Logudice, Steven Luton, Dana Hensley, and Ruth Ebert. Bob Richter arranged for and delivered a great deal of field equipment. Nat Murphy and Bill Wagner provided information on the occurrence and history of MOR 980. Peter Larson was an endless source of information and enthusiasm concerning *Tyrannosaurus rex*. Finally, Larry and Joyce Tuss, Winifred, MT, hosted us while preparing this chapter in the extended aftermath of Hurricane Katrina.

Reference Cited

Rigby, J. K., Linford, C. B., and Rigby, J. K., Jr. 2001. Geology of the McRae Springs Quadrangle, McCone County, northeastern Montana. *Geology Studies, Brigham Young University* 46: 75–91.

Figure 6.1. *BMR P2002.4.1 (Jane)* on display at the Burpee Museum of Natural History, Rockford, IL.

TAPHONOMY AND ENVIRONMENT OF DEPOSITION OF A JUVENILE TYRANNOSAURID SKELETON FROM THE HELL CREEK FORMATION (LATEST MAASTRICHTIAN) OF SOUTHEASTERN MONTANA

6

Michael D. Henderson and William H. Harrison

Extensive outcrops of the Hell Creek Formation (uppermost Maastrichtian) occur in eastern Montana. The formation is widely regarded as having been deposited in a lowland fluviolacustrine system; these sections consist of a stacked series of fining-upward sedimentary sequences formed by channels meandering across a lowland floodplain (Kirk Johnson, personal communication). The soft texture of the Hell Creek strata, combined with the sporadic rainfall in the northern Great Plains, has resulted in the development of extensive badlands throughout its outcrop area. These badlands contain an abundant and diverse fossil flora and fauna, including dinosaurs.

In 2001, an expedition from the Burpee Museum of Natural History, with permission from the Bureau of Land Management (BLM) to survey for vertebrate fossils, discovered several foot and lower limb elements of a theropod dinosaur weathering from an exposure of the Hell Creek Formation in Carter County, MT. After initial evaluation, the site was winterized and an application made to the BLM to open a quarry the following year. In the summer of 2002, a field crew returned to excavate the specimen. During the course of the excavation, a sizable quarry was created, which yielded major portions of the skeleton of a tyrannosaur (BMR P2002.4.1), nicknamed Jane, approximately 7 m in length (Fig. 6.1). In addition to Jane, many associated plant, invertebrate, and vertebrate fossils were recovered from the quarry.

Lack of fusion of the neural arches to their respective centra in the vertebral column indicates that Jane is a juvenile animal. Examination of histological sections from a rib, fibula, and metatarsal indicates that at death, Jane was 11 years old and still in a phase of rapid growth (G. Erickson, personal communication 2003). Recovered skull bones and teeth of Jane (BMR P2002.4.1) bear a close resemblance to those of CMNH 7541, a controversial tyrannosaurid skull that has been interpreted as belonging to either a juvenile *Tyrannosaurus rex* (Carr 1999; Carr and Williamson 2004) or a separate taxon, *Nanotyrannus lancensis* (Bakker et al. 1988;

Introduction

Currie 2003a, 20003b; Currie et al. 2003). Research is currently ongoing into the systematic position of both specimens.

In spite of many taxa of dinosaurs known from the Hell Creek Formation, most finds consist of isolated bones (Pearson et al. 2002). The discovery of a substantial portion of a juvenile tyrannosaurid skeleton in the formation is so unusual that the circumstances of its burial and preservation merit careful analysis so that like environments can be explored.

Stratigraphy

Extensive exposures of the Hell Creek Formation occur in southeastern Montana. Widely regarded as a prograding, clastic wedge associated with the retreat of the Western Interior Sea, the Hell Creek primarily consists of poorly cemented channel and crevasse splay sandstones, overbank mudstones and siltstones, paleosols, carbonaceous claystones, and thin and sparse lignite beds deposited during the last years of the Cretaceous Period (Murphy et al. 2002; Johnson this volume).

Near its type section in Garfield County, north-central Montana, the Hell Creek Formation is 170 m thick (Johnson et al. 2002). Its thickness in southeastern Montana is estimated to be 150 m; however, no complete sections of the formation are exposed. The nearest complete sections are in southwestern North Dakota (about 100 km from the quarry), where the formation is approximately 100 m thick (Murphy et al. 2002). Unfortunately, no identifiable marker beds occur in the Hell Creek, with the exception of the top and bottom contacts. Within the formation, there is little lateral continuity of beds and bentonitic surface weathering obscures

Figure 6.2. Jane Quarry in the Hell Creek Formation (latest Maastrichtian) in northwestern Carter County, MT.

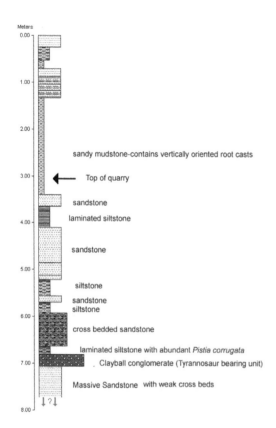

Figure 6.3. *Stratigraphic section of the Hell Creek Formation exposed in the Jane Quarry, Carter County, MT.*

bedding (Johnson et al. 2002). Consequently, the exact stratigraphic placement of Jane within the formation is complicated. Pollen and plant megafossils from the Jane Quarry correlate with a stratigraphic level in southwestern North Dakota that is 28 to 35 m below the top of the formation.

The Jane Quarry is located on the northern side of an elongate east-west-trending ridge in northwestern Carter County, MT (Fig. 6.2); the exact locality is available from us. The tyrannosaur was discovered weathering out near the base of the butte. The Jane Quarry section exposes a fining-upward sequence of clastic sediments approximately 8 m thick (Fig. 6.3).

At the base of the section is massive, poorly cemented, dirty, tan-brown, crossbedded sandstone. Its total thickness is unknown because it is incompletely exposed. Abundant wood and coniferous and deciduous leaf impressions are present on or near the upper surface of the sandstone. Deciduous leaves recovered from this unit are identified as belonging to *Dryophyllum subfalcatum* and "*Vitis*" *stantoni*. The only vertebrate fossil encountered in the unit was a single midseries cervical vertebrae of a large azhdarchid pterosaur (Henderson and Peterson in press).

A clay-ball conglomerate composed of poorly sorted sand, silt, and rounded greenish-colored clay clasts overlies the sandstone (Fig. 6.4). This unit is lenticular, showing rapid lateral variation in thickness (12–40 cm). The tyrannosaur skeleton was discovered in the lower part of this conglomerate, at its contact with the underlying sandstone. Diagenetically pro-

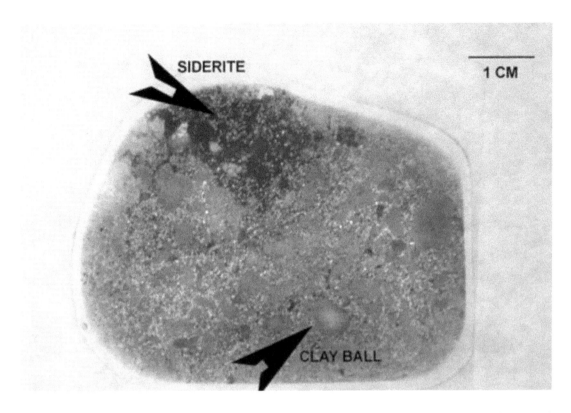

Figure 6.4. Thin section of the clay-ball conglomerate that contained the juvenile tyrannosaur Jane (BMR P2002.4.1). Arrows indicate siderite and a clay ball. Scale bar = 1 cm.

duced siderite nodules occur within the conglomerate and partly encased several bones of BMR P2002.4.1. Plant fossils recovered from this unit include wood and bark impressions, conifer cones and needles, and numerous small, round to oval seeds, preserved as internal casts. In addition, an abundant, diverse, and well-preserved palynoflora occurs in siderite nodules and clay balls within the conglomerate. To date, 51 genera of pollen, spores, and cysts have been recovered. These indicate the presence of a diverse flora of flowering plants, conifers, ferns, cycads, and palms. Pollen of *Gunnera*, a herbaceous plant, is especially common. Recovered pollen and spores are typical of the *Aquilapollenites* palynofloral province, which is found in Upper Cretaceous rocks from western North America westward into northeastern China. Invertebrates are represented by poorly preserved internal casts of unionid bivalves (2 species) and high-spired gastropods (1 species). Remains of vertebrates (dinosaurs, lizards, freshwater fish, crocodilians, champsosaurs, turtles) are common and represented by disarticulated skeletal elements randomly distributed within the unit. Preservation of these bones and teeth range from pristine to significantly worn. A 2- to 4-cm-thick layer of siderite caps the conglomerate.

Above the siderite cap is a siltstone. The basal 30 to 35 cm of the siltstone is finely laminated and contains extremely abundant fossils of aquatic monocots, principally a kind of water lettuce, *Pistia corrugata*, many preserved as whole plants (Fig. 6.5). A second 2- to 4-cm stratum of siderite caps the *Pistia*-bearing layers. Higher in the unit, sporadic sandstone lenses (up to 1.5 m thick) occur. Crossbedding within the sand units indicates water flow from

south to north. The top 0.5 m of the quarry is sandy siltstone, which is not laminated, and contains vertically oriented root casts.

Environment of Deposition

The Jane Quarry section appears to be a typical floodplain sequence. We interpret the massive, crossbedded sandstone on which the tyrannosaur lay as a point bar sand. The clay-ball conglomerate that contained the juvenile tyrannosaur records a mudflow from a flood event, or a bank collapse. Above Jane, the laminated siltstone containing reeds and *Pistia* indicate a stream avulsion and the subsequent development of an oxbow lake on the site, while sandstone lenses higher in the section are thought to have been produced by underwater dunes migrating through the lake during times of high water when the abandoned channel was temporarily reconnected to the river system. Vertically oriented root casts present at the top of the quarry indicate that the lake eventually silted up and vegetation was established.

Fossils of plants and animals associated with Jane (BMR P2002.4.1) correspond closely with those collected in association with a *Tyrannosaurus rex* (Peck's Rex) from the upper Hell Creek Formation of McCone County, MT (Derstler and Myers this volume), and another *T. rex* (known as Scotty) from the contemporaneous Frenchman Formation of southwestern Saskatchewan, Canada (Tokaryk and Bryant 2004). The environment of deposition of all these specimens indicates burial took place on a warm, wet, lowland floodplain.

Figure 6.5. Leaves of *Pistia corrugata* from the laminated siltstones above the tyrannosaur-bearing unit.

Taphonomy

All tyrannosaur skeletal elements recovered from the Jane Quarry are consistent with derivation from a single individual. The 145 bones, representing approximately 52% of the skeleton, were collected from a 4 m² area (Fig. 6.6).

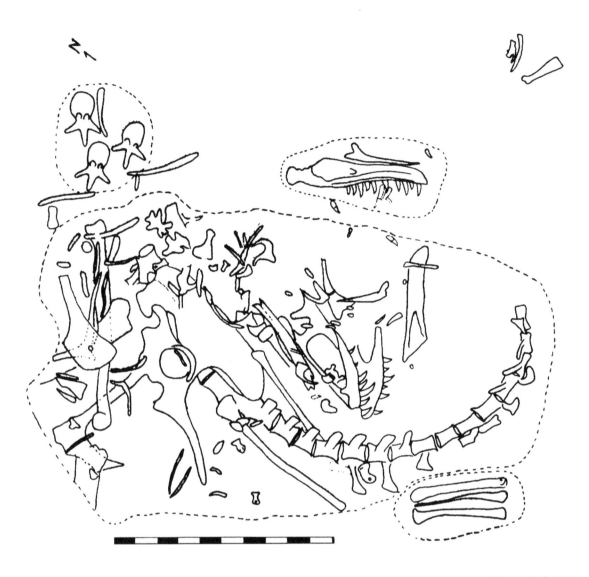

Figure 6.6. *Quarry map showing distribution of skeletal elements of BMR P2002.4.1 (Jane). Dotted lines indicate field jackets. Scale bar = 1 m.*

The tyrannosaur lay on its right side when buried. Portions of the right foot remained in articulation. Skull bones were disarticulated but concentrated in a limited area over the hips. A segment of 16 proximal caudal vertebrae, with their associated hemal arches, was found arcing over the back. This distribution of bones indicates that after death, shrinkage of muscles and ligaments along the vertebral column contorted the carcass into the classic dinosaur-avian death pose. Loose teeth from the right dentary were found north of it in a pattern consistent with movement by water or mud from south to north. Many of Jane's ribs and presacral vertebrae were scattered or missing. This could be the result of movement by water or mud, bloating then bursting of the carcass, or scavenging of the carcass. The right humerus was found about a meter from the main bone concentration, upstream from inferred paleocurrent direction and in direct contact with a shed tyrannosaur tooth. The context suggests scavenging, but the completeness of the skeleton and concentration of bones indicate scavenging was not extensive. No tooth marks were observed on preserved bones.

The bone preservation of Jane is excellent, with no signs of postmortem weathering. This indicates that the skeleton was not exposed on the point bar for an extended period of time. Rapid burial of Jane's skeleton by a mudflow appears to be the key event responsible for its completeness. The finely laminated siltstones of the oxbow lake deposited on top of the remains provided further protection from disturbance.

Summary

The sequence of events leading to burial may be summarized as follows:

1. A juvenile tyrannosaur died. At the time of death, or shortly thereafter, its body came to rest on the point bar of a channel on a forested, lowland floodplain.
2. Shrinking muscles and ligaments contorted the tyrannosaur's corpse into the classic dinosaur-avian death pose concurrent with, or soon after, minor scavenging of the corpse occurred.
3. As decay proceeded, disarticulation of the skeleton reached an advanced stage. Ligaments remaining along the hips, base of the tail, and feet kept these elements in place.
4. Burial was accomplished by a viscous mudflow composed of poorly sorted sand, silt, and clay balls. Entrained in these sediments were pieces of wood, seeds, leaves, and a variety of vertebrate bones and teeth. Possibly, burial was related to a cutbank failure triggered by a flood event. Some skeletal elements were moved by the mudflow.
5. After deposition, diagenetic siderite nodules formed around some of the dinosaur's bones and in the conglomerate.
6. The meander channel was abandoned and became a deep oxbow lake. Under quiet water conditions, aquatic plants flourished and laminated silts were deposited.

Acknowledgments

Peter Larson (Black Hills Institute) generously provided advice and assistance in collecting the tyrannosaur. Kirk Johnson (Denver Museum of Nature & Science) visited the tyrannosaur quarry site and provided assistance in interpreting the strata exposed there and helped identify plant remains. Doug Nichols (USGS Denver) provided assistance in identifying the palynoflora associated with the tyrannosaur. Reed Scherer and J. Michael Parrish (Northern Illinois University) read and commented on a draft of this chapter. John Warnock (Northern Illinois University) prepared the stratigraphic column and Molly Holman (Burpee Museum) produced the skeletal drawing of Jane. A number of staff and volunteers from Burpee Museum assisted in the excavation and preparation of Jane's skeleton, including: Melissa Birks, Dave Carlson, Lew Crampton, Joseph De La Morte, Shannon Farley-Maconaghy, Chris Garnhart, Jill and Richard Hertzing, Lisa Johnson, Jim Keller, Chrissy Majerowisz, Miriam Michaelis, Deborah Moauro, Brian Ostberg, Holli Palmer, Joseph E. Peterson, Sheila, Richard, and Nancy Rawlings, Scott Santoyo, Melissa Schrock, Ernie and Sue Smith, Mike Spiachello, Mindy Thompson, Katie Tremaine, Carol and Hazen Tuck, and Scott Williams.

References Cited

Bakker, R. T., Williams, M., and Currie, P. 1988. *Nanotyrannus*, a new genus of pygmy tyrannosaur, from the latest Cretaceous of Montana. *Hunteria* 1: 1–30.

Carr, T. D. 1999. Craniofacial ontogeny in Tyrannosauridae (Dinosauria, Coelurosauria). *Journal of Vertebrate Paleontology* 19: 497–520.

Carr, T. D., and Williamson, T. E. 2004. Diversity of late Maastrichtian Tyrannosauridae (Dinosauria: Theropoda) from western North America. *Zoological Journal of the Linnean Society* 142: 479–523.

Currie, P. 2003a. Allometric growth in tyrannosaurids (Dinosauria: Theropoda) from the Upper Cretaceous of North America and Asia. *Canadian Journal of Earth Science* 40: 651–665.

———. 2003b. Cranial anatomy of tyrannosaurid dinosaurs from the late Cretaceous Alberta, Canada. *Acta Palaeontologica Polonica* 48: 191–226.

Currie, P., Hurum, J. H., and Sabath, K. 2003. Skull structure and evolution in tyrannosaurid dinosaurs. *Acta Palaeontologica Polonica* 48: 227–234.

Henderson, M. D., and Peterson. J. E. In press. An Azhdarchid Pterosaur cervical vertebra from the Hell Creek Formation (Latest Maastrichtian) of southeastern Montana. *Journal of Vertebrate Paleontology*.

Johnson, K. R., Nichols, D. J., and Hartman, J. H. 2002. Hell Creek Formation: a 2001 synthesis. P. 503–510 in Hartman, J. H., Johnson, K. R., and Nichols, D. J. (eds.). *The Hell Creek Formation and the Cretaceous-Tertiary Boundary in the Northern Great Plains: An Integrated Continental Record of the End of the Cretaceous*. Geological Society of America Special Paper 361.

Murphy, E. C., Hoganson, J. W., and Johnson, K. R. 2002. Lithostratigraphy of the Hell Creek Formation in North Dakota. P. 503–510 in Hartman, J. H., Johnson, K. R., and Nichols, D. J. (eds.). *The Hell Creek Formation and the Cretaceous-Tertiary Boundary in the Northern Great Plains: An Integrated Continental Record of the End of the Cretaceous*. Geological Society of America Special Paper 361.

Pearson, D. A., Schaefer, T., Johnson, K. R., Nichols, D. J., and Hunter, J. P. 2002. Vertebrate biostratigraphy of the Hell Creek Formation in southwestern North Dakota and northwestern South Dakota. P. 503–510 in Hartman, J. H., Johnson, K. R., and Nichols, D. J. (eds.). *The Hell Creek Formation and the Cretaceous-Tertiary Boundary in the Northern Great Plains: An Integrated Continental Record of the End of the Cretaceous*. Geological Society of America Special Paper 361.

Tokaryk, T. T., and Bryant, H. N. 2004. The fauna from the *Tyrannosaurus rex* excavation, Frenchman Formation (Late Maastrichtian), *Saskatchewan*. *Saskatchwan Geological Survey, Summary of Investigations* 1: 1–12.

Figure 7.1. *Medullary bone in extant laying hen. (A) Gross cross section of femur of actively laying hen shows extensive medullary bone formation. New bone is randomly oriented and much more porous than overlying cortical bone. (B) Low magnification and (C) high magnification of histological section of demineralized bone from laying hen. Chemical differences between cortical and medullary bone are indicated by differential response of each bone type to hematoxylin and eosin staining. In (C), separation of the medullary bone from cortical bone is seen as sectioning artifact. Large, multinucleated osteoclasts are visible around bone spicules, and small osteoblasts align along preexisting bone spicules, active in deposition of new bone. Abbreviations: CB, cortical bone; MB, medullary bone; ELB, endosteal laminar bone; OCL, osteocyte lacunae; OC, osteoclast; OB, osteoblast. Scales as indicated. See color version of this figure online at https://www.iupress.indiana.edu/media/tyrannosaurusrex/*

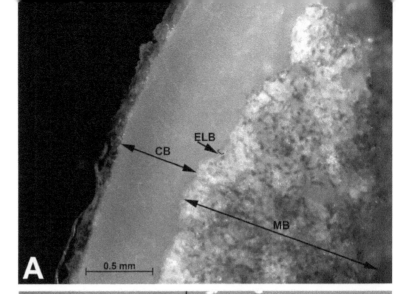

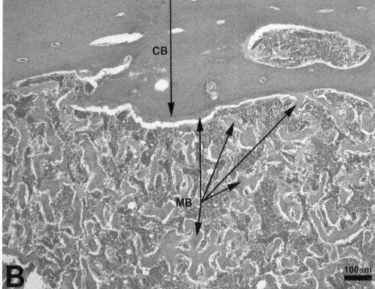

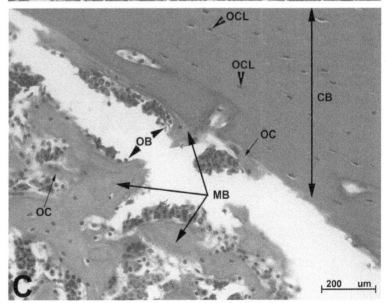

ONE PRETTY AMAZING *T. REX*

Mary Higby Schweitzer, Jennifer L. Wittmeyer, and
John R. Horner

Introduction

Determining sex in extinct animals is difficult because most features commonly used to assign sex are lost in the process of fossilization. Despite this difficulty, many bony features of dinosaurs have been interpreted to be evidence of sexual dimorphism, including degree of robustness in sauropods and their close relatives (Weishampel and Chapman 1990; Galton 1997; Benton et al. 2000), theropods (Carpenter 1990; Larson 1994; Smith 1998) and protoceratopsids (Tereschenko and Alifanov 2003); horn core size in ceratopsids (Godfrey and Holmes 1995); or presence or absence of the first caudal chevron (Larson and Frey 1992; Larson 1994), to name a few. However, even if such features could definitively be shown to be products of sexual differentiation, it remains impossible to assign a particular feature unambiguously to a specific sex (e.g., the robust morph being female; Carpenter 1990; Larson 1994). At best, assigning sex to a specific morphotype of dinosaurs has fallen within the realm of speculation. What is needed is an unambiguous means of assigning a particular sex to male and female morphs. One possibility is the identification of medullary bone in dinosaurs.

Medullary bone is an ephemeral reproductive tissue that is present in living taxa and that is found exclusively in female, actively reproducing birds. This bony tissue lines the medullary cavities of the long bones of extant birds; it is chemically and morphologically distinct from other bone types. Special characteristics of composition and structure contribute to the high metabolic rates of medullary bone. In fact, it is capable of being metabolized 10 to 15 times faster than cortical bone (Simkiss 1961; Dacke et al. 1993), and it serves as an easily mobilized calcium storage tissue for the production of calcareous eggshell (Sugiyama and Kusuhara 2001). Its presence in dinosaurs would indicate sex, support phylogenetic proximity, suggest shared reproductive physiological strategies with extant birds, and indicate reproductive phase at the time of death.

Comparison of Medullary and Cortical Bone Characteristics

In addition to protection and support of vital internal organs, bone plays an important role in calcium metabolism in vertebrates, including all avian taxa (Miller and Bowman 1981). Long bone formation in extant birds procedes much the same as in other vertebrate taxa through endochondral ossification of preexisting cartilage models (Whitehead 2004; Taylor et al. 1971). Bone elongation involves periosteal deposition, and concurrent endosteal osteoclastic resorption at the metaphyseal region, resulting in overall maintenance of bone morphology and thickness during longitudinal growth (Taylor et al. 1971).

In both formation and elongation, bone production involves 2 phases, which reflect the composite nature of bone material. In the first, the bone-forming cells (osteoblasts) secrete organic matrix (osteoid) (Taylor et al. 1971; McKee et al. 1993). This matrix primarily consists of the fibrous helical protein collagen I and the accessory collagen V; the noncollagenous proteins osteocalcin, osteopontin, and osteonectin (Bonucci and Gherardi 1975; McKee et al. 1993; Gerstenfeld et al. 1994; Sugiyama and Kasuhura 2001; Wang et al. 2005), and bone sialoprotein (Gerstenfeld et al. 1994; Robey 1996 and references therein); serum proteins, including hemoglobin and albumin (McKee et al. 1993); and various glycosaminoglycans (Bonucci and Gherardi 1975; Dacke et al. 1993; Arias and Fernandez 2001; Wang et al. 2005). Therefore, cortical and trabecular bone have specific, characteristic, and defineable chemical and molecular profiles.

However, in female birds, a unique bone type is formed as the result of a surge in blood estrogen levels at the onset of sexual maturity (Bonucci and Gherardi 1975; Knott and Bailey 1999; Dacke et al. 1993; Whitehead 2004). Medullary bone does not occur naturally in any other taxon (Elsey and Wink 1986; Dacke et al. 1993), and it is present only during the reproductive period in all living female birds, filling the marrow cavities of many skeletal elements (Wilson and Thorpe 1998; Van Neer et al. 2002). It is produced by specialized osteoblasts that lie within the endosteum, a thin connective tissue layer that lines the marrow surfaces of the bones (Van Neer et al. 2002). Medullary bone exists only to offset the effects of bone resorption during shelling by serving as an easily mobilized source of calcium, and it has no direct biomechanical function (Bonucci and Gherardi 1975; Wilson and Thorp 1998). It is chemically and morphologically distinct from other bone types. Although medullary bone has been assumed to be present in extant paleognaths, it has not been previously imaged or studied, and no data exist regarding the morphology or chemistry of this bone type in ratites.

The mineral phase of both medullary and cortical bone is primarily hydroxyapatite ($Ca_{10}(PO_4)_6(OH)_2$), but the ratio of mineral to organics is measurably higher in medullary bone (Ascenzi et al. 1963; Taylor et al. 1971; Dacke et al. 1993), and medullary bone incorporates a higher proportion of calcium carbonate (Pelligrino and Blitz 1970) than other bone types. Medullary bone is not only more highly mineralized than cortical bone, but also the distribution of minerals is different between the 2 bone types. In cortical bone, the mineral crystals are regularly distributed at the head of the A bands of collagen molecules (Taylor et al. 1971), but in medullary bone, mineral distribution and orientation is much more random, with mineral crystals additionally deposited in intrafibrillar spaces (Ascenzi et al. 1963; Taylor et al. 1971). In addition, medullary bone does not exhibit birefringence because of the random arrangement of both collagen fibrils and mineral, whereas other bone types are anisotropic in polarized light (Miller and Bowman 1981; Wilson and Thorp 1998). Finally, the mineral crystals incorporated into medullary bone are somewhat larger than the microcrystalline apatite of other bone types (Ascenzi et al. 1963), producing a greater crystallinity index.

The organic phase of medullary bone differs significantly from that of cortical and trabecular bone. Collagen makes up a greater proportion of the organic matrix of cortical bone, whereas the percentage of noncollagenous proteins to collagen is far greater in medullary bone, comprising approximately 40% of the total organics (Knott and Bailey 1999). The concentration of various glycosaminoglycans is greater in medullary than cortical bone, and it incorporates different amino sugars (Bonucci and Gherardi 1975). Hexosamine and keratan sulfate are much more prevalent in medullary than cortical bone (Taylor et al. 1971; Wang et al. 2005), which incorporates chondroitin sulfate instead. In addition, relatively high concentrations of tartrate-resistant acid phosphatase (TRAP), an enzyme involved in digestion of bone (Sugiyama and Kusuhara 2001), are found in medullary bone. These chemical differences are reflected in the differential response of the 2 bone types to various histochemical stains (Fig. 7.1; Taylor et al. 1971; Sugiyama and Kusuhara 2001; Wang et al. 2005).

Function of Medullary Bone

Unlike other bone types, medullary bone has no biomechanical or other supportive function. It exists solely as a calcium storage tissue that aids in mineral mobilization to the shell gland during lay (Dacke et al. 1993; Wilson and Thorp 1998; Whitehead 2004). As mentioned previously, medullary bone formation in birds is triggered by increased levels of both estrogen and androgens that accompany ovulation, activating osteoblasts to begin secretion of osteoid while inhibiting osteoclast activity (Dacke et al. 1993; Whitehead 2004). The formation of medullary bone begins approximately 1 or 2 weeks before lay. It is maintained during the full laying cycle, and it may persist up to 1 week after lay before resorption is complete (Reynolds 2003). Medullary bone osteoclasts in female birds are specialized to contain estrogen receptors in their cell membranes, which, when triggered by rising reproductive hormones, increases the efficiency of mobilizing stored calcium (Miller 1981). Although evidence of medullary bone may be found in virtually all skeletal elements of extant birds, it is most abundant in the femur and tibiotarsus of most birds studied (Reynolds 2003), and, consistent with its function as a source of rapid calcium mobilization, it is infused with abundant vessels and blood sinuses. In fact, up to 40% of the calcium used in eggshell formation comes directly from the resorption of medullary bone (Mueller et al. 1969; Dacke et al. 1993). Although it is not known to serve a direct mechanical function, in reducing the resorption of cortical and trabecular bone, it may aid in maintaining integrity and strength of structurally important bone (Whitehead 2004), and indeed, the presence of medullary bone in long bones of laying birds has been shown to increase fracture resistance of these elements (Fleming et al. 1998).

Like birds, most reptiles, including crocodiles and alligators, also produce calcareous eggshell, but apparently do not produce medullary bone (Elsey and Wink 1986; Dacke et al. 1993). This may be because of different mechanisms of shelling (Jackson et al. 2002) and overall greater bone density that can offset the calcium draw without requiring additional bone storage sources. Thus, extant nonavian archosaurs undergo bone resorption during lay, but the structural integrity and biomechanical function of these organisms is apparently not compromised during shelling.

Although medullary bone has not been previously observed or described in dinosaurs, it was proposed that reproducing dinosaurs, at least in the theropod lineage most closely related to avian dinosaurs, would possess this ephemeral tissue (Martill et al. 1996; Chinsamy and Barrett 1997). The failure to observe or identify these fragile reproductive tissues in dinosaurs previously may be due to a number of taphonomic and/or biological factors, or observational bias. First, we do not have any way of estimating the length of the reproductive cycle in theropods. There is a wide range of reproductive strategies among living birds, and the extent and distribution of medullary bone in these taxa differ correspondingly (Schraer and Hunter 1985). If theropods reproduce seasonally, they may only possess the tissue for a maximum of a month or less. Second, because of the relatively thick, dense cortex, the need for medullary bone may be less in these animals, so the medullary layer may be quite thin. Third, in extant birds, the tissue is quite fragile, and separates easily from the overlying cortex (Fig. 7.1B, C). It may be that the tissues are lost, either during fossilization or during subsequent recovery and preparation. Third, it may be that medullary bone differs sufficiently from that of extant derived birds so that it is not recognized. A fourth factor may be the failure to examine bones for its presence because of collection techniques requiring bone to be conserved, and not broken to expose interior fragments.

Medullary Bone in T. Rex

At the end of field season in 2002, a well-preserved specimen of *Tyrannosaurus rex* (MOR 1125) was found as an association of disarticulated elements. The site was located at the base of the Hell Creek Formation (Lancian), about 8 m above the Fox Hills Sandstone. Soft, well-sorted sandstones derived from an estuarine or fluvial setting surrounded the skeletal elements. Some of the elements evidenced slight crushing, but overall preservation was excellent. MOR 1125, nicknamed Bob-rex after its discoverer, Bob Harmon, is a relatively small but fully adult *T. rex*. In comparison with another *T. rex*, FMNH PR2081, with a femur length of about 131 cm, the femur of MOR 1125 is only 107 cm in length. By use of lines of arrested growth, MOR 1125 was calculated to be about 18 years old at the time of death (Horner and Padian 2004).

The remote region where MOR 1125 was recovered had no roads into the site, so a helicopter was required to transport field jackets to the MOR laboratories. However, the jacket containing the femur and other elements was too heavy to be airlifted out, and the jacket and bones they contained were broken and rejacketed for removal. In the process, many internal fragments that were visually free of preservative or consolidants were collected for analyses.

When these fragments were examined in hand sample, a bony tissue lining the endosteal surface of the bone could be seen that was distinct in texture, appearance, and distribution from other described dinosaur bone types. The morphological similarity of the new tissues to avian medullary bone was immediately apparent (Schweitzer et al. 2005b). Figure 7.2 shows fresh-fracture images of *Tyrannosaurus rex* endosteal tissues (Fig. 7.2A, B),

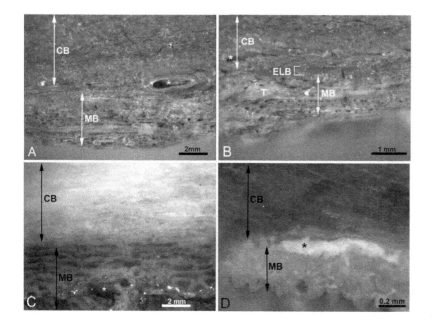

Figure 7.2. Fresh fracture of tibiae of MOR 1125 (A, B), ostrich (C), and emu (D). Morphology and microstructure of medullary bone is observed in all cases as distinct from overlying cortical bone. Medullary bone is less organized and more vascular. Large vascular sinuses can be seen in the medullary bone, and in some cases, large erosion cavities (*) are visible at the interface between medullary and cortical bone and in the medullary bone itself, indicating some resorption of bone has occurred. In emu (D), a large elongate erosion room is infilled with new medullary bone with characteristic corrugated texture. Abbreviations as in Figure 7.1. T, trabecular spicule. Scales as indicated. See color version of this figure online at https://www.iupress.indiana.edu/media/tyrannosaurusrex/

compared with medullary bone tissues in reproducing ostrich (Fig. 7.2C) and emu (Fig. 7.2D). The hallmark traits of medullary bone—dense vascularity and random, woven bone pattern—are clearly visible in all samples. Large erosion cavities are visible in all medullary tissues (indicated with an asterisk in figures), indicating that calcium mobilization has occurred.

Demineralization of extant bony tissues is commonly used to more clearly observe microstructural characteristics, such as fibril orientation; and when mineral is removed, the primarily collagenous protein matrix is exposed. Conventional wisdom has held that when fossilized dinosaur bone is subjected to the same treatment, the bone would dissolve completely because no proteinaceous material would persist over the course of geological time (Hoss 2000).

In order to determine characteristics of presumed medullary tissues, we prepared a partial demineralization designed to etch mineral enough to expose underlying patterns. At this point, we discovered an unexpected and novel characteristic to this bony tissue. As minerals were dissolved from the medullary bone, the sample did not disintegrate, but, similar to extant bone, tissues remained (Schweitzer et al. 2005a). Furthermore, these dinosaur tissues exhibited apparent original flexibility, comparable to that seen in extant ratites. However, these characteristics are not germane to this chapter and will be discussed elsewhere. The retention of a pliable and fibrous matrix after demineralization speaks to unusual preservation in this dinosaur material and suggests that perhaps theorized modes of fossilization may need to be reevaluated. Figure 7.3 demonstrates the persistence of fibrous tissues after demineralization. Small fragments of emu and dinosaur demineralized medullary bone tissues show random fiber orientation, and large open spaces for vessels and vascular sinuses permeate the tissues. The morphological similarity between extant and fossil samples is clearly visible and supports the hypothesis of a common origin to the tissues.

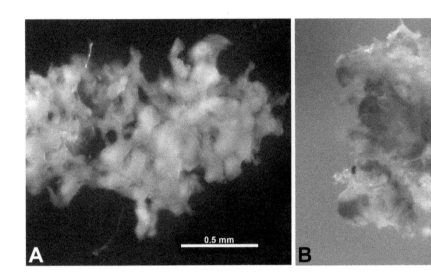

Figure 7.3. Demineralized fragments of medullary bone from emu (A) and MOR 1125 (B). The fibrous, woven pattern of bone matrix is visible in both cases, and the relatively lacy appearance results from penetration of the bone by blood vessels. Scales as indicated. For methods on demineralization, see Schweitzer et al. (2005a), supplemental online information. See color version of this figure online at https://www.iupress.indiana.edu/media/tyrannosaurusrex/

Conclusion

The endosteally derived bone tissues observed in this specimen of *Tyrannosaurus rex* (MOR 1125) have all of the morphological characteristics of medullary bone, a distinctive avian reproductive tissue. Although not identical to published accounts of extant neognaths, the dinosaur tissues fall within the range of morphological variation observed in ratites. This bone tissue is derived from the endosteum, is highly vascular, and exhibits the random, woven bone arrangement consistent with very rapidly deposited bone. In addition, it has been identified on the endosteal surfaces of both femora and one tibia, the only bones examined for the presence of this tissue. The distribution is consistent with that seen in extant birds and suggests an organismal, rather than pathological, response.

Pathologies of the endosteum are relatively rare and localized, and they are usually accompanied by cortical bone anomalies in the affected regions. No anomalies were observed, either grossly or microscopically, in MOR 1125. In light of the fact that the relationship between theropod dinosaurs and birds is robustly supported (e.g., Gauthier 1986; Sereno 1997; Holtz 2004), it is most parsimonious to conclude that this novel tissue seen in MOR 1125 is medullary bone, and its presence in theropods not only adds independent support to the robustly demonstrated relationship between theropods and birds, but also suggests that these organisms had similar reproductive physiological strategies. In addition, its presence provides a means for unambiguous assignment of sex in dinosaurs. With careful examination, other, less ephemeral morphological traits may be identified in this specimen that can be applied to differentiate nonreproducing females in this lineage.

References Cited

Arias, J. L., and Fernandez, M. S. 2001. Role of extracellular matrix molecules in shell formation and structure. *World Poultry Science Journal* 57: 349–355.

Ascenzi, A., Francois, C., and Bocciarelli, D. S. 1963. On the bone induced by estrogens in birds. *Journal of Ultrastructure Research* 8: 491–505.

Benton, M. J., Juul, L., Storrs, G. W., and Galton, P. M. 2000. Anatomy and systematics of the prosauropod dinosaur *Thecodontosaurus antiquus* from the Upper Triassic of southwest England. *Journal of Vertebrate Paleontology* 20(1): 77–108.

Bonucci, E., and Gherardi, G. 1975. Histochemical and electron microscope investigations on medullary bone. *Cell and Tissue Research* 163: 81–97.

Carpenter, K. 1990. Variations in *Tyrannosaurus rex*. P. 141–146 in Carpenter, K., and Currie, P. J. (eds.). *Dinosaur Systematics: Approaches and Perspectives*. Cambridge University Press, Cambridge.

Chinsamy, A., and Barrett, P. M. 1997. Sex and old bones? *Journal of Vertebrate Paleontology* 17(2): 450–450.

Dacke, C. G., Arkle, S., Cook, D. J., Wormstone, I. M., Jones, S., Zaidi, M., and Bascal, Z. A. 1993. Medullary bone and avian calcium regulation. *Journal of Experimental Biology* 184: 63–88.

Elsey, R. M., and Wink, C. S. 1986. The effects of estradiol on plasma calcium and femoral bone structure in alligators (*Alligator mississippiensis*). *Comparative Biochemistry and Physiology* 84A: 107–110.

Fleming, R. H., McCormack, H. A., McTeir, L., and Whitehead, C. C. 1998. Medullary bone and humeral breaking strength in laying hens. *Research in Veterinary Science* 64: 63–67.

Galton, P. M. 1997. Comments on sexual dimorphism in the prosauropod dinosaur *Plateosaurus engelhardti* (Upper Triassic, Trossingen). *Neues Jahrbuch für Geologie und Palaeontologie, Monatshefte* 1997: 674–682.

Gauthier, J. 1986. Saurischian monophyly and the origin of birds. P. 1–55 in Padian, K. (ed.). *The Origin of Birds and the Evolution of Flight*. California Academy of Science Memoir 8.

Gerstenfeld, L. C. 1994. Selective extractability of noncollagenous proteins from chicken bone. *Calcified Tissue International* 55 (3): 230–235

Godfrey, S. J., and Holmes, R. 1995. Cranial morphology and systematics of *Chasmosaurus* (Dinosauria: Ceratopsidae) from the Upper Cretaceous of western Canada. *Journal of Vertebrate Paleontology* 15: 726–742.

Holtz, T. R., Jr. 2004. Tyrannosauroidea. P. 111–136 in Weishampel, D. B., Dodson, P., and Osmálska, H. *The Dinosauria*. 2nd ed. University of California Press, Berkeley.

Horner, J. R., and Padian, K. 2004. Age and growth dynamics of *Tyrannosaurus rex*. *Proceedings of the Royal Society of London Series B* 271: 1875–1880.

Hoss, M. 2000. Neanderthal population genetics. *Nature* 404: 453–454

Jackson, F. D., Schweitzer, M. H., and Schmitt, J. G. 2002. Dinosaur eggshell study using scanning electron microscopy. *Scanning* 24: 217–223.

Knott, L., and Bailey, A. J. 1999. Collagen biochemistry of avian bone: comparison of bone type and skeletal site. *British Poultry Science* 40: 371–379.

Larson, P. L. 1994. *Tyrannosaurus* sex. P. 139–155 in Rosenberg, G. D., and Wolberg, D. L. (eds.). *Dinofest*. Paleontological Society Special Publication 7.

Larson, P. L., and Frey, E. 1992. Sexual dimorphism in the abundant Upper Cretaceous theropod, *Tyrannosaurus rex*. *Journal of Vertebrate Paleontology* 12(Suppl. to 3): 38a.

Martill, D. M., Barker, M. J., and Dacke, C. G. 1996. Dinosaur nesting or preying? *Nature* 379: 778–778.

McKee, M. D., Farachcarson, M. C., Butler, W. T., Hauschka, P. V., and Nanci, A. 1993. Ultrastructural immunolocalization of noncollagenous (osteopontin and osteocalcin) and plasma (albumin and alpha-2HS glycoprotein) proteins in rat bone. *Journal of Bone and Mineral Research* 8(4): 485–496

Miller, S. C. 1981. Osteoclast cell-surface specializations and nuclear kinetics during egglaying in Japanese quail. *American Journal of Anatomy* 162: 35–43.

Miller, S. C., and Bowman, B. M. 1981. Medullary bone osteogenesis following estrogen administration to mature male Japanese quail. *Developmental Biology* 87: 52–63.

Mueller, W. J., Brubaker, R. L., and Caplan, M. D. 1969. Eggshell formation and bone resorption in egg-laying hens. *Federation of the American Society of Experimental Biologists* 28: 1851–1855.

Pelligrino, E. D., and Blitz, R. M. 1970. Calcium carbonate in medullary bone. *Calcified Tissue Research* 6: 168–171.

Reynolds, S. J. 2003. Mineral retention, medullary bone formation and reproduction in the white-tailed ptarmigan (*Lagopus leucurus*): a critique of Larson et al. (2001). *Auk* 120(1): 224–228.

Robey, P. G. 1996. Vertebrate mineralized matrix proteins: structure and function. *Connective Tissue Research* 35(1–4): 185–190.

Schraer, H., and Hunter, S. J. 1985. The development of medullary bone: a model for osteogenesis. *Comparative Biochemistry and Biophysiology* 82A(1): 13–17.

Schweitzer, M. H., Wittmeyer, J. L., and Horner, J. R., and Toporski, J. 2005a. Soft tissue vessels and cellular preservation in *Tyrannosaurus rex*. *Science* 307: 1952–1955.

———. 2005b. Gender specific reproductive tissue in ratites and *Tyrannosaurus rex*. *Science* 308: 1456–1460.

Sereno, P. C. 1997. The origin and evolution of dinosaurs. *Annual Reviews of Earth and Planetary Science* 25: 435–489 1997

Simkiss, K. 1961. Calcium metabolism and avian reproduction. *Biological Review* 36: 321–367.

Smith, D. K. 1998. A morphometric analysis of *Allosaurus*. *Journal of Vertebrate Paleontology* 18(1): 126–142.

Sugiyama, T., and Kusuhara, S. 2001. Avian calcium metabolism and bone function. *Asian-Australian Journal of Animal Science* 14: 82–90.

Taylor, T. G., Simkiss, K., and Stringer, D. A. 1971. The skeleton: its structure and metabolism. P. 621–640 in Freeman, B. M. (ed.). *Physiology and Biochemistry of the Domestic Fowl*. Vol. 2. Academic Press, New York.

Tereschenko, V. S., and Alifanov, V. R. 2003. *Bainoceratops efremovi*, a new protoceratopsid dinosaur (Protoceratopidae, Neoceratopsia) from the Bain-Dzak locality (south Mongolia). *Paleontology Journal* 37: 293–302.

Van Neer, W., Noyen, K., and DeCupere, B. 2002. On the use of endosteal layers and medullary bone from domestic fowl in archaeozoological studies. *Journal of Archeological Science* 29: 123–134.

Wang, X., Ford, B. C., Praul, C. A., and Leach, R. M. 2005. Characterization of the non-collagenous proteins in avian cortical and medullary bone. *Comparative Biochemistry and Physiology B* 140: 665–572.

Weishampel, D. B., and Chapman, R. E. 1990. Morphometric study of *Plateosaurus* from Trossingen (Baden-Wurttemberg, Federal Republic of Germany). P. 43–51 in Carpenter, K., and Currie, P. J. (eds.). *Dinosaur Systematics: Approaches and Perspectives*. Cambridge University Press, Cambridge.

Whitehead, C. C. 2004. Overview of bone biology in the egg-laying hen. *Poultry Science* 83: 193–199.

Wilson, S., and Thorp, B. H. 1998. Estrogen and cancellous bone loss in the fowl. *Calcified Tissue International* 62: 506–511.

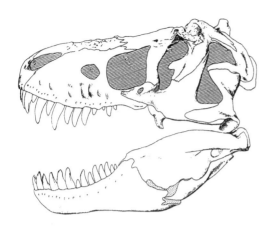
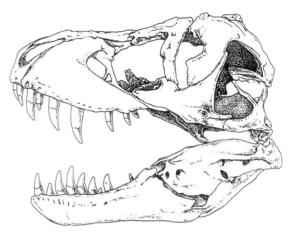

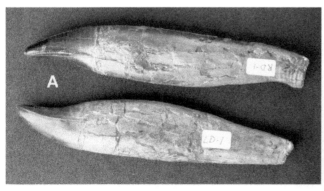
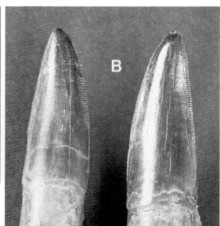

VARIATION AND SEXUAL DIMORPHISM IN *TYRANNOSAURUS REX*

Peter Larson

Introduction

The science of paleontology has often been accused of being more art than science. This assessment stems from the problems encountered when dealing with the paucity and incompleteness of the fossil record. Not the least of the problems confronting paleontologists is the scarcity of specimens. To date, 46 specimens (N. L. Larson this volume) consisting of more than a few associated bones have been assigned to *Tyrannosaurus rex* Osborn (1905, 1906). Although this is a robust representation for extinct theropods, when compared with extant populations, this number seems extremely inadequate. For example, Buss (1990) reported a 1973 count of 14,309 African elephants (*Loxodonta africana*) in the 3840 km^2 (1483 mi^2) Kabalega Falls National Park in Uganda. On its face, 46 specimens seems a paltry number from which to define a species, let alone attempt to identify males and females. Yet that is exactly what this study attempts. The use of modern taxonomic methods may be used to identify anomolous morphological characters and to remove questionable specimens from a taxon to which they have been unnaturally joined (more below). Taken even further, morphometrics, physiology, and pathology can be used to help separate and define sex morphotypes.

For this study, 34 specimens attributed to *Tyrannosaurus rex*, including specimens listed as *Tyrannosaurus* "x" and *Nanotyrannus* (considered as specimens of *T. rex* by Carr 1999), were examined. In addition, 2 specimens assigned to *Tarbosaurus bataar*, one assigned to *Gorgosaurus* and another to *Albertosaurus*, were examined as outgroups. These specimens are listed in Table 8.1.

Figure 8.1. *(Left)* **Tyrannosaurus "x"** *(AMNH 5027). (Right)* **Tyrannosaurus rex** *(BHI 3033).*

Figure 8.2. *Medial view of right dentary of the type* **Tyrannosaurus rex** *CM 9380. Note the incisiform first dentary tooth.*

Figure 8.3. *Left and right first dentary teeth of* **Tyrannosaurus rex** *BHI 3033. (A) Lateral view. (B) Posterior view. Note that both serrations are exposed in the posterior view, creating the typical tyrannosaurid D-shaped cross section.*

Variation

In any population, individual variation within a species will occur. This variation is due to ontogeny, nutrition, genetic variance, pathology, and, of course, sexual dimorphism. Thus, it is imperative that these factors be excluded when examining the question: "Have researchers included specimens within the species *T. rex*, with variation beyond that expected within a living population?" Extant phylogenetic bracketing techniques (Witmer 1995) were used to evaluate the characters used in this study for the purpose of isolating those attributable to intraspecific variation.

Table 8.1. *Specimens Used in the Study*

Tyrannosaurus rex	Tyrannosaurus "x"	Nanotyrannus	Outgroups
CM 9380	AMNH 5027	BMR P2002.4.1	*Tarbosaurus*
CM 1400	MOR 008	CMNH 7541	BHI 6236
LACM 23844	SDSM 12047	BHI 6235	ZPAL-MgD-I/4
LACM 2345	Samson	LACM 28471	*Gorgosaurus*
MOR 009			TCM2001.89.1
MOR 1128			*Albertosaurus*
MOR 1125			BHI 6234
MOR 555			
MOR 980			
FMNH PR2081			
BHI 3033			
BHI 4100			
BHI 4182			
BHI 6232			
BHI 6231			
BHI 6233			
BHI 6230			
BHI 6242			
TCM2001.90.1			
RTMP 81.12.1			
RTMP 81.6.1			
UCMP118742			
BMNH R7994			
NHM R8001			
USNM V6183			
LL.12823			

Ontogenetic variation may include aspects other than the obvious increase in size. For example, it may also include an increase in the number of alveoli, or tooth positions (e.g., *Edmontosaurus annectens*; personal observation). In certain groups (i.e., mammals), growth to adulthood may also include modification of tooth morphology, along with an increase in the number of tooth positions (Romer 1966). For many vertebrates, ontogeny also includes an increase in body size at a faster rate than for the brain, eyes, and skull (Lockley et al. this volume). Nutritional variation may manifest itself as smaller body size and smaller body mass—differences that are not generally confused with taxonomic characters. Genetic variation may be monitored by using extant populations as examples (Darwin 1868). Pathologic specimens showing evidence of disease or healed injury are relatively easily recognized, and are generally not reproducible from specimen to specimen in a form that would be noted as a taxonomic character. Finally, sexual dimorphism will be discussed in depth near the end of this chapter.

The Case for Tyrannosaurus "x"

More than 25 years ago, Robert Bakker (personal communication) made the case for dividing the North American genus *Tyrannosaurus* into 2 species, *T. rex* and what he refers to as *Tyrannosaurus* "x" (Fig. 8.1). Bakker's reasoning was based on a peculiar variation in the anterior dentition of the dentary. The type of *Tyrannosaurus rex* (AMNH 973 = CM 9380) possesses a single incisiform tooth occupying the anterior position in the dentary. This tooth is morphologically reminiscent of the teeth of the premaxilla, is D shaped in cross section, and is substantially smaller than those directly posterior to it (Figs. 8.2 and 8.3). Bakker also noted that AMNH 5027 appears to possess 2 incisors in each dentary. For lack of specimens, his views were never published. Paul (1988) and Molnar (1991) have both also considered the possibility of a second species of *Tyrannosaurus*.

A quarter of a century later, there now exist at least 15 reasonably complete *Tyrannosaurus* skulls. Three of these specimens (MOR 008, SDSM 12047, and Samson) share certain characters, including the double lower incisors, with AMNH 5027 (Figs. 8.4 and 8.5). Because these "incisors" are either missing or were restored on all 4 specimens, without computed tomographic scans to look at unerupted teeth, the D-shaped morphology of these "incisors" is in question. The apparent differences seem to be best expressed by comparing the size of the second dentary tooth with that of the third, and because the teeth themselves were not always available to measure, the length of the second and third alveoli were measured and compared. The results of these measurements are found in Table 8.2.

Although all 4 skulls seem short when compared with full-grown individuals (i.e., BHI 3033 and FMNH PR2081 = BHI 2033), ontogenetic variation may be ruled out because other individuals of approximately the same skull length do not share this character. One of the specimens, Samson, has a femur (length, 129 cm) of comparable length to Stan (BHI 3033; length, 131 cm), but whose skull is less than 80% as long (104 cm). A shorter skull and variation in lower jaw dentition is unlikely to be caused by differences in nutrition. Pathology may be ruled out because of the lack of any associated manifestation of healed injury. Genetic variance also seems improbable because no modern correlates exist. A case could be made for the differences in the dentition being attributable to sexual dimorphism.

Table 8.2. Comparison of Lengths of Second (DT2-L) and Third Dentary Tooth or Alveolus (DT3-L)

* From the holotype.

Specimen	DT2-L (mm)	DT3-L (mm)	Ratio of DT3 to DT2
*T. rex**			
CM 9380	55	54	1.0
MOR 555	52	56	1.1
MOR 980	51	51	1.0
BHI 3033	56	60	1.1
T. "x"			
BHI 4182	33	34	1.0
MOR 008	48	64	1.3
SDSM12047	35	55	1.6
Samson	33	54	1.6

Figure 8.4. Dorsal view of the anterior portion of the left dentary of *Tyrannosaurus rex* CM 9380 preserving small first dentary tooth DT1 and large second dentary tooth DT2.

Although there are modern examples of sexual dimorphism in the canines of some primates (Martin et al. 1994) and in the canines or incisors of walrus, elephants, bush pig, and hippopotamus (Lincoln 1994), sexual dimorphism expressed as differences in dentition in extant taxa seems to be restricted to mammals. Any dental expression of sexual dimorphism remains undocumented for crocodilians, extinct toothed birds (extant phylogenetic bracketing), or other extant reptiles.

Can the difference in the teeth be attributable to speciation? Although stratigraphic information for the 4 specimens is unavailable, there are good records available for *Tyrannosaurus rex*. BHI 2033 was collected 16 m below the K-T boundary in the Hell Creek Formation (the Hell Creek in the area, near Buffalo, SD, is approximately 150 m thick). A second indisputable specimen of *Tyrannosaurus rex* (BHI 4182) was collected nearby, from within 10 m of the base of the formation, and it represents perhaps the oldest known record of *Tyrannosaurus* from North America (Kirk Johnson, personal communication). Geographic distribution is also not a factor, because *T. rex* co-occurs with *T.* "x."

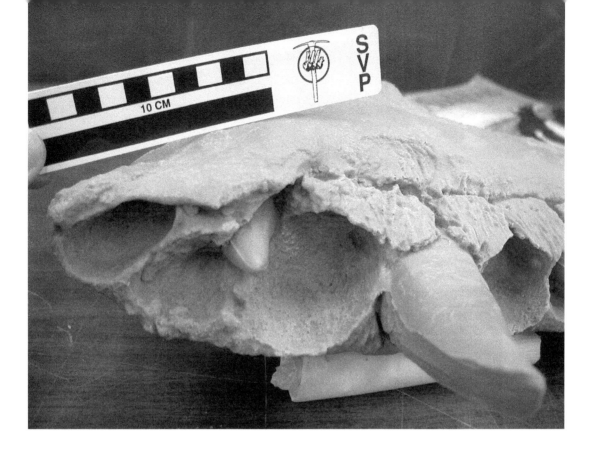

Figure 8.5. Dorsal view of the anterior portion of the left dentary of **Tyrannosaurus "x"** (Samson) preserving small alveoli for DT1 and DT2 and a large alveolus for the third dentary tooth DT3.

Dentary and maxillary tooth (alveoli) counts also seem to vary between the 2 "species." This is particularly evident in the dentary, with 13 or 14 for *Tyrannosaurus rex* and 14 or 15 for *T.* "x." The distribution of all of these characters, with *Tarbosaurus bataar* as an outgroup, are listed in Table 8.3.

A fourth character separating the 2 forms is the relative size of the lateral pneumatic lachrymal foramen. Specimens referable to *T.* "x" have relatively smaller lateral pneumatic lachrymal foramina than those of *Tyrannosaurus rex* (Fig. 8.6). When measured and plotted as lachrymal foramina length vs. lachrymal length (Fig. 8.7), *Tyrannosaurus* "x" clusters separately from *T. rex* (as do *Gorgosaurus* and *Nanotyrannus*). However, it should be noted that the size of the lachrymal foramina in *Allosaurus* is extremely variable, and this difference between *T. rex* and *T.* "x" may not be statistically significant, especially given the sample size (Kenneth Carpenter, personal communication).

Table 8.3. Comparison of Skull Characters

Skull Character	*Tyrannosaurus rex*	*Tyrannosaurus* "x"	*Tarbosaurus*
Lateral lachrymal pneumatic foramina	Small	Very small	Small
Maxillary tooth count	11 or 12	12	12 or 13
Dentary tooth count	13 or 14	14 or 15	15
Dentary incisor count	1	2	1
L3DT/L2DT DT3-L/DT2-L	1.0–1.1	1.3–1.6	1.2

Figure 8.6. *Lateral view of the left lachrymals of (A)* **Tyrannosaurus "x"** *AMNH 5027 and (B)* **Tyrannosaurus rex** *BHI 3033. Note the larger lateral pneumatic foramen on* **T. rex**.

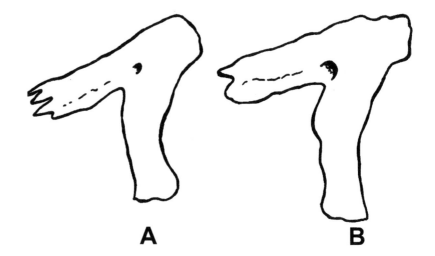

Are these 4 cranial characters enough to erect a new species? (No significant postcranial characters were noted.) Because we are dealing with an extinct group, doing so at this time might be premature. Although it is likely that a second North American Latest Cretaceous species of *Tyrannosaurus* exists, all of the specimens in questions are in need of further preparation that will permit a more thorough comparison with the type (AMNH 973 = CM 9380) and other referred specimens. Fortunately, preparation of 2 of the specimens (SDSM 12047 and Samson) is already underway. The ultimate disposition of *Tyrannosaurus* "x" may soon be resolved.

Is *Nanotyrannus lancensis* a Juvenile *Tyrannosaurus rex*?

The genus *Nanotyrannus* was erected by Bakker et al. (1988) for the type specimen (CMNH 7541) of *Gorgosaurus lancensis* Gilmore (1946). This specimen (Fig. 8.8) consists of a relatively complete skull preserved with the jaws in occlusion, with very little distortion and no associated postcra-

Figure 8.7. *Lachrymal length vs. lachrymal foramina length.*

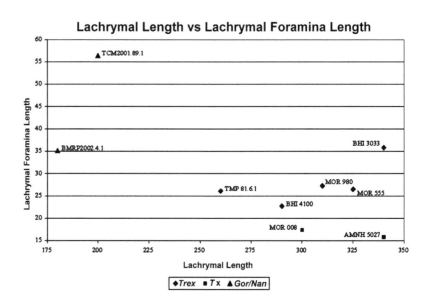

Figure 8.8. *Type specimen of* **Nanotyrannus lancensis** *CMNH 7541.*

nial material. Bakker et al. (1988) argued that certain derived characters, including the construction of the basicranium, the angle of the occipital condyle, the maxillary tooth count, the overall tooth morphology, the relative narrowness of the snout, and the expansion of the temporal region of the skull clearly separated this specimen from other tyrannosaur clades (*Gorgosaurus, Albertosaurus, Daspletosaurus,* and *Tyrannosaurus*).

Although the characters discussed by Bakker et al. (1988) clearly separated this specimen from its earlier assignment to *Gorgosaurus,* its distance from the *Tyrannosaurus* clade seemed less defined. They both "achieved the highest degree of potential stereoscopy known among large theropods," and they agree in characters, including the orientation of the occipital condyle (Bakker et al. 1988, p. 25). They also address the question of the skull being that of a juvenile: "The sutures between the lachrymal and prefrontal have thoroughly coalesced in *Nanotyrannus,* as have the sutures between frontals and prefrontals. . . . Without question, the type of *Nanotyrannus* was fully adult and had reached the maximum size the individual would have attained if it had lived longer" (Bakker et al. 1988, p. 17).

Carpenter (1992, pp. 259, 260) disagreed with Bakker et al. (1988) when he noted that "the coalescence of cranial bones is known to be variable in dinosaurs" bringing under suspicion "its usability to 'age'" dinosaurs. Carpenter further noted that "the oval shape of the orbit" may well

Variation and Sexual Dimorphism

be a juvenile character. He concluded that *Nanotyrannus lancensis* could be a juvenile *T. rex*.

Carr (1999) expanded this possibility. On the basis of 17 specimens referred to *Albertosaurus libratus*, Carr erected an ontogenetic series of growth stages (1–3). From bone texture, lack of fusion, shape of the orbit, and overall skull morphology, Carr placed CMNH 7541 into his stage 1, the youngest in his ontogenetic series. Carr then declared *Nanotyrannus lancensis* to be a juvenile *Tyrannosaurus rex*. In later arguments (Carr and Williamson 2004; Carr et al. 2005), this designation was used to establish a growth series for *T. rex*, establishing a sequence of changes from the small juvenile LACM 28471, followed by the juvenile CMNH 7541 (stage 1), through subadults LACM 23845 and AMNH 5027, to the fully grown adults LACM 23844 and FMNH PR2081 (BHI 2033).

Although Carr (1999) presented a compelling and thoughtful argument, not all paleontologists agree with his assessment. Currie (2003, p. 223) pointed out that "most of the characters used to demonstrate that *Nanotyrannus* and *Tyrannosaurus* are synonymous are also characters of *Tarbosaurus* and *Daspletosaurus*." Bakker et al. (1988; personal communication) noted the discrepancy in tooth counts—15 maxillary teeth in *Nanotyrannus* and 11 or 12 in *Tyrannosaurus rex*—and the lack of tooth reduction ontogenetically in the maxilla of any extant species. The primitive compressed nature of *Nanotyrannus* teeth (Bakker et al. 1988) as compared with the derived inflated teeth seen in *T. rex* and evidence of feeding behavior differences also argue for the uniqueness of CMNH 7541 (Larson 1999). Because the growth series argument of Carr is rooted in the assumption that *Nanotyrannus* is a juvenile *T. rex*, much of Carr's concept of ontogenetic change and ontogenetic stages in *Tyrannosaurus rex* is in question (Jorn Hurum, personal communication). I agree with Carr and Williamson's (2004) assessment of LACM 28471 (the so-called Jordan theropod) with CMNH 7541 (the type of *Nanotyrannus*), and with the designation of the subadult LACM 23845 as *Tyrannosaurus rex*. However, I disagree with the subadult designation of AMNH 5027, which groups as a full adult with *Tyrannosaurus* "x" and with *Nanotyrannus* as a juvenile *T. rex*.

An isolated left lachrymal (BHI 6235) comparable in size and morphology to CMNH 7541 was found associated with Sue (FMNH PR2081) and erroneously identified as a juvenile *T. rex* (Larson 1997). It, too, should be referred to *Nanotyrannus*. Finally, the recent discovery of a fourth specimen (BMR P2002.4.1) is clearly referable to *Nanotyrannus*. This specimen, nicknamed Jane, in addition to many uncrushed and well-preserved skull elements with a nearly complete dentition, also preserves much of the postcranial skeleton.

Although this subject is discussed in detail elsewhere (Currie 2003; Currie et al. 2003; Larson in press), a list of characters separating *Nanotyrannus* from *Tyrannosaurus* is presented in Table 8.4. For purposes of comparison as outgroups, those characters are also listed for *Tyrannosaurus* "x," *Tarbosaurus*, *Gorgosaurus*, and *Albertosaurus*.

Table 8.4. Comparison of Tyrannosaurid Skull Characters

Skull Character	Tyrannosaurus rex	Tyrannosaurus "x"	Tarbosaurus bataar	Tarbosaurus bataar Juvenile	Nanotyrannus lancensis	Gorgosaurus sp.	Albertosaurus sp.
Antorbital fossa	Very deep	Deep	Deep	Deep	Very shallow	Shallow	Shallow
Ventral antorbital maxillary ridge meets jugal	No	No	No	No	Yes	Yes	Yes
Maxillary fenestra reaches rostral margin of antorbital fossa (Carr et al. 2005)	Yes	Yes	Yes	Yes	No	No	No
Maxillary fenestra reaches ventral margin of antorbital fossa (Carr et al. 2005)	Yes	Approaches	Yes	Yes	No	No	Approaches
Vomer expansion	Lateral	Lateral	Lateral	?	Dorsoventral	Dorsoventral	?
Posterior dorsal Quadratojugal notch	No	No	No	No	Yes	Yes	No
Central dorsal quadratojugal notch	No	No	No	No	Yes	Yes	No
Anterior dorsal medial notch in quadratojugal (Carr and Williamson 2004)	Yes	?	No	?	No	No	No
Lachrymal horn (cornual process)	Absent	Absent	Absent	Absent	Present	Present	Present
Lachrymal shape	Inverted L	Inverted L	Inverted L	Inverted L	T	T	T
Quadrate—squamosal articulation	Double	Double	?	?	Single	Double	Single
Cranial nerve V-2 bounded by:	Maxilla and premaxilla	Maxilla and premaxilla	Maxilla and premaxilla	Maxilla and premaxilla	Maxilla only	Maxilla only	?
Anterior maxilla fossa at cranial nerve V-2	Maxilla and premaxilla	Maxilla and premaxilla	Maxilla and premaxilla	Maxilla and premaxilla	Maxilla only	Maxilla only	?

Table 8.4 (continued).

Skull Character	Tyrannosaurus rex	Tyrannosaurus "x"	Tarbosaurus bataar	Tarbosaurus bataar Juvenile	Nanotyrannus lancensis	Gorgosaurus sp.	Albertosaurus sp.
Tooth cross section at base of crown	Ovate	Ovate	Compressed	Compressed	Compressed	Ovate	Compressed
Fourth maxillary tooth L/W (at base of crown)	1.76	1.23	1.68	?	2.12	1.36	?
Fourth dentary tooth L/W (at base of crown)	1.38	1.34	1.39	?	1.66	1.23	?
First maxillary tooth small and incisoform	No	No	No	No	Yes	Yes	?
D-shaped first dentary tooth	Yes	?	?	?	No	No	?
First dentary tooth reduced	Yes	Slightly	Slightly	?	Greatly	No	?
Maxillary tooth count	11–12	12–13	12–13	13	15–16	13–15	13–15
Dentary tooth count	13–14	14–15	15	?	17	15–17	13–15
Medial postorbital fossa	No	No	No	?	Yes	Yes	Yes
Foramina on lateral aspect (center) of quadratojugal	Absent	Absent	Absent	Absent	Large pneumatic	Small	Absent
Anterior squamosal pneumatic foramina	Very large	Very large	Present	?	Absent	Small	Absent
Lateral lachrymal pneumatic foramina	Small	Very small	Small	Small	Multiple, large	Large	Large
Medial lachrymal pneumatic foramina	Present, large	Present, large	Present	?	Absent	Small	Absent
Ectopterygoid pneumatic foramina bounded by thick lip (Carr et al. 2005)	Yes	Yes	Yes	?	No	No	No
Jugal pneumatic foramina	Anterolateral facing	Anterolateral facing	Anterolateral facing	Anterolateral facing	Dorsolateral facing	Anterolateral facing	?

Sexual Dimorphism in *Tyrannosaurus rex*

Is it possible to recognize sexual dimorphism in *Tyrannosaurus rex*? The subject of sexual dimorphism in nonavian theropods has been examined by a number of authors over the years (e.g., Paul 1988; Colbert 1989, 1990; Raath 1990; Chinsamy 1990; Gay 2005). The subject of sexual dimorphism in *Tyrannosaurus rex* has surfaced repeatedly since Carpenter first broached the subject in 1990 (Molnar 1991; Larson and Frey 1992; Larson 1994, 1995, 2001; Horner and Lessem 1993; Carpenter and Smith 2001; Larson and Donnan 2002; Brochu 2003; Molnar 2005). These authors have also explored the possibilities of identifying, or at least separating, the sexes of various theropod species on the basis of differences in cranial ornamentation (Larson 1994; Molnar 2005), pelvic construction (Carpenter 1990; Larson 1994, 1995, 2001; Larson and Donnan 2002), erosion of the femur to liberate calcium for egg production (Chinsamy 1990), preservation of medullary bone (Schweitzer et al. 2005), differences in hemal arch (chevron) morphology (Larson and Frey 1992; Larson 1994, 1995; Erickson et al. 2005), the presence of eggs within the pelvic arc (Sato et al. 2005), and skeletal morph (i.e., gracile vs. robust morphs) (Paul 1988; Carpenter 1990; Raath 1990; Chinsamy 1990; Larson and Frey 1992; Larson 1994, 1995, 2001; Larson and Donnan 2002; Carpenter and Smith 2001).

Sexual dimorphism in extant animals is well documented. We recognize this in mammals as the presence of antlers in male cervids; longer and more massive tusks in male elephants, suids, and walrus; larger horns in male bovids; the presence of canines in male equids; and a generally larger male body size (e.g., Macdonald 1984). This sexual size dimorphism can be quite impressive, reaching as much as a 7:1 (3500 kg : 500 kg) ratio of male to female body mass in the southern elephant seal, *Mirounga leonina* (Lindenfors et al. 2002). Interestingly, for many mammals, the only obvious sexual dimorphism, excluding genitalia, is expressed in adult size, with males outweighing females (Macdonald 1984).

Many reptile groups (e.g., crocodilians; Bellairs 1970) seem to follow this mammalian pattern of sexual size dimorphism. However, it is not always the males who outweigh the females. In turtles and snakes (Fitch 1981), and even in a few mammal groups like baleen whales (Minasian et al. 1984) and hyenas (Estes 1991), sexual size dimorphism is expressed by females being larger than males. Species of invertebrates, to offer other examples, are often quite sexually size dimorphic, with the female, almost without exception, being the larger. In fact, the world record holder for the most sexually size-dimorphic animal is the blanket octopus, *Tremoctopus violaceus*, where females may outweigh males by as much as 40,000 to 1 (Norman et al. 2002).

Birds, the closest living relatives to nonavian theropods, are often quite sexually dimorphic. This dimorphism may be expressed as differences in coloration (the ostrich, *Struthio camelus*), plumage (the common peafowl, *Pavo cristatus*), keratinous structures (the rhinoceros hornbill, *Buceros rhinoceros*), fleshy head ornamentation (the common turkey, *Meleagris gallopavo*), or even inflatable fleshy structures (the greater prairie chicken, *Tympanuchus cupido*). Unfortunately, because none of these features is likely to be preserved in the fossil record, they are not much use in recognizing sexual dimorphism in extinct theropods. Sexual size dimorphism, however, is effective in separating males from females in some bird species (Brad Livezey, personal communication). Sexual size dimorphism may also prove recognizable in nonavian theropods like *Tyrannosaurus rex*.

For many birds, sexual size dimorphism is measurable. It manifests itself as males larger than females in gulls (Ingolfsson 1969; Schnell et al. 1985; Bosch 1996), steamer ducks (Livezey and Humphrey 1984), sparrows (McGillivray and Johnston 1987), and skimmers and terns (Coulter 1986; Quinn 1990), among others. Sexual size dimorphism also occurs with females larger than males in spotted owls (Blakesley et al. 1990), ospreys (Schaadt and Bird 1993), sandpipers (Sandercock 1998), emus (Maloney and Dawson 1993), and so forth. Morphometric analysis, performed by skeletal measurements, has proven effective in separating sex when the difference in mass is over 6% (Schnell et al. 1985). It has even been possible to separate the sexes of mature individuals through morphometric examination (by using bill, wing, and tail measurements) when mass differences between the sexes was insignificant or even indiscernible (Winker et al. 1994).

Although researchers have referred to the presence of robust and gracile morphotypes, Molnar (2005) points out that to date, these morphotypes have not been adequately quantified, but rather are generally based on visual assessments. Is it possible to recognize and quantify sexual size dimorphism, and clearly classify individual *Tyrannosaurus* specimens as robust or gracile morphs? To answer this, I have taken measurements of select elements from 25 specimens of *Tyrannosaurus rex*. Measurements were also taken for 2 outgroup specimens assigned by this study to *Nanotyrannus lancensis* (CMNH 7541 and BMR P2002.4.1) and one to *Gorgosaurus* sp. (TCM2001.89.1). Even though this study considers *Tyrannosaurus* "x" to be the same genus as *T. rex* and hence should be separable in a consistent manner, 3 of these specimens (AMNH 5027, Samson, and MOR 008) also appear as outgroups. Measurements varied from element to element and consisted of lengths, widths, heights, and/or circumference, as shown in Figure 8.9; the values are found in Tables 8.5 and 8.6. Clustering on graphs is assumed to separate sexual size dimorphs. The results were then compared with a visual analysis that divided robust morphs from gracile. Some elements failed to provided significant results (e.g., dentary length vs. tooth row length). For other elements, there was simply not enough data to yield meaningful results (e.g., metatarsal II length vs. circumference, Fig. 8.11; ilium length vs. height, Fig. 8.12; and humerus length vs. circumference, Fig. 8.13), although visual examination was able to separate them, indicating that the human eye can see apparent differences (as in Fig. 8.10). Elements that provided too few data may yet prove useful for quantifiable analysis when additional specimens are discovered. Elements that were abundant, such as the femur (Fig. 8.14) and humerus (Fig. 8.13), yielded clear results, which confirmed their separation by visual inspection: robust plotted individuals look more robust.

From the results of the analysis, 2 morphs of *Tyrannosaurus* are apparent, a robust and a gracile morph. Neither geographic nor stratigraphic distribution can explain these differences. Therefore, because both crocodiles and birds show sexual size dimorphism, extant phylogenetic bracketing tells

Table 8.5. *Measurements Used in Morphometric Analysis*

Specimen	Humerus		Ulna		Scapula		Ilium			Sacrum			Metatarsal			Femur	
	L	C	L	C	L	C	L	H	L	S-1	S-5	S-6	L II	C II	L IV	C IV	L	C
T. rex																		
CM 9380	350	165															1200	545
MOR 009							1160	400										
MOR 1128																	1260	580
MOR 1125			200	99		180											1150	510
MOR 555	375	162	198	106	980	185	1490	565	1010	219	202	170	585	295	605	253	1275	514
MOR 980	362	165			940	146	1397	483	851				597	232	655		1232	483
FMNH PR 2081	390	185	220	121	1140	205	1480	590	980	285	210	240		280	600	247	1340	580
BHI 3033							1550	590	1060				595				1310	500
BHI 4100					800	142												
BHI 6232	360	172															1180	527
BHI 6231																	1110	515
BHI 6233	330	145	185	82			1470						600	272	625	238	1190	494
BHI 6230																	1180	512
BHI 6242			176	96	940	196	1275	490	895	240	185	204	550	267	565	263	1200	560
TCM 2001.90.1	302	150						535	980	220	200	210					1210	470
RTMP 81.12.1																		490
RTMP 81.6.1																		490
BMNH R7994																	990	425
NHM R8001																	1200	467
USNM 6183																		
LL 12823																		
T. "x"																		
Samson													610	305	635	280	1295	560
Nanotyrannus																		
BMR P2002.4.1	280	118	180	74	675	109	720	220	500	85	64	63	510	149	513	136	720	250
Gorgosaurus																		
TCM 2001.89.1	305	130				120	865	305					490	195	500	184	825	270

Abbreviations:—C, circumference; L, length; S, sacrals; II, IV, metatarsal II, IV.

Table 8.6. Measurements (mm) Used in Morphometric Analysis

Specimen	Maxilla							Dentary					Lachrymal			
	H	LA	LB	LA/B	LTR	TP	L	HA	H	LTR	TP	H	L	LF	H/LF
T. rex																
CM 9380	380	695	620	1.12	535	12	860	180	290	510	13				
CM 1400	390	760			620	12									
LACM 23844								175	260	545	13				
MOR 1125	360	680	590	1.15	520	12	920	140	230	485	14				
MOR 555	395	798			620	12	760	170	310	630	13	365	325	26.5	13.8
MOR 980	340	770			560	11	990	175	280	545	13	310	310	27.3	11.4
FMNH PR2081	400	855	720	1.19	645	12	900	200	320	620	13				
BHI 3033	395	775	650	1.19	605	11	1010	175	280	570	13	340	340	35.8	9.5
BHI 4100		730					915	160		530	13	290	290	22.7	12.8
BHI 4182							770	170	270	530	14				
RTMP 81.6.1	318				540		910	143	248	530		280	260	26.1	10.7
UCMP 118742	390	810	690	1.17	625	12	770								
AMNH 5027	360	710	710	1	530	12	850	135	190	520	14	360	340	15.8	22.8
MOR 008	350	720			580	12	880	180	280	560	13	340	300	17.4	19.5
T. "x"																
Samson							870	170	270	540					
Nanotyrannus																
BMR P2002.4.1	160	470	385	1.22	355	15	505	69	107	318	17	163	180	35.1	4.6
CMNH 7541	150	385	309	1.25	278	15	375								
BHI 6235												120		18	6.7
Gorgosaurus																
TCM 2001.89.1	230	568	420	1.35	405	14	580	95	197	360	15	235	200	56.4	4.2

Note.—See Figure 8.9.

Abbreviations:—A, B refer to points in Figure 8.9. F, foramen (=lacrymal foramen); H, height; L, length; LTP, number of tooth positions; TR, tooth row.

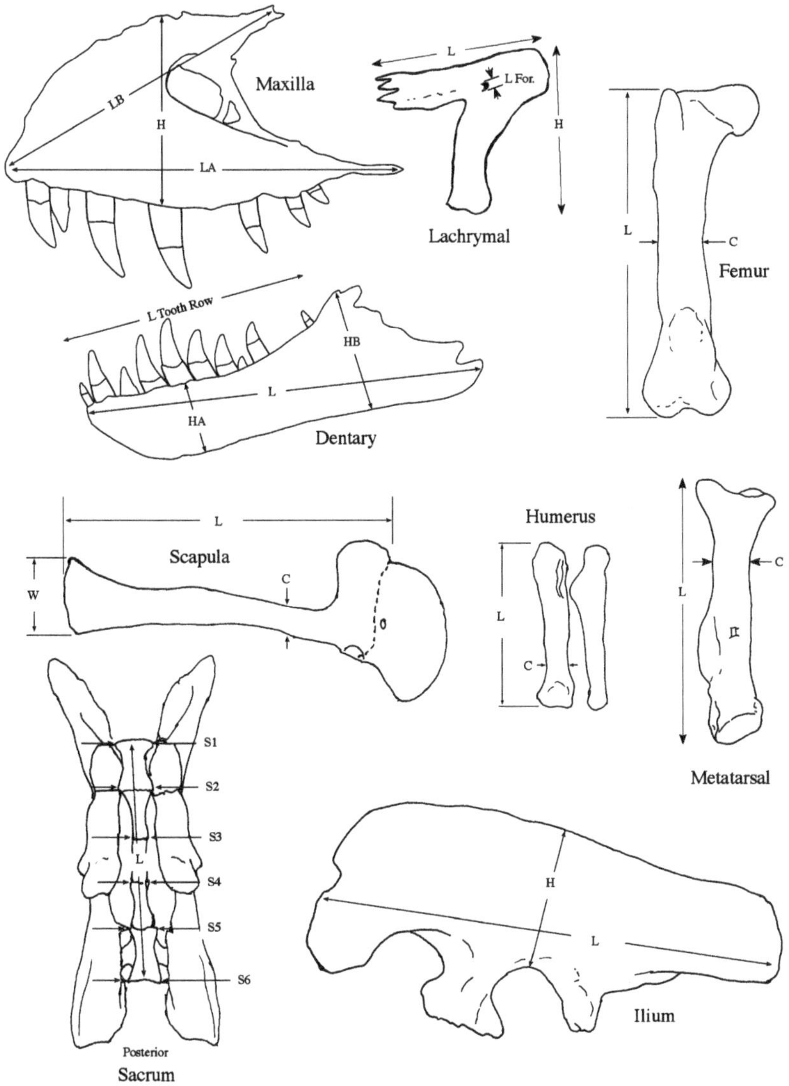

Figure 8.9. *Examples of measurement techniques used in this study.*

us that the most parsimonious explanation for the presence of these 2 morphs is sexual size dimorphism. The formula developed by Anderson et al. (1985) was used to estimate the mass of the robust and gracile morphs from femur diameter. The weight estimates (Table 8.7) show a maximum weight for the gracile morphs of 4.0 metric tonnes, with a mean of 3.5 metric tonnes (6 individuals); and a maximum weight of 5.6 metric tonnes and a mean of 4.7 metric tonnes for robust morphotypes (9 individuals).

Male or Female

Given that the presence of 2 morphs has been established for *Tyrannosaurus rex*, can we determine the sex of the morphotypes? Carpenter (1990) suggested that, on the basis of the greater divergence of ischium (Fig. 8.17), the robust form was female. Larson and Frey (1992) agreed with Carpenter, and they further suggested that the location and morphology

Variation and Sexual Dimorphism

Figure 8.10. *Anterior view of right metatarsal II of (A) gracile (BHI 3033) and (B) robust (TCM 2001.90.1) morphotypes.*

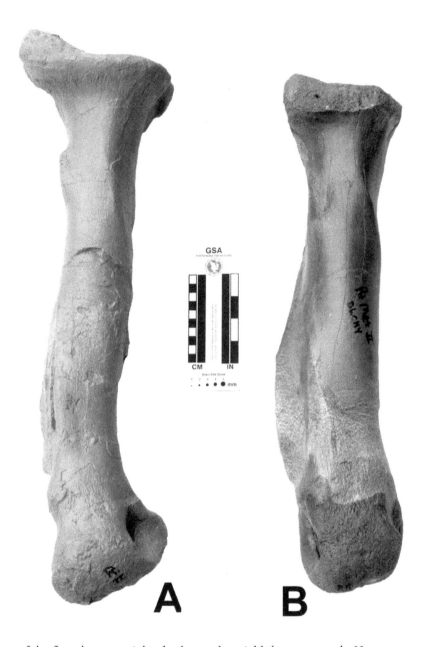

of the first chevron might also be used to yield the same result. However, this method has proven unreliable (Erickson et al. 2005). Elsewhere, I (Larson 1994, 1995) have suggested that the wider pelvic arch and healed injuries of the proximal caudal vertebrae (consistant with injuries potentially inflicted by a mounting male during copulation) were restricted to robust morphotypes. I (Larson 2001; Larson and Donnan 2002) supported Carpenter's (1990) conclusion that robust individuals were female. But because of the tenuous nature of these conclusions, I have speculated that one way to positively recognize a female is to locate medullary bone within the skeleton (Larson and Donnan 2002). Medullary bone is only deposited within the medullary cavity in the long bones of female birds during ovulation, as an aid to the quick mobilization of calcium for egg

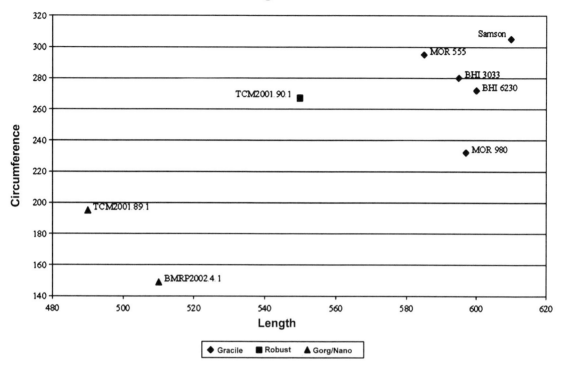

Figure 8.11. *Metatarsal II length vs. circumference.*

Figure 8.12. *Ilium length vs. height.*

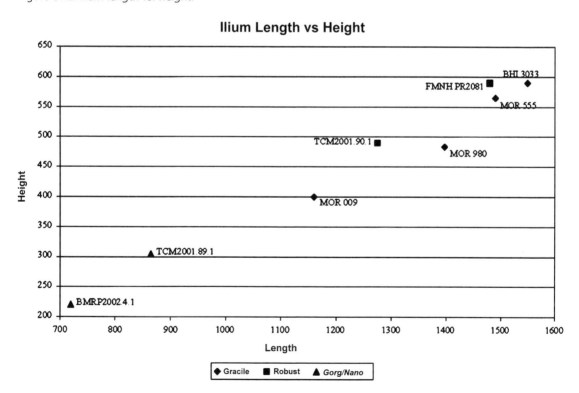

Variation and Sexual Dimorphism   119

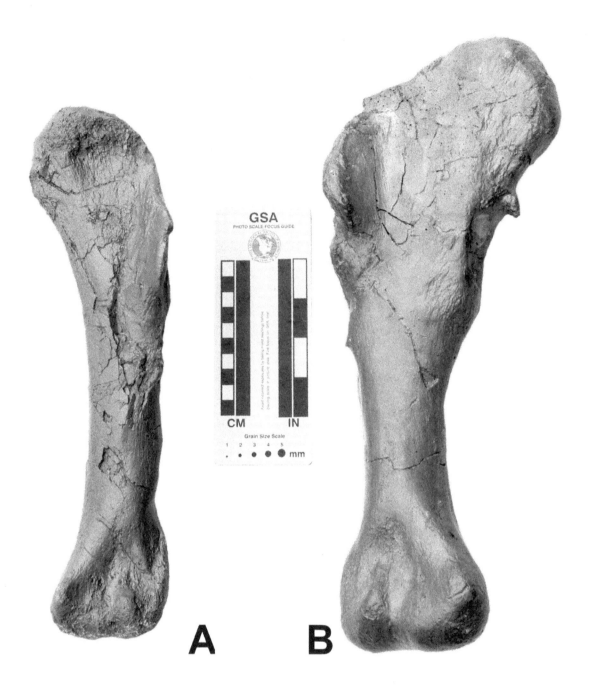

Figure 8.13. Anterior view of (A) left humerus of gracile (BHI 6230) and (B) right humerus of robust (FMNH PR2081) morphotypes.

production (Taylor 1970; Welty and Baptista 1988; Schweitzer et al. this volume). Although the absence of medullary bone is inconclusive (it is not found in males and nonovulating females), its presence unequivocally identifies a female.

Medullary bone has not been documented in ovulating female crocodilians. Although ovulating birds have medullary bone, there were no guarantees that ancestral nonavian theropods shared this character. What would be the chances of finding the fossil of an ovulating female *Tyrannosaurus rex*, preserving the medullary bone, exposing the inside of the medullary cavity, and recognizing and then verifying that the tissue is medullary bone? Unbelievably, that is exactly what Schweitzer et al. (2005, this

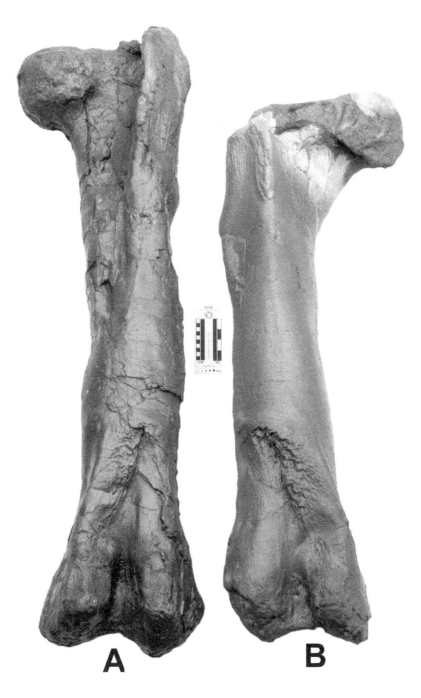

Figure 8.14. *Anterior view of (A) left femur of gracile (BHI 3033) and (B) right femur of robust (TCM2001.90.1) morphotypes.*

volume) did. Schweitzer et al. have verified the presence of medullary bone within the femur of a specimen of *Tyrannosaurus rex* by comparison with medullary bone extracted from laying chickens (*Gallus gallus*) and ostriches (*Struthio camelus*). By plotting information from the femur from which the medullary bone was found (MOR 1125), it was found that the specimen clusters with robust morphotypes (Fig. 8.18), thereby providing independent supporting evidence that the robust morphotypes are most

Variation and Sexual Dimorphism

Table 8.7. Calculated Mass for **Tyrannosaurus** Specimens and Outgroups, **Nanotyrannus** and **Gorgosaurus**

Specimen	Type	Femur Length (mm)	Femur Circumference (mm)	Mass (kg)	Mass (tonne)	Morph
T. rex						
CM 9380		1200	545	4726	4.7	R
BMNH R7994			490	3535	3.5	
MOR 1128		1260	580	5601	5.6	R
MOR 1125	B-rex	1150	510	3943	3.9	R
MOR 555	Wankel rex	1275	514	4028	4.0	G
MOR 980	Peck's Rex	1232	483	3399	3.4	G
FMNH PR2081	Sue	1340	580	5601	5.6	R
BHI 3033	Stan	1310	500	3735	3.7	G
BHI 6232		1180	527	4312	4.3	R
BHI 6233		1110	515	4049	4.1	R
BHI 6230	Wyrex	1190	494	3614	3.6	G
BHI 6242	Henry	1180	512	3985	4.0	R
RTMP 81.12.1	Huxley	1200	560	5090	5.1	R
RTMP 81.6.1	Back Beauty	1210	470	3155	3.2	G
USNM V6183		990	425	2397	2.4	
LL 12823		1200	467	3100	3.1	G
T. "x"						
Samson	Z-rex	1295	560	5090	5.1*	R
BMR P2002.4.1	*Nanotyrannus*	720	250	563	0.6	
TCM2001.89.1	*Gorgosaurus*	825	270	695	0.7	

Abbreviations:—
G, gracile; R, robust.
* Mean, 4.1.

certainly females. We may therefore assume that the gracile morphotypes are males.

Conclusion

This study examined 34 specimens that have been assigned by various authors to *Tyrannosaurus rex*. This list also included specimens ascribed by some authors to *Nanotyrannus lancensis* but synonymized by others with *T. rex*. By use of shared and derived characters, these specimens (CMNH 7541, LACM 28471, BMR P2002.4.1, and BHI 6235) may clearly be removed from the clade, thus validating the work of Gilmore (1946) and Bakker et al. (1988). Also of contention is a group of 4 specimens (AMNH 5027, MOR 008, SDSM 12047, and Samson) that have been referred to as *Tyrannosaurus* "x." Again, by use of taxonomic characters, there is ample evidence to remove them from the species *rex*, but maintain them within the genus *Tyrannosaurus*.

By use of morphometric analysis, gracile and robust morphs are confirmed to be present within the clade *Tyrannosaurus rex*. Extant phylogenetic bracketing (comparison with living crocodiles and birds) leads us to conclude that the existence of these 2 morphs most parsimoniously repre-

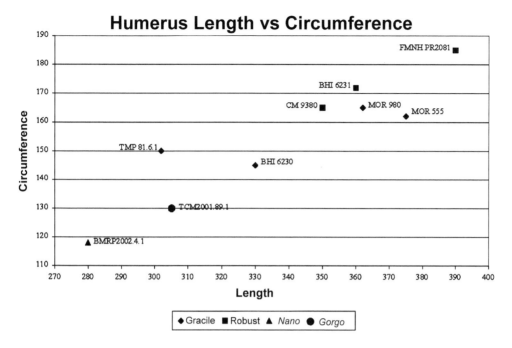

Figure 8.15. *Humerus length vs. circumference.*

Figure 8.16. *Femur length vs. circumference.*

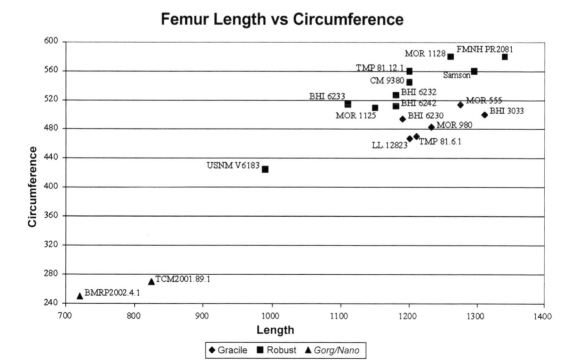

Variation and Sexual Dimorphism 123

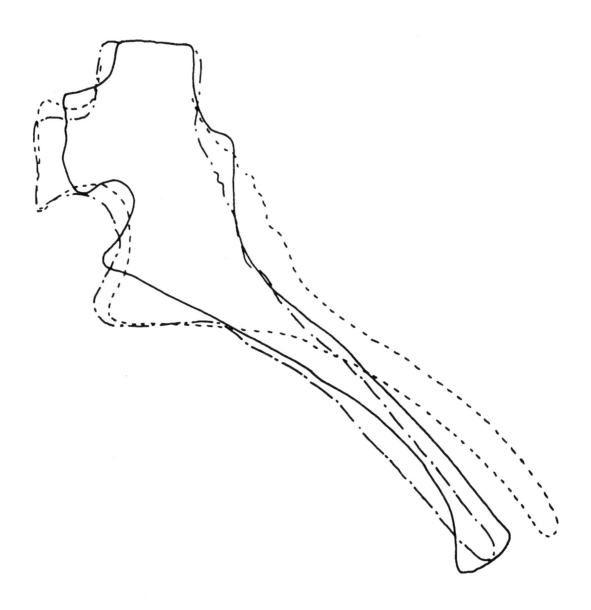

Figure 8.17. Overlay of the ischia of (A) CM 9380; (B) RTMP 81.61; and (C) AMNH 5027 (after Carpenter 1990).

sents sexual dimorphism. The discovery of medullary bone within the medullary cavity of a robust specimen of *T. rex* established MOR 1125 as female (Schweitzer et al. 2005), and therefore all other robust *T. rex* specimens are, in all probability, also female.

Acknowledgments

I thank Larry Shaffer of Black Hills Institute for preparation of the figures and tables, and Neal Larson for some of the photography. I am extremely grateful to Phil Currie, Bill Simpson, Tim Tokaryk, Phil Fraley, Tom Williamson, Chris Morrow, Phil Manning, Kenneth Carpenter, Mike Henderson, and Scott Williams, who grabbed their tape measures and supplyed missing data at a moment's notice. Conversations with Kenneth Carpenter, Ralph Molnar, Thomas Carr, Mike Henderson, Phil Currie, Jorn Hurum, Greg Erickson, Bob Bakker, and a host of others have provided insight. The

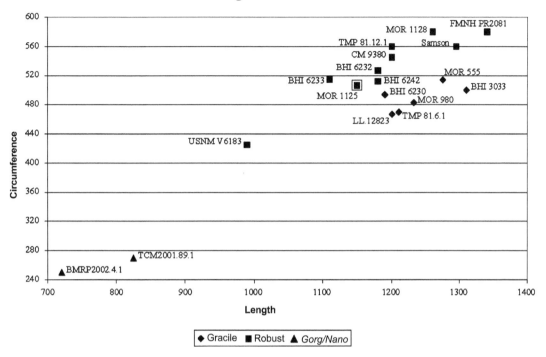

Figure 8.18. MOR 1125, a female **Tyrannosaurus rex,** clusters with other robust individuals.

compilation of data would not have been possible without the access, help, and patience provided by the curators, collection managers, and preparators at all the institutions I worked with, especially Bill Simpson, Jack Horner, Carrie Herbel, Mark Norell, Luis Chiappe, and Matt Lamanna. Last, but certainly not least, thanks to all the discoverers and collectors who saved the specimens that provided my data.

References Cited

Anderson, J. F., Hall-Martin, A., and Russell, D. A. 1985. Long-bone circumference and weight in mammals, birds, and dinosaures. *Journal of the Zoological Society of London* A 207: 53–61.
Bakker, R. T., Williams, M., and Currie, P. 1988. *Nanotyrannus,* a new genus of pygmy tyrannosaur, from the Latest Cretaceous of Montana. *Hunteria* 1(5): 1–26.
Bellairs, A. 1970. *The Life of Reptiles.* Universal Books, New York.
Blakesley, J. A., Franklin, A. B., and Gutierrez, R. J. 1990. Sexual dimorphism in northern spotted owls from northwest California. *Journal of Field Ornithology* 61(3): 320–327.
Bosch, M. 1996. Sexual size dimorphism and determination of sex in yellow-legged gulls. *Journal of Field Ornithology* 67(4): 534–541.
Brochu, C. A. 2003. Osteology of *Tyrannosaurus rex*: insights from a nearly complete skeleton and high-resolution computed tomography analysis of the skull. *Journal of Vertebrate Paleontology* Memoir 7.
Buss, I. O. 1990. *Elephant Life: Fifteen Years of High Population Density.* Iowa University Press, Ames.
Carpenter, K. 1990. Variations in *Tyrannosaurus rex.* P. 141–145 in Carpenter,

K., and Currie, P. J. (eds.). *Dinosaur Systematics: Approaches and Perspectives*. Cambridge University Press, Cambridge.

———. 1992. Tyrannosaurids (Dinosauria) of Asia and North America. P. 250–268 in Mateer, N., and Chen, P. J. *Aspects of Nonmarine Cretaceous Geology*. China Ocean Press, Beijing.

Carpenter, K., and Smith, M. 2001. Forelimb osteology and biomechanics of *Tyrannosaurus rex*. P. 90–116 in Tanke, D. H., and Carpenter, K. (eds.). *Mesozoic Vertebrate Life*. Indiana University Press, Bloomington.

Carr, T. D. 1999. Craniofacial ontogeny in tyrannosauridae (Dinosauria, Coelurosauria). *Journal of Vertebrate Paleontology* 19(3): 497–520.

Carr, T. D., and Williamson, T. E. 2004. Diversity of Late Maastrichtian tryrannosauridae (Dinosauria: Theropoda) from western North America. *Zoological Journal of the Linnean Society* 142: 479–523.

Carr, T. D., Williamson, T. E., and Schwimmer, D. R. 2005. A new genus of tyrannosaurid from the Late Cretaceous (Middle Campanian) Demopolis Formation of Alabama. *Journal of Vertebrate Paleontology* 25(1): 119–143.

Chinsamy, A. 1990. Physiological implications of the bone histology of *Syntarus rhodesiensis* (Saurischia: Theropoda). *Palaeontologia Africana* 27: 77–82.

Colbert, E. H. 1989. *The Triassic Dinosaur Coelophysis*. Museum of Northern Arizona Bulletin 57.

———. 1990. Variation in *Coelophysis bauri*. P. 81–90 in Carpenter, K., and Currie, P. J. (eds.). *Dinosaur Systematics: Approaches and Perspectives*. Cambridge University Press, Cambridge.

Coulter, M. C. 1986. Assertive mating and sexual dimorphism in the common tern. Savannah River Ecology Lab, South Carolina. *Wilson Bulletin* 98(1): 93–100.

Currie, P. J. 2003. Cranial anatomy of tyrannosaurid dinosaurs from the Late Cretaceous of Alberta, Canada. *Acta Palaeontologica Polonica* 48(2): 191–226.

Currie, P. J., Hurum, J. H., and Sabath, K. 2003. Skull structure and evolution in tyrannosaurid dinosaurs. *Acta Palaeontoligica Polonica* 48(2): 227–234.

Darwin, C. 1868. *The Variation of Animals and Plants under Domestication*. Vol. 1. John Murray, London.

Erickson, G. M., Lappin, A. K., and Larson, P. L. 2005. Androgynous *rex*—the utility of chevrons for determining the sex of crocodilians and non-avain dinosaurs. *Zoology* 108: 277–286.

Estes, R. D. 1991. *The Behavior Guide to African Mammals*. University of California Press, Berkeley.

Fitch, H. S. 1981. *Sexual Size Differences in Reptiles*. University of Kansas Museum of Natural History Miscellaneous Publication 70.

Gay, R. 2005. Sexual dimorphism in the Early Jurassic theropod dinosaur *Dilophosaurus* and a comparison with other related forms. P. 277–283 in Carpenter, K. (ed.). *The Carnivorous Dinosaurs*. Indiana University Press, Bloomington.

Gilmore, C. W. 1946. A new carnivorous dinosaur from the Lance Formation of Montana. *Smithsonian Miscellaneous Collection* 106: 1–19.

Horner, J. R., and Lessem, D. 1993. *The Complete T. rex*. Simon & Schuster, New York.

Ingolfsson, A. 1969. Sexual dimorphism of large gulls. *Auk* 86: 732–737.

Larson, P. L. 1994. *Tyrannosaurus* sex. P. 139–155 in Rosenberg, G. D., and Wolberg, D. L. (eds.). *Dino Fest Proceedings*. Paleontological Society Special Publication 7.

———. 1995. To sex a *rex*. *Nature Australia* 25(2): 46–53.

———. 1997. The king's new clothes: a fresh look at *Tyrannosaurus rex*. P. 65–71 in Wolberg, D. L., Stump, E., and Rosenberg, G. D. *Dinofest International Proceedings*. Academy of Natural Sciences, Philadelphia.

———. 1999. Guess who's coming to dinner; *Tyrannosaurus* vs. *Nanotyrannus*: variance in feeding habits (abstract). *Journal of Vertebrate Paleontology*, Abstract of Papers, 58a.

———. 2001. Paleopathologies in *Tyrannosaurus rex* (in Japanese). *Dino Press* 5: 26–35.

———. In press. *The Case for Nanotyrannus*. Black Hills Institute of Geological Research.

Larson, P. L., and Donnan, K. 2002. *Rex Appeal: The Amazing Story of Sue, the Dinosaur that Changed Science, the Law and My Life*. Invisible Cities Press, Montpelier, VT.

Larson, P. L., and Frey, E. 1992. Sexual dimorphism in the abundant Upper Cretaceous theropod, *Tyrannosaurus rex* (abstract). *Journal of Vertebrate Paleontology*, Abstract of Papers, 38a.

Lincoln, G. A. 1994. Teeth, horns and antlers: the weapons of sex. P. 131–159 in Short, R. V., and Balaban, E. (eds.). *The Differences between the Sexes*. Cambridge University Press, Cambridge.

Lindenfors, P., Tullberg, B. S., and Biuw, M. 2002. Phylogenic analyses of sexual selection and sexual size dimorphism in pinnipeds. *Behavior, Ecology, Sociobiology* 52: 188–193.

Livezey, B. C., and Humphrey, P. S. 1984. Sexual dimorphism in continental steamer-ducks. The Cooper Ornithological Society. *Condor* vezey(86): 368–377.

McGillivray, W. B., and Johnston, R. F. 1987. Differences in sexual size dimorphism, and body proportions between adult and subadult house sparrows in North America. 1987. *Auk* 104: 681–687.

Macdonald, D. W. 1984. *The Encyclopedia of Mammals*. Facts on File, New York.

Maloney, S. K., and Dawson, T. J. 1993. Sexual dimorphism in basal metabolism and body temperatures of a large bird, the emu. *Condor* (95): 1034–1037.

Martin, L. A., Willner, L. A., and Dettling, A. 1994. The evolution of sexual size dimorphism in primates. P. 159–202, in Short, R. V., and Balaban, E. (eds.). *The Differences between the Sexes*. Cambridge University Press, Cambridge.

Minasian, S. M., Balcomb, K. C., III, and Foster, L. 1984. *The World's Whales: The Complete Illustrated Guide*. Smithsonian Books, Washington, DC.

Molnar, R. E. 1991. The cranial morphology of *Tyrannosaurus rex*. *Palaeontographicia*. Abteilung A 217: 137–176.

———. 2005. Sexual selection and sexual dimorphism in theropods. P. 277–283 in Carpenter, K. (ed.). *The Carnivorous Dinosaurs*. Indiana University Press, Bloomington.

Norman, M. D., Paul, D., Finn, J., and Tregenza, T. 2002. First encounter with a live male blanket octopus: the world's most sexually size-dimorphic large animal. *New Zealand Journal of Marine and Freshwater Research* 36: 733–736.

Osborn, H. F. 1905. *Tyrannosaurus rex* and other Cretaceous carnivorous dinosaurs. *American Museum of Natural History Bulletin* 21: 259–296.

———. 1906. *Tyrannosaurus rex*, Upper Cretaceous carnivores dinosaur (second communication). *American Museum of Natural History Bulletin* 22: 281–296.

Paul, G. S. 1988. *Predatory Dinosaurs of the World: A Complete Illustrated Guide*. Simon & Schuster, New York.

Quinn, J. S. 1990. Sexual size dimorphism and parental care patterns in a monomorphic and a dimorphic Larid. *Auk* 107: 260–274.

Raath, M. A. 1990. Morphological variation in small theropods and it's meaning in systematics: evidence from *Syntarsus rhodesiensis*. P. 91–105 in Carpenter, K., and Currie, P. J. (eds.). *Dinosaur Systematics: Approaches and Perspectives*. Cambridge University Press, Cambridge.

Romer, A. S. 1966. *Vertebrate Paleontology*. University of Chicago Press, Chicago.

Sandercock, B. K. 1998. Assortive mating and sexual size dimorphism in western and semipalmated sandpipers. *Auk* 115(3): 786–791.

Sato, T., Cheng, Y., Wu, X., Zelenitsky, D. K., and Hsiao, Y. 2005. A pair of shelled eggs inside a female dinosaur. *Science* 308: 375.

Schnell, G. D., Worthen, G. L., and Douglas, M. E. 1985. Morphometric assessment of sexual dimorphism in skeletal elements of California gulls. *Condor* 87: 484–493.

Schweitzer, M. H., Wittmeyer, J. L., and Horner, J. R. 2005. Gender-specific reproductive tissue in ratites and *Tyrannosaurus rex*. *Science* 308: 1456–1460.

Schaadt, C. P., and Bird, D. M. 1993. Sex-specific growth in ospreys: the role of sexual size dimorphism. *Auk* 110: 900–910.

Taylor, T. G. 1970. How an egg shell is made. *Scientific American* 222(3): 88–95.

Welty, T. C., and Baptista, L. 1988. *The Life of Birds*. 4th ed. Saunders College Publications, Fort Worth, TX.

Winker, K., Voelker, G. A., and Klicka, J. T. 1994. A morphometric examination of sexual dimorphism in the *Hylophilus*, *Xenops*, and an *Automolus* from southern Veracruz, Mexico. *Journal of Field Ornithology* 65(3): 307–323.

Witmer, L. M. 1995. The extant phylogenetic bracket and the importance of reconstructing soft tissue in fossils. P. 19–33 in Thomason, J. J. (ed.). *Functional Morphology in Vertebrate Paleontology*. Cambridge University Press, Cambridge.

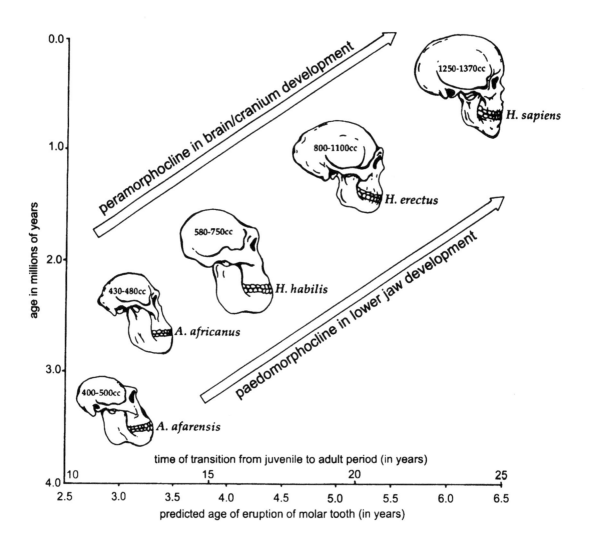

Figure 9.1. The compensation principle is well illustrated by the hominid skull, which demonstrates a reciprocal relationship between cranium and jaw if one is peramorphic, the other is paedomorphic, and vice versa. Note that the main evolutionary trend is toward cranial enlargement (encephalization or anteriorization). After McNamara (1997).

9

WHY *TYRANNOSAURUS REX* HAD PUNY ARMS: AN INTEGRAL MORPHODYNAMIC SOLUTION TO A SIMPLE PUZZLE IN THEROPOD PALEOBIOLOGY

Martin Lockley, Reiji Kukihara, and Laura Mitchell

The purpose of this chapter is to show that the reason *Tyrannosaurus* (and *Carnotaurus*) had miniature arms may be surprisingly simple as well as consistent with the broad morphological context of theropod and saurischian growth dynamics and heterochrony. Inherent, or formal, morphodynamic growth trends (sensu Gould 2002) lead to strong anteriorization (of the head) in derived, large (mainly peramorphic) members of various theropod, saurischian, and dinosaur clades, as well as in other vertebrates. These seem to be of no obvious functional significance (sensu Gould 2002), leading to the inference that we should pay more attention to these morphodynamic trends as part of the inherent structure of vertebrate organization. For example, in amphibians (salamanders versus frogs), pterosaurs (rhamphoryhnchoids versus pterodactyloids), sauropodomorphs (prosauropods versus brachiosaurs), ornithischians (primitive thyreophorans versus derived ceratopsians), and primates (monkeys versus hominids), the latter (derived) groups always show more anteriorization (encephalization) than their primitive relatives. There is also a compensatory reduction or loss of the tail.

Likewise, in derived forms, the inner (proximal) portions of the limb (femora and humeri) are typically more developed than the distal portions. This differentiation of proximal and distal limbs and feet or hands gives rise to diverse morphologies that have been interpreted as being of functional utility—for example, grasping theropod hands or slashing raptor claws. However, such developmental exaggeration or emphasis in one organ or region of the body inevitably results in underdevelopment in adjacent organs, as required by the principle of compensation, also known as heterochronic trade-offs (sensu McNamara 1997). Mounting evidence from evolution of development studies (Carroll 2005)—popularly known as evo-devo—suggests that organs must be looked at holistically (i.e., in the context of the whole body) and that formal developmental patterns and morphologies reiterate fractally and convergently throughout the vertebrate world, and indeed in the biosphere in general.

T. rex is undoubtedly one of the most popular, fascinating, and controversial dinosaurs. Although its large head and ferocious teeth have obvious

Introduction

*If grasping hands and long arms generally typify theropods, how does one explain the outrageously diminutive forelimbs of **Tyrannosaurus** and **Carnotaurus**. . . . It is not at all clear why these animals independently miniaturized their arms.*

Fastovsky and Weishampel 2005, p. 283

functional use in fostering its reputation as the king of the predators (or an undiscriminating bone-crunching scavenger or scavenger-predator), its tiny arms stand out as an unusual or anomalous morphological feature. Indeed, the forelimbs are described as "outrageously diminutive" by leading experts in the latest textbooks (e.g., Fastovsky and Weishampel 2005, p. 283). Given the prevalence of functional explanations generated by the Darwinian paradigm (Gould 2002), it is not surprising that paleontologists have pondered the use of such small and apparently vestigial organs. Although Darwinism typically demands or at least prefers functional explanations, biology generally accepts that some organs are more functional than others, and the use of terms like *vestigial* acknowledges the fact that function may have been lost or diminished in some organs, as Paul (1988) inferred in the case of tyrannosaurids. We will return to the specific case of *T. rex* forelimbs once we have presented a broader context in which to develop our understanding of theropod limbs.

Functional arguments can only take us so far because they ignore inherent, intrinsic, or formal growth and development. Evo-devo promises to greatly subsume and reorient the functional Darwinian paradigm. But evo-devo is not new; it is merely the rediscovery of the subdisciplines of heterochrony and morphodynamics (Gould 1997; McNamara 1995, 1997) by biologists. This intrinsic, formal approach has been known for decades, although underutilized, and can be traced as far back as Wolfgang Goethe and Geoffroy Saint-Hilaire in the late 1700s and early 1800s (LeGuyader 2004). These approaches are holistic and internally consistent, and, as current excitement about evo-devo shows, they help demonstrate that we can read developmental patterns in the fossil record, even though some biological (genetic and molecular) evidence is not directly available. We nevertheless see the same patterns of morphological expression in ancient and modern forms and can thus unify biology and paleontology more thoroughly. Increasingly, it is possible to demonstrate that these organizational trends or patterns repeat in many groups in an ordered, lawful way.

The main thrust of this chapter is therefore 2-fold. First, we make the case that morphology is not necessarily completely or wholly explicable in terms of function because intrinsic growth dynamics, which Gould (2002) describes as formal (and related to intrinsic generation of form), play an important role, as recognized by students of heterochrony (McKinney and McNamara 1997; McNamara 1997) and evo-devo (Carroll 2005). Second, an understanding of these formal dynamics reveals that shifts in the timing or morphological development produces a cascade of compensating effects throughout the body so that if one organ grows large, one or more adjacent organs will be reduced in size and vice versa.

We argue that *T. rex* and other tyrannosaurids had small forelimbs because they had such large heads—or more accurately, we stress the morphodynamic compensation between head and forelimbs. Thus, anterior growth bypassed other anterior organs and concentrated in the head. This is in contrast to the patterns observed in other related theropod dinosaurs (coelurosaurs), such as the ornithomimids, which developed long front limbs and necks but had small heads. In such animals, anterior growth concentrated in the forelimbs and neck (as well as the anterior organs), but never became exaggerated in the head. Support for such a hypothesis is derived from an overview of recurrent morphological trends in the Theropoda as a whole, and in the Saurischian clade to which they belong. Fractal or recursive trends mean that they repeat at different levels or scales of organization with similar but not identical patterns (see Bird 2004 for definitions). For example, as noted below, in many theropods, including *T. rex*, short forelimbs are associated with relatively long hind limbs.

As pointed out by Lockley and Kukihara (2005) and Lockley (in press), dimorphism in the well-known dinosaur *Coelophysis* reveals similar compensations (small forelimbs = large head, and large forelimbs = small head) at the species level. (There is also a pattern of compensation between large forelimbs and small hind limb, or vice versa, and corresponding compensations throughout the whole body.) These are lawful in the sense that they can be shown to recur fractally at many different taxonomic levels within the vertebrates, and so must represent some inherent pattern of biological organization. For example, just as there is a polarity between the 2 *Coelophysis* dimorphs or the primitive and derived coelurosaurs (ornithomimids versus derived tyrannosaurs), so too there is a similar polarity between primitive and derived ceratosaurs (e.g., *Coelophysis* versus *Carnotaurus*). In the latter case, the convergence between *Carnotaurus* and *T. rex* in respect to forelimb-head compensations is striking.

We argue that the implications of this morphodynamic approach are far-reaching. For example, it has long been known that theropods had their hands free, unlike other dinosaurs, and so were not overspecialized or committed to quadrupedal locomotion, like most other dinosaurs and the majority of terrestrial vertebrates. However, we take this argument further by suggesting that not only did the anterior limbs of theropods develop physical flexibility, which in turn helps support a physical or biomechanical connection with wing development in birds, but also a close relationship existed between the respiratory system as an anterior organ and the anterior limbs. Thus, it is important to think in terms of how the aforementioned morphodynamic development in limbs, neck, or head is closely integrated with development of physiological systems like respiration. We know that many theropods (coelurosaurs and oviraptosaurs) and birds had a fundamentally important relationship with the air (i.e., air sacs, feathers, wings, and flight), and we argue that this represents an emphasis on the anterior part of the middle system (pulmonary system, anterior torso, and forelimbs). For example, Carrano and O'Connor (2005) describe the vertebrae and small ribs of the neck region of *Ornithomimus* as being perforated by the pulmonary system, although in larger, more peramorphic theropods, like *Tyrannosaurus* and *Carnotaurus*, somatic growth evidently bypassed the forelimbs and was concentrated in the head. These large forms also frequently developed extensive pneumaticity—not just in the neck region, but often throughout the whole body (Xu et al. 2004). The same pattern is seen most modern birds. Thus, small birds may show less pneumaticity than large ones. This is not directly correlated with flight because some large birds, such as the ostrich, are flightless. Therefore, one might argue that in the larger theropods, as is the case in large birds, the whole body had become like an enlarged lung. Feathers are also a means of incorporating air within the boundaries of the physical body. One might argue, therefore, that feathered forms with little skeletal pneumaticity have a more outward relationship with the air than large forms with well-developed internal pneumaticity but no feathers. Again, this is an example of compensation.

Before presenting the data that support these morphological comparisons and inferences, it is necessary to outline the morphodynamic paradigm and its strongly heterochronic flavor.

The Morphodynamic Paradigm

Two generations before Darwin introduced the concept of natural selection, with all its functional implications, the German founders of modern biology, including Goethe, who introduced the term *morphology*, thought in terms of dynamic processes. The recent introduction into the English language of the verb *to morph* emphasizes both this dynamic connotation and the resurgence of such process thinking. (The *Oxford English Dictionary* defines this verb as to "change smoothly and gradually from one image to another.") Interestingly, this usage is intimately associated with the dynamic field of animation. This dynamism contrasts with the mind-set of spatial coordinates, measurements, and discrete character attributes associated with many taxonomic, anatomical, morphometric, and cladistic approaches that followed. Although these are all useful to varying degrees, many of the contributions of the early German biologists (e.g., Goethe and Ernest Haeckel) were often overlooked while attention was directed to the well-established tradition of monographic description of anatomy for classification purposes. In effect, our concept of morphology became frozen, and species were mostly described on the basis of adult forms that represented the final manifestation of the dynamic process of ontogeny. Such cataloging, although important, has sometimes been characterized as mere stamp collecting, and it shifts our mind-set away from the dynamic, organic process. In the context of the dynamic nature of the growth process, this charge is not wholly unjustified. For example, as noted by Arthur (2006), only a few 20th-century biologists working between 1900 and 1975 were really focused on the dynamic relationships between evolution and development; among these were Julian Huxley, Gavin de Beer (1940), and D'Arcy Thompson, whose classic *On Growth and Form* (1917) has been cited as an example of "the theory of transformations" (Arthur 2006, p. 401) dealing with the types of macromorphological dynamics we discuss here.

However, despite the freezing of the concept of morphology, which was originally closely allied to the dynamic concepts of metamorphosis and transmutation (the latter an early 19th-century synonym for evolution), German biologists such as Haeckel (who coined the term *biology*) were at the forefront of embryological research, thus maintaining a focus on dynamic processes. Haeckel (1866) also coined the term *heterochrony*, and he developed the famous biogenetic law that "ontogeny recapitulates phylogeny." Thus, an anteriorization trend (P-A) in ontogeny may parallel one in phylogeny (as seen in *Tyrannosaurus*; see below). Despite the shortcomings in this law if taken too literally, its general relevance in linking individual development and evolution has merit. Indeed, the recent emergence of the evo-devo field is in itself strong validation of the renewed interest in this approach (Arthur 2006). There are even claims of a reverse biogenetic law (Suchantke 1995), which can be characterized as a reverse morphodynamic movement (i.e., an anterior-posterior [A-P] trend in one organ may be compensated for by a P-A trend in another).

What is of fundamental importance here is to note that heterochrony is not some specialized branch of biology. Indeed, a compelling case can be made that it is essentially a synonym of the original dynamic concept of morphological development as process, which emphasizes the highly organic nature of changing or morphing of anatomy through time.

Moreover, the dynamic processes are driven by internal ontogenetic forces. Although such dynamics were recognized in the biometric sense by such concepts as allometry or nonlinear growth, they have been overlooked by the Darwinian notion that the organism is too much under the spell of external influences that force it to passively adapt to the environment. Gould (2002) clearly recognized the difference between the former intrinsic Goethean (or formal) perspective of the Germanic school and the latter extrinsic Darwinian (or functional) perspective of the English school.

It is only now that the Darwinian and neo-Darwinian (genetic) paradigm has been explored to the point where deficiencies in functional explanations are evident that the intrinsic or formal paradigm is coming back into vogue among mainstream biologists. Genetics, in its early days, aimed to support the selectionist Darwinian paradigm by using mathematical population-based statistical models (harking back to 19th-century social Darwinism and Malthus's ideas of populations competing for scare resources). Ironically, modern developmental genetics and evo-devo now inform us that form and species diversity is generated by internal processes that are most dynamic and complex in the very early formative stages of ontogeny. Put simply, focus has shifted from the paradigm of a passive Darwinian organism, pushed around by the external environment and competing to survive by functioning properly, to an active model of dynamic organisms generating form through complex internal or formal organization.

Despite the previous strong focus of mainstream biology and evolutionary studies on Darwinian selection paradigms, a number of workers have kept alternative perspectives alive. Among the most relevant studies are those that have explored the dynamics of heterochrony (Gould 1977; McKinney and McNamara 1991; McNamara 1997). The latter study introduced the important concept of heterochronic trade-offs that are, in fact, basically a synonym of the compensation principle introduced by Goethe (1795). This principle essentially tells us that no organ can develop without a reciprocal effect on an adjacent organ. McNamara (1997) gives the excellent example of the trade-off between the large human cranium and small jaw and contrasts it with the opposite or reciprocal case of the small cranium and large jaw, as seen in the chimp or certain primitive hominids (Fig. 9.1). In such a case, the larger organ is peramorphic (exhibiting more growth) and the smaller one is paedomorphic (exhibiting less growth). It seems that such compensations are found universally throughout the organic world and that they are an integral factor in the evolutionary process.

In this regard, we may note the recent resurgence of interest in the work of Geoffroy Saint-Hilaire (1822). Endorsements by Thom (1975), Gould (1985), DeRobertis and Sasai (1996), and LeGuyader (2004), among others, clearly demonstrate Saint-Hilaire's profound understanding of integrated organization in organic systems as early as the 1790s. Thus, Saint-Hilaire recognized and endorsed Goethe's compensation principle, which he called the "law of balancement of organs," and tried to encourage French anatomists such as George Cuvier to understand its fundamental importance and pay more attention to the higher level of biological thinking going on in Germany (Lenoir 1987). Despite the rejection of this system of thinking by Cuvier and later by most Darwinians, who labeled the school "transcendental Nature Philosophy," Saint-Hilaire has been proved right in his thesis that all organisms display a fundamental unity of composition and unity of organization. Thus, he realized that differences in morphology between organisms were only superficial and the result of different emphasis of organs during development. This is essentially the central message of heterochrony, which is still paid too little attention today. Saint-Hilaire give explicit examples of differential development, noting that if a bone did not develop in one species, it could be shown to have been arrested at an early stage, when it was still tissue or cartilage.

More pertinent to the present study, Saint-Hilaire noted that body plans repeat again and again in what today we would call a recursive or fractal pattern, which he variously called "unity of organization," "unity of plan," and "unity of composition." Thus, when "developmental genetics . . . arrived on the scene like a thunderbolt" (LeGuyader 2004, p. 244), it became clear that homeotic genes (*Hox* genes) of the homeobox showed the same A-P organization as the macromorphology—exactly as Saint-Hilaire frequently noted in his principle of connections (or theory of analogues). This is now called *colinearity* (Duboule and Morata 1994). It is outside the scope of this chapter to delve further into Saint-Hilaire's prescient observations, except to note that his claim that arthropods were organized like upside-down vertebrates, although long scorned by Cuvier and many subsequent generations of anatomists, has been shown to be genetically correct: that is, "two major genes have been discovered which intervene in antagonistic directions . . . the insect's gene with ventral expression is the same as that with dorsal expression in the vertebrate and vice versa!" (LeGuyader 2004, p. 252). The message is that it is dangerous to dismiss prescient holistic thinkers because their ideas are perceived to be too general, too complex, not obviously applicable to specific cases, or otherwise hard to follow. Saint-Hilaire, like his contemporary Goethe, would have been quite at home discussing the compensation principle, antagonistic genes, and the balancement of organs with anyone versed in modern heterochronic and developmental studies. Like modern students of heterochrony (McNamara, personal communication), Goethe and Saint-Hilaire would likely have been surprised to learn that developmental genetics claims that many of these insights are new in principle, when in fact such insights were around at the birth of biology, and since then, developmental genetics has been mostly concerned with rediscovery of these principles and explanation of details and mechanisms.

The fuller potential of such approaches is brilliantly realized in the work of Schad (1977) on *Man and Mammals*, which was briefly summarized by Riegner (1985, 1998) and Lockley (1999a) and applied to dinosaurs (Lockley 1999a, 1999b, 2004, in press). In these and a few other related studies, the term *morphodynamics* has been introduced partly as a means of redynamizing the concept of morphology so as to reflect its original process meaning, but also as a convenient way of stressing that there are morphodynamic trends in the evolution of various clades, such as increasing anterior or posterior developmental emphasis. These cephalo-caudal or caudo-cephalic trends have been known for many years, and a substantial literature on the topic exists, especially in the field of physical anthropology (Kingsbury 1924; Verhulst 2003). In recent years, developmental microbiology has been routinely preoccupied with how A-P axes and polarities develop (e.g., Wallenfand and Seydoux 2000).

Possibly the best-known and most obvious example of a recursive anteriorization trend is that referred to as *cephalization*, which is seen in vertebrate clades and in vertebrates in general (Fig. 9.1). Other examples include the polarity between primitive amphibian organization, as seen in the salamander morphotype (characterized by a long tail and small head), and frog morphology (large head and no tail). Every schoolchild knows the story of tadpole-frog metamorphosis. Likewise, we see the same trends in primates (from monkeys to hominids). The same trend is also obvious in many dinosaur clades such as ceratopsians, and here we point to the reiteration of the trend in theropods. Such anteriorization trends can generally be classed as correlated progressions, which refer to the reiteration of directional trends in evolution that follow repeated or fractal variations on a theme (sensu Kemp 1999). This concept is essentially similar to the concept of colinearity (LeGuyader 2004).

Schad's genius was to recognize that the expression of physical exaggeration of anterior or posterior organs in any animal (mammals in his 1977 study) must be seen as only one side of a compensatory relationship with physiological processes. Physiologically, for example, ungulates (which are predominantly large, placid, long-lived, derived, and evolutionarily specialized) have well-developed posterior digestive (or metabolic) systems (multichambered stomach) and limbs designed for sustained (long distance) locomotor efficiency. However, the main physical exaggeration is seen in the anterior regions (horns, shoulders, manes, beards, front limbs often larger than hind). In striking contrast, the predominantly small and evolutionarily unspecialized rodents show the opposite, or reciprocal physical and physiological organization. They are physiologically dominated (anteriorly) by overactive sensory and nervous systems (sense-nerve emphasis) and are behaviorally frenetic and short-lived, with weak, low-endurance limbs, but they express their maximum or exaggerated physical development in the posterior part of the body (long tail and hind limbs longer than front limbs). The carnivores represent a middle or central group (typically intermediate in size), in which the anterior and posterior physical and physiological systems are

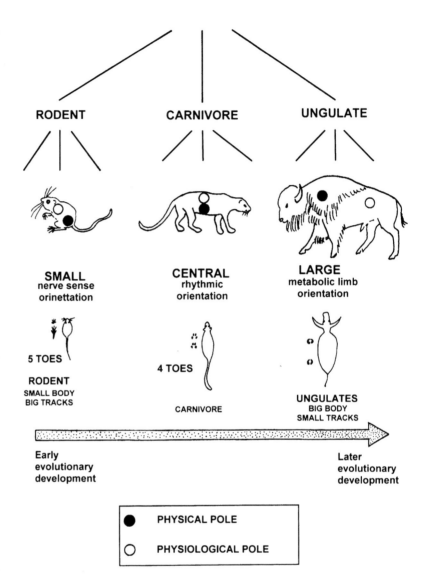

Figure 9.2. *A holistic view of 3 major groups of placental mammals (after Schad 1997; Lockley 1999a). The rodents, carnivores, and ungulates express different emphasis of the physiological organ systems, i.e., sense-nerve, central-rhythmic (or respiratory-circulatory), and metabolic (digestive)–limb systems, respectively. Note that these anterior, central, and posterior centers of gravity (emphasis) of these physiological systems are the reciprocal of the posterior, central, and anterior physical (somatic) emphasis. Such organization reveals many fractally repeated (iterative) gradients such as, small (paedomorphic–short lived–short gestation–local range) and large (peramorphic–long lived–long gestation–wide range). These gradients also correspond to evolutionary trends from primitive to derived. Moreover, the morphology of the tracks also show the posterior to anterior shift, with progressive loss of heel in larger forms. Thus, the track-body size relationship cannot be interpreted functionally, but instead corresponds to relative shifts in the development of proximal and distal portions of the limb.*

less exaggerated (more balanced) and the developmental emphasis is more centered in the circulatory and respiratory systems (heart and lung rhythmic athleticism) (Fig. 9.2). Thus, they rhythmically alternate between intense frenetic and placid behaviors as an expression of the strong influence of their rhythmic organs.

The distribution of incisor, canine, and molar teeth also reiterates the outward-inward and A-P polarity of mammal organization. Because the jaws are analogous to the limbs of the head, the molar teeth correspond to a greater development of the inner or proximal jaw, paralleling the inner limbs of the ungulate, whereas the incisors correspond to the distal or outer region.

As previously noted, the essential message of Schad's work has been summarized by Riegner (1985, 1998) and Seamon, and Zajonc (1998), and has been applied to dinosaurs (Lockley 1999a, 1999b). Although this morphodynamic approach is not yet widely appreciated, is highly holistic in that it accommodates morphology, physiology, heterochrony, fractal organization, and even color patterns. All major animal groups are revealed to be complex systems with the type of organized and compensatory polarities outlined above. The rodent pole (i.e., small, predominantly paedomorphic forms with posterior orientations: long tails and hind limbs) is more outward and connected to the environment and its stimuli, whereas the ungulates are more inward and emancipated from the environment. This is literally true in terms of size, but it also affects behavior such as home range, aggression, social structure, and need for nesting protection (see Schad 1977, 1992, for detailed explanations). There is also a polarity of organization between small, predominantly paedomorphic clades, such as rodents, which are very diverse (in species richness) but not disparate in form, and large peramorphic clades, such as ungulates, which are less diverse in species but much more disparate in form. Thus, in the same way that the individual ungulate takes up more physical space than the rodent, the ungulate is also further separated from other ungulates in terms of its morphospace characteristics (sensu Foote 1991).

Again, these characteristics indicate an organization that is fractal and recursive, in that the general structure of taxonomic groups repeats at all hierarchical levels with an almost endless, staggered sequence of minor variations on the general theme (see Bird 2004 and Carroll 2005 for discussion of the inherently fractal and recursive organization of organic systems, and the use and definition of terms like *iteration, recursion,* and *staggered sequences*). Conway (2003) also stresses the rampant convergence in nature that points to inherent organization rather than randomness.

An excellent example of such fractal organization was demonstrated by Portmann (1964), who showed that in mammals, ungulates have large forebrains relative to hind brains, in comparison with rodents, which have the reverse organization. He called this the "neo-pallial index," which is a measure of the relative size of the forebrain or neocortex. It is small in rodents, intermediate in carnivores, and large in ungulates. So ungulates (e.g., ruminants) manifest more complex social behavior and are perceived as being inclined to ruminate inwardly. Thus, the distribution of physical brain mass in these major mammal groups is a fractal reiteration of whole body A-P organization. This general trend in vertebrate brains has been confirmed by subsequent studies of reptiles (Hopson 1977, 1979; Jerison 1969), nonavian theropods (Chatterjee 1997; Larsson et al. 2000), and mammals (Kaas 2000). The same is true of feet and footprints (large posterior emphasis in rodents and more anterior emphasis in large ungulates; Lockley 1999a). In the case of brain mass distribution, the inward-anterior characteristics of the ungulate neocortex mirror the inward or proximal characteristics of the molars.

With this background, we can turn to the evo-devo paradigm, which is highly convergent with the Schadian paradigm. It shows that the organization and influence of *Hox* genes lead to similar morphological expressions in related organisms that are highly recursive or fractal, leading to endless forms that are variations on a universal or common theme or organization. (*Organization* literally means "an ordered sequence of organs.") However, before being completely seduced by the compelling evidence for the organizational role played by *Hox* genes in controlling, driving, or explaining morphology, we should note that Schad (1977) and others, such as Portmann (1964), must be given credit as pioneering biologists who recognized these organizational patterns in macromorphology, physiology, behavior, and color patterns before the same organizational patterns and structure were demonstrated in the microscopic world of molecules and *Hox* genes. It is now time for a synthesis, and it is in paleontology, where macromorphology is available but genes are not, that we can effect a marriage of these macro and micro (evo-devo) fields (Carroll 2005).

The Evolution of Development, or the Evo-Devo Paradigm

In many respects, the evo-devo paradigm (Carroll 2005) is similar to that outlined above as the morphodynamic paradigm. Molecular and cellular biologists have in recent years studied the homeobox (sequences of homeotic genes in regular repeated segments or boxes) and have shown that diverse invertebrate and vertebrate species share an orderly A-P arrangement of these developmental genes known as *Hox* genes within the homeobox. Moreover, the arrangement of these *Hox* genes is in the same, or similar, A-P order as they appear in the organs of the body (Fig. 9.3). Thus, the sequence or order of developmental *Hox* genes is like a microcosm of the whole body, implying a highly fractal or recursive form of organization.

Such work shows that not only do individual *Hox* genes develop from anterior to posterior in early embryogenesis, but this same A-P directionality recurs in the development of the 7 rhombomeres of the hind brain, and

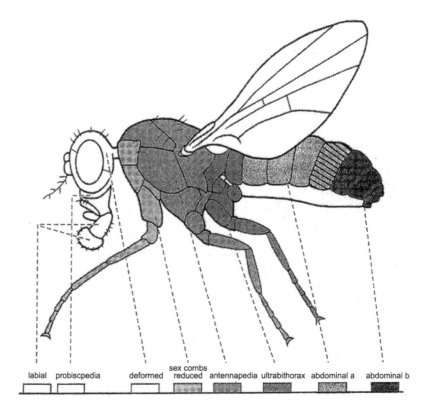

Figure 9.3. *The homeobox demonstrates a parallelism between the A-P organization of* **Hox** *genes and the A-P organization of actual body organs. Modified after Carroll (2005).*

in the development of somites throughout the whole body and in limb formation (Fig. 9.3). As summarized by Carroll (2005, p. 100), we know exactly which *Hox* genes are involved in the expression of these stages of cellular and somatic development and can characterize them as "the staggered expression of the *Hox* genes" (see Fig. 9.4 for details). This clearly implies a sequence of development that is heterochronically controlled or timed. Moreover, in the case of the development of body somites, *Hoxc6* is associated with the boundary between cervical and thoracic vertebrae and is therefore closely associated with the point of origin of anterior limbs. Carroll (2005), like other evo-devo biologists, stresses that all development is strongly controlled by the timing of *Hox* gene expression, which in turn is controlled by complex genetic switches. Again, the importance of heterochronic timing in generating morphology is emphasized. Slight differential changes at early stages in development can have profound effects and have unequivocally been shown to develop different races (morphotypes). Carroll (2005) goes on to stress that all evolution can be understood through the application of this paradigm.

Such work shows that although individual ontogeny is a complex heterochronic process, the general arrangement of *Hox* genes is repeated fractally or recursively throughout the organic world. Thus, it appears that all organisms have the same general patterns of organization, but they are different in detail owing to subtle shifts in the timing of development and morphology from cell to cell, organ to organ, individual to individual, species to species, and clade to clade. (For a discussion of terms like *iteration* and *recursion* and the application of complexity and chaos concepts to evolutionary theory, see Bird 2004.) In short, we can now be increasingly confident that the evolutionary relationships we perceive in related clades (e.g., ceratosaurs, coelurosaurs, and other groups within the theropods, or among related clades such as modern birds) can be understood in terms of these A-P morphodynamics and heterochrony, which have been shown to apply consistently to a wide range of extant vertebrate clades.

Limb Construction and Morphological Organization in Theropods and Related Vertebrates

As indicated above, there is growing evidence to suggest that organism growth (morphogenesis) is much more ordered than previously thought. For example, Shubin and Alberch (1986) demonstrated that the limbs of most higher vertebrates follow similar growth patterns or programs. Similar observations have been made by Verhulst (2003) in his study of primates. These studies show that during early ontogeny, the distal limb develops first (A-P direction; Fig. 9.4). Later, the proximal limb develops at an accelerated rate, relative to the foot, especially in large peramorphic species (McNamara 1995; Long and McNamara 1995, 1997; Lockley in press). This is, therefore, a posterior-anterior (P-A; distal-proximal or caudo-cephalic) developmental direction for the limb as a whole. In different species, as differential (heterochronic) growth retards or prolongs development, the ratio of distal to proximal limb proportions will vary considerably and will show a strong relationship to size.

Figure 9.4. *(Top to bottom)* Dynamic A-P (or west-east) flow directions in *Hox* genes, hind brain rhombomere, whole-body somite, and limb development. Proximal-distal orientations in the limb correspond to A and P, respectively. Modified after Carroll (2005).

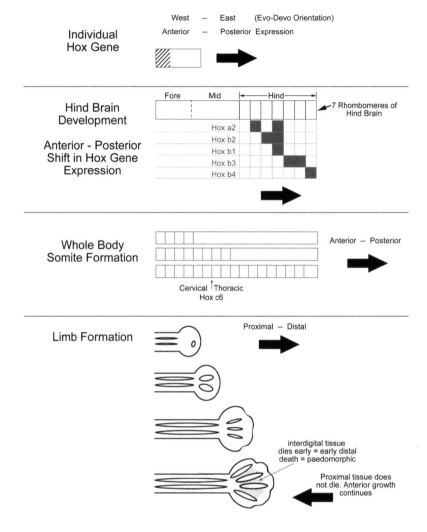

Some Hind Limb Proportions

The distal-proximal differential is particularly obvious in the saurischian clade. If we compare small, primitive theropods with large, derived sauropods, we find that the former have relatively large feet (distal organs) and short inner (proximal) limb bones (femora and humeri), whereas the reverse is true in the sauropods (Figs. 9.5, 9.6). In general, the theropod-sauropod polarity is also a P-A polarity: that is, theropods are bipedal, hind limb walkers, whereas sauropods moved anteriorly into quadrupedal postures with long anterior necks (Lockley 1999a, in press). The same posterior (biped)–anterior (quadruped) polarity repeats in many other dinosaur clades and is related to many factors, including size and phylogeny (primitive versus derived characters).

If we narrow our focus to theropod limbs, we can look at both primitive and derived clades. Among the ceratosaurs, it appears that small, gracile forms like *Coelophysis* and *Elaphrosaurus* have relatively short femora compared with larger robust forms like *Ceratosaurus* (Fig. 9.7).

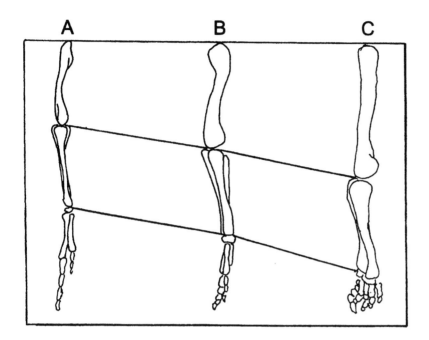

Figure 9.5. *The principle of distal versus proximal emphasis in saurischian limb elements (after Lockley and Jackson in press; Lockley in press). From left to right, each row shows a theropod, prosauropod, and a sauropod. (Top)* **Dilophosaurus** *(A),* **Anchisaurus** *(B), and* **Apatosaurus** *(C) are all relatively primitive representatives of their respective groups. (Bottom)* **Tyrannosaurus** *(D),* **Plateosaurus** *(E), and* **Brachiosaurus** *(F) are more derived representatives of their respective groups. Cf. Figure 9.6.*

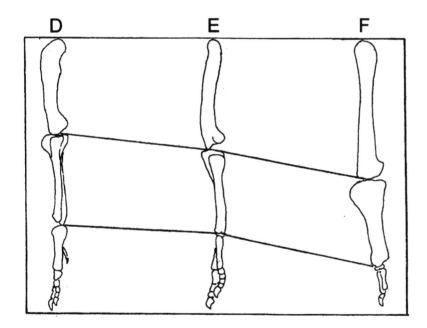

In the more derived coelurosaurs, which include the tyrannosaurs and ornithomimids, we find similar patterns. The gracile ornithomimids suggest a contrast with the robust tyrannosaurids. The relative proximal limb (femora) lengths are somewhat shorter than in the tyrannosaurids and their distal limb elements (metatarsals) are longer, thus fitting the general trend toward distal emphasis in smaller, predominantly paedomorphic forms (Fig. 9.8).

Within the Tyrannosauroidea, taxa like *Dilong* and *Albertosaurus* juveniles tend to have shorter femora than large forms like *Tyrannosaurus*. In

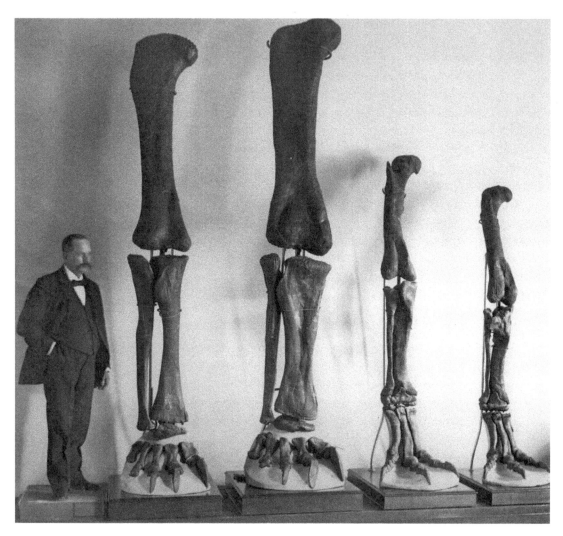

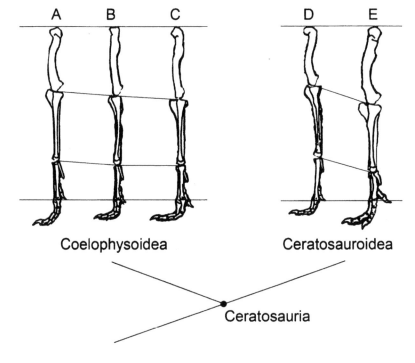

Figure 9.6. The principle of distal versus proximal emphasis in saurischian limb elements illustrated by a display in the American Museum (after Norell et al. 1995). Genera from left to right are **Diplodocus**, **Apatosaurus**, and 2 **Allosaurus**. Cf. Figure 9.5.

Figure 9.7. Hind limb proportions in Ceratosaurs. (Left) Coelophisoidea **Coelophysis** (A), **Liliensternus** (B), and **Dilophosaurus** (C). (Right) Ceratosauroidea **Elaphrosaurus** (D) and **Ceratosaurus** (E).

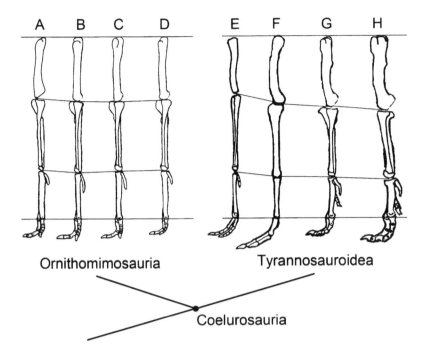

Figure 9.8. *(Left)* Hind limb proportions of Coleurosaurs. Ornithomimisaurs: juvenile **Gallimimus** (A), **Ornithomimus** (B), **Struthiomimus** (C), and **Gallimimus** (D). *(Right)* Tyrannosaurids: hypothetical juvenile **Albertosaurus** (E), **Dilong** (F), adult **Albertosaurus** (G), and adult **Tyrannosaurus** (H). Shown are their proportions when scaled to the same limb size.

small forms like *Dilong* (Xu et al. 2004), the feet (phalanges) are significantly longer than in the larger taxa.

Long and McNamara (1995) inferred important developmental shifts during tyrannosaur ontogeny. Thus, juvenile *Albertosaurus* purportedly has a relatively long foot (distal limb) and short femur compared with an adult. (Note, however, that the reconstruction of the juvenile *Albertosaurus* was hypothetical [Russell 1970, figs. 8 and 9 therein], and so must be further tested before inferences are drawn.) It is nevertheless interesting that Russell (1970) and Long and McNamara (1995) depicted the juvenile with strong posterior emphasis, with long tail and relatively small head compared with the adult (Fig. 9.9). This whole-body A-P polarity appears to mirror the pattern seen in other vertebrate groups discussed above, and is further comparable to the proximal-distal polarity seen in limb elements. To clarify this point, we stress that vectors of A-P (or cephalo-caudal) growth for the whole body follow a head-trunk-tail direction along the vertebral column. However, appendicular organs such as the limbs and the jaw are better described in terms of their proximal-distal elements or orientations. So the hind foot is posterior relative to the whole body, but also distal relative to the more proximal (anterior) location of the femur. The same proximal-distal orientation is evident in forelimbs and the jaws, although it is clear that these organs are more closely connected to anterior organs of the axial body. In the case of the jaw, especially in mammals where incisor-canine-molar heterodonty is well developed, these teeth are situated distally, medially, and proximally relative to the cranium. However, because jaws protrude anteriorly relative to the axial body, the terms *anterior medial* and *posterior* could equally well be used. As noted below, the respective A-P or proximal-distal polarity between cranium and jaw is proving important in studies of tyrannosaur

Figure 9.9. *Whole-body proportions of hypothetical juvenile* **Albertosaurus** *(A) (after Russell 1970),* **Dilong** *(B), adult* **Albertosaurus** *(C), and adult* **Tyrannosaurus** *(D) scaled to the same body size. Modified after McNamara (1995).*

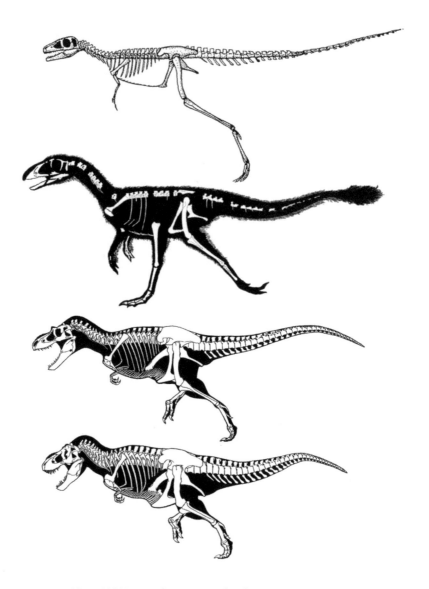

ontogeny (Carr 1999). It is also noteworthy that upper jaw maintains a fixed cranial (anterior-proximal) orientation while the lower jaw can move in a posterior (distal) direction.

We provide a few further theropod hind limb examples to help stress the widespread nature of this pattern. First, among the large allosaurids and related forms (Fig. 9.10), we see a similar emphasis on the development of the inner limb (femora). By contrast, in the small dromaeosaurids, we see a much greater emphasis of the distal limb (tibia and fibula: Fig. 9.11). This, of course, can be partially explained by the effects of scaling (Pike et al. 2002), but such single explanations do not negate the inherent proximal-distal limb compensations discussed herein. For example, the repetition of the large proximal versus small distal element polarity in large species and its opposite or reciprocal (small proximal versus large distal) elements repeats not only in dinosaur limbs used for walking, but also, as discussed

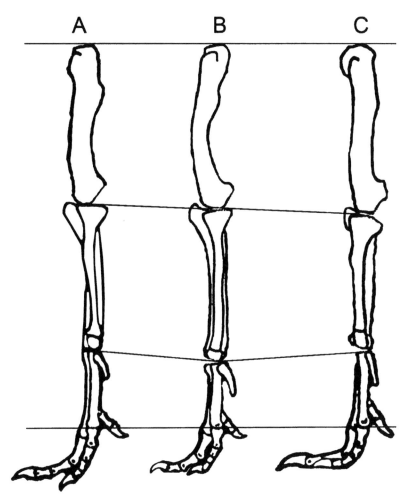

Figure 9.10. *Hind limbs of allosaurids and selected large theropods* **Piatnizkysaurus** *(A),* **Allosaurus** *(B), and* **Gigantosaurus** *(C).*

below, in wings of pterosaurs and birds with entirely different functions (Kellner 2003; Middleton and Gatesy 2000, respectively).

Some Forelimb Proportions

As noted by Middleton and Gatesy (2000), in comparison with *Tyrannosaurus* and *Carnotaurus*, the forelimbs of most theropods are well developed (cf. Fastovsky and Weishampel 2005). We have already established that the proximal portion of the hind limbs is more developed in large tyrannosaurids than in juveniles, but the same appears to be true of the forelimbs. The humerus makes up about 50% of the length of the limb. The humeral proportion is even more exaggerated in *Carnotaurus*, comprising more than 60% of the limb length. Compare this with the humerus proportion of less than 40% for *Struthiomimus* and *Ornithomimus* (Fig. 9.12), and the comment of Hutt et al. (2001, p. 229) that the relatively small tyrannosaur *Eotyrannus* has a "manus proportionally long (digit II c. 95% of humerus length)." It is clear that the same dynamics affect the front and hind limbs in all these forms. The consistency in developmental dynamics

Figure 9.11. *Deinonychosaur* hind limbs ***Bambiraptor*** *(A),* ***Velocoraptor*** *(B), and* ***Deinonychus*** *(C). Note short proximal limb in* ***Bambiraptor****. Cf. Figures 9.7 and 9.8.*

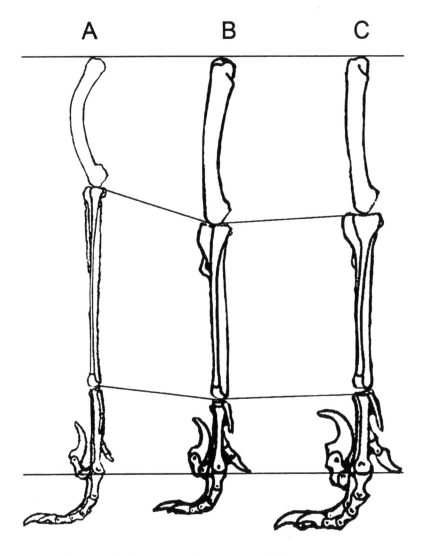

across these multiple organs and across multiple theropod clades can hardly be attributed to solely to function.

Middleton and Gatesy (2000) have shown similar patterns in the forelimbs (wings) of birds. For example, small, predominantly paedomorphic flying birds such as passerines have short proximal humeri and long distal wings, whereas large (mostly peramorphic) ground birds have long humeri and short distal wings (Fig. 9.13). The same trends in proximal-distal limb proportions are seen in comparing small, primitive (paedomorphic) rhamphorhynchoid pterosaurs and large, derived pterodactyloids (Kellner 2003; Mitchell and Lockley, in preparation). Here we stress the point made by all heterochronic studies (e.g., McNamara 1997) that compensations or trade-offs make most organisms a mixture of paedomorphic and peramorphic characteristics. Thus, a small or large animal may only be paedomorphic or peramorphic in the general sense of body size (less or more growth, respectively). Thus, we see the same or similar patterns in small (primitive) and large (derived) mammals (Schad 1977) and in primates (Verhulst

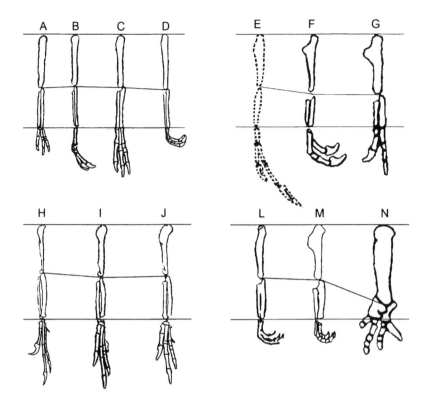

Figure 9.12. *Dinosaur forelimbs.* (Top left) Juvenile **Gallimimus** (A), **Ornithomimus** (B), **Struthiomimus** (C), and **Gallimimus** (D). (Top right) Tyrannosauroidea: hypothetical juvenile Albertosaurus (E), adult **Albertosaurus** (F), and **Tyrannosaurus** (G). (Lower left) Deinonychosaurs: **Bambiraptor** (H), **Velociraptor** (I), and **Deinonychus** (J). (Lower right) Ceratosauroidea: **Elaphrosaurus** (L), **Ceratosaurus** (M), and **Carnotaurus** (N). Cf. Figures 9.7, 9.8, and 9.11 for corresponding hind limbs and cf. Figure 9.13 for bird forelimbs.

2003). In short, it is a characteristic of all these major groups of vertebrates that the more derived and more obviously peramorphic forms have proximal rather than distal limb emphasis. Put another way, the inner organs are more developed.

Relationships between Head, Neck, Front Limbs, and Other Anterior Organs in Theropods

It is important to stress that compensation principles, balancement of organs (or heterochronic trade-offs) evidently apply at many different levels of morphological organization. Thus, we can move from comparing proximal versus distal organs within the hind limb or forelimb in different species (cf. Middleton and Gatsey 2000), or to comparison of limb proportions at different stages of development (Reisz et al. 2005). This inevitably leads to comparisons of adjacent organs such as head, neck, and forelimbs. Why, for example, among many saurischians and some birds do we get very long-necked forms with relatively small heads? Why are such morphologies less well developed in the vast majority of ornithischian dinosaurs and mammals? Put another way, why do some saurischian (notably theropods like *T. rex*) have an opposite or reciprocal organization (large heads and relatively short necks)? And why do some forms have such diminutive forelimbs, yet well-developed hind limbs? Given that developmentally the forelimb bud originates at the junction between the neck and trunk (cervicals and dorsals; Carroll 2005), the possibility of a developmental explanation must be explored.

According to Russell (1970) and Long and McNamara (1997), the anterior part of the juvenile tyrannosaur body is relatively small. The juvenile head is relatively small compared with the adult head, and the front limb is

Figure 9.13. *Relative proportions of proximal and distal wings in modern birds; after Middleton and Gatesy (2000). Note polarity between small (paedomorphic) forms with distal emphasis and large (peramorphic) forms with proximal emphasis. Cf. Figure 9.12.*

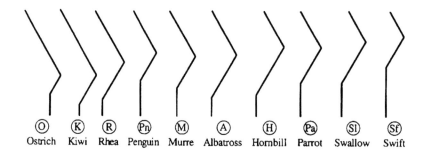

relatively small compared with the hind limb. Assuming this reconstruction has some merit, the situation is the reverse or adult organization, showing how there was a progressive anteriorization of growth, or shunting of growth from more posterior to more anterior organs (Fig. 9.9). However, this reconstruction of a juvenile is entirely hypothetical, and we can gain better insight into tyrannosaur ontogeny from recent studies by Carr (1999) and Currie (2003). For example, the lightly built albertosaurines have "longer, tibiae, longer metatarsals and longer toes" (Currie 2003, p. 663). These all point to emphasis of distal limb elements in the smaller, lighter, more gracile group, as compared with proximal limb elements in the robust tyrannosaurines. Carr (1999) studied only the cranial morphology but nevertheless showed the shift from a narrow, elongate gracile skull in early ontogeny (stage 1) to one that is robust, shorter, and deeper in later stages (stages 3 and 4). This shift is accompanied by the loss of rostral (distal) end of the maxilla and a widening of the proximal portion of the skull, all of which have far-reaching implications for tyrannosaurid taxonomy (Carr 1999).

As noted previously, such shifts fall under the general label of anteriorization (or emphasis of proximal organs) typical of large derived forms in most vertebrate clades (Lockley 1999a, in press). Indeed, Carr (1999) goes so far as to hint a recapitulation by comparing juvenile tyrannosaurs with other more distantly related primitive theropods, which had long and narrow (rather than broad and deep) skulls. This general pattern is clearly supported by the discovery of small, gracile, long-snouted basal tyrannosauroids like *Dilong* (Xu et al. 2004) and *Guanlong* (Xu et al. 2006). Thus, it is perhaps no surprise that as derived tyrannosaurs, ceratosaurs, brachiosaurs, ornithopods, ceratopsians, and pterosaurs became larger than their ancestors, they developed large heads and exaggerated headgear as their evolutionary cycles culminated. Observations by Hone et al.(2004) have revived the idea of Cope's rule and the increase in evolutionary tendency toward larger size through time. Indeed, it appears that this tendency is prevalent in many different dinosaur groups, and is "not limited to any particular subclade" (Hone et al., 2004, p. 590). These authors note that the tendency is particularly pronounced in the Saurischia, and they single out the tyrannosaurids as a pronounced example. We infer that the correlation between large size and large heads is noteworthy, so there is likely a connection between Cope's role and pronounced anteriorization.

As we follow these whole-body anteriorization trends, which have the most obvious manifestations in the heads of the large derived species, it is necessary to keep track of similar fractal patterns that affect the individual organs such as limbs, and their inner (proximal) and outer (distal) elements (Figs. 9.5–9.13). In this regard, it is no coincidence that there is a distalless gene that controls or affects the development of distal limbs in a wide range of animals species, and that it is so named because it inhibits the growth of distal limb elements when mutated (Carroll 2005). This is the opposite of the malformations caused by the fertility drug thalidomide, which often attacked the proximal portion of the limb during a critical growth stage of the fetus—that is, 20 to 36 days after conception (Gilbert 2000). Because the relative growth of distal portions of the limb is dependent on whole-body size—as shown in the case of the saurischians—we can infer that the relative importance of a distalless organizing principle (or its equivalent gene) is correspondingly expressed. It is assumed that the manifestation of more or less distal growth is linked to heterochronic developmental dynamics. Such dynamics are directly and indirectly explored in a recent study of prosauropod embryos by Reisz et al. (2005). They concluded that the juveniles had large heads and forelimbs, but that negative allometry during postnatal development reduced forelimb growth and growth of the head (relative to the neck). Indeed, they showed that relative to the femur (taken as the standard), the tibia and dorsal vertebrae grew isometrically, but that there was strong positive allometry in cervical vertebrae and negative allometry in the skull. This stresses the dynamic growth compensation between the neck and the skull, as discussed herein for coelurosaurians. Such dynamic compensations have far-reaching implications. Thus, Reisz et al. (2005) inferred that giant sauropods likely evolved as a result of the opposite process in which forelimb growth was not suppressed—that is, through paedomorphic retention of early ontogenetic features in the adult (cf. Bonaparte and Vince 1979). Thus, the sauropod may show many features of an overgrown prosauropod embryo. The polarity between long-necked sauropods with small heads and long forelimbs, and shorter-necked prosauropods with shorter forelimbs may be developmentally convergent with the polarities seen in theropods.

We have already argued that the head is the most anterior organ of the body and is characteristically largest in highly derived forms that show extreme encephalization. Here some qualification is necessary to distinguish between absolute and relative head and/or brain size. For example, absolute head encephalization is generally less pronounced in the saurischian than in the ornithischian dinosaurs (Lockley 1999a), as seen in a comparison between sauropods and ceratopsians. However, as shown by the famous example of encephalization in *Troodon* (Russell and Seguin 1982), a small animal may have a relatively large head and/or brain relative to body size, even if the absolute size is not large in comparison with other, much larger dinosaurs. As noted above, there is a polarity in head-neck relationships in many dinosaur groups (within the theropods, within saurischians, and within the Dinosauria as a whole). This suggests a fractal recursion of growth dynamics and/or subsequent functional adaptations. For example, among certain groups of Saurischia, (e.g., the smaller, gracile theropods, ornithomimids, and sauropodomorphs), the neck can be very developed (elongate) while the

LARGE HEAD

Short neck

Short Arms

Small head

LONG NECK

LONG ARMS

Figure 9.14. *The polarity between small (paedomorphic) taxa with small heads and long necks and forelimbs and large (peramorphic) taxa with reciprocal or compensatory organization (large head and short neck and forelimbs) is well demonstrated by the respective ceratosaurians (**Coelophysis** and **Carnotaurus**) and the respective coleurosaurians (**Ornithomimus** and **Tyrannosaurus**).*

head remains relatively small. This characteristic is also typical of many bird groups (as discussed below in relation to *Hoxc6*). By contrast, the neck is never hyperdeveloped in the large ornithischians, although it is more developed in large hadrosaurs and iguanodontids than in other groups, and even in smaller forms, it does not reach exaggerated proportions. Thus, the whole tendency toward anteriorization is less developed in the saurischian head than in that of ornithischians (Lockley 1999a), although in compensation, it is more developed in the saurischian neck.

If we just limit our view to the theropods, we note that this same polarity is clear within various theropod groups. For example, the Ceratosauria include smaller, gracile (primitive) forms such as *Coelophysis* with longer necks, smaller heads, and longer forelimbs, and also include larger, robust (derived) forms like *Carnotaurus* with larger heads, shorter necks, and shorter forelimbs. The same pattern of polarity is reiterated in the younger coelurosaurian clade, where the ornithomimid and tyrannosaurid distinction is an example morphological convergence with the ceratosaurian polarity just outlined (Fig. 9.14). One could even argue that in *Carnotaurus* the distal part of the head is quite short, with the exaggeration proximally expressed in the horns.

If we limit our view to the morphological variation within the genus *Coelophysis* (Colbert 1989), where 2 dimorphs are recognized, we find that the large-headed, robust form has relatively shorter forelimbs, whereas the smaller, gracile form has relatively longer forelimbs (Lockley and Kukihara 2005; Fig. 9.15 herein). Note here that the differences between 2 dimorphs of the same species are much less than between different genera within larger clades. Also, given the similarity in size (Table 9.1), actual sizes are given rather than relative proportions, except in the case of the cervical/dorsal ratio. These compensatory relationships between adjacent organs in *Coelophysis* extend throughout the whole body so that the robust form has longer hind legs and a shorter body, whereas the gracile form has a longer body and shorter hind limbs.

As pointed out by Lockley (1999a), smaller, more primitive dinosaurs with posterior emphasis have narrow, elongate bodies and short limbs, whereas larger forms with anterior emphasis typically have foreshortened bodies and longer limbs. This generalization corresponds to the case of the *Coelophysis* dimorphs and appears to be supported by the observations of Bakker and Bir (2004, p. 303), who noted exactly the same trends in "progressive shortening of the torso" in the sequence *Ceratosaurus-Allosaurus-*

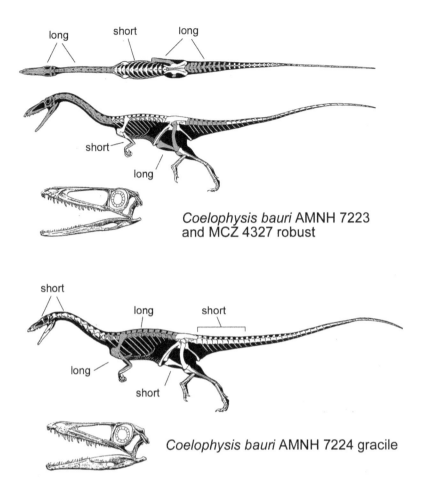

Figure 9.15. Morphodynamics of *Coelophysis* dimorphism based on comparison of the robust from (AMNH 7223) and the gracile form (AMNH 7224). Note the alternation of long and short organs (i.e., natural body segments: head, neck, trunk, tail, limb, foot) showing an organized, reciprocal pattern of heterochronic compensation. Ultimately the robust form has more anterior emphasis and longer limbs, and the gracile form has posterior emphasis and short limbs. Thus, one is the perfect reciprocal of the other (after Lockley in press).

Gorgosaurus and correlated it with "increasingly sharp S flexure of the neck," which is also a manifestation of foreshortening.

Clearly, changes in morphological development in any organ have a cascade effect throughout the body, and they certainly have a compensatory effect in adjacent organs. This is the dynamic essence of the morphodynamic paradigm. It has been shown that specific *Hox* genes play a role in these changes or shifts. The anterior expression of *Hoxc6*, for example, plays a pivotal role in defining the transition from cervical to thoracic vertebrae. This boundary is closely associated with the growth of forelimbs. As summarized by Carroll (2005), the mouse, chicken, and goose have short, medium, and long necks with 7, 14, and 17 vertebrae, respectively. Each, it can be argued, has increasingly well-developed fore limbs. By contrast, according to Carroll (2005), the snake has no cervical vertebrae and hence no neck or forelimbs. It is all head and exaggerated thorax (Fig. 9.16). These shifts in organization in the central and anterior part of the vertebrate body may give us clues to the morphodynamic organization in saurischian dinosaurs and the likely relationship to *Hox* gene expression. Thus, the polarity between the theropods with short necks, large heads, and small forelimbs is contrasted with the sauropods that have long necks and forelimbs but small heads. *T. rex*, like most theropods, has 10 cervical

Skeletal Element	AMNH 7223 (Big Head)	AMNH 7224 (Small Head)
Skull-spine to caudal 17	1838 mm	1761.5 mm
Skull length (mm)	265 long	222 short
Neck (cervicals)	485 long	405 short
Trunk (dorsals)	425 short	455 long
Cervical/dorsal ratio	Long neck (1.14), short body	Short neck (0.89), long body
Sacrum	120 short	148 long
Proximal tail (caudals 1<N>9)	291 long	279 short
Midtail (caudals 11–17)	252	252.5
Distal tail	822.5	Missing
Forelimb total	354.9 short	412.9 long
Scapula-coracoid	134 short	156 long
Humerus	120 short	134 long
Radius	65 short	89 long
Metacarpal 3	35.9 short	40.9 long
Hind limb total	559 long	549 short
Femur	209 long	203 short
Tibia	224 long	221 short
Metatarsal 3	126 long	125 short

Table 9.1. *Measurements for Major Organs of 2 Coelophysis Dimorphs*

Note.—Data from Colbert (1989).

and 13 thoracic vertebrae, whereas in the sauropods, there are 12 to 19 cervicals and 9 to 14 thoracics (Fig. 9.16; Weishampel et al. 2004).

This polarity (long neck–long forelimb–small head versus short neck–short forelimb–large head) recurs fractally within the coelurosaurian theropods (tyrannosaur versus ornithomimids) and within the sauropods (diplodocids versus brachiosaurids). However, it is important to note that within the nonavian theropods, the number of cervical and dorsal vertebrae do not vary much. On the basis of analogy with modern reptiles, birds, and mammals, we could assume that the expression of *Hoxc6* is constant within the theropods (although it is known to vary in birds). However, among the sauropods, we can infer that *Hoxc6* expression is more variable because of the greater variability (disparity) in number of neck vertebrae (Fig. 9.16). We speculate that this is consistent with the model of greater disparity of form in large, derived clades such as sauropods, as compared with less disparate but more diverse clades such as theropods. For example, Weishampel et al. (2004) list 2 well-known theropod genera for each known sauropod genus. Recent finds have also tended to emphasize sauropod disparity, as in the case of *Amargosaurus*, which has an extraordinary exaggeration of the neck in the form of bifurcated neural spines (Salgado and Bonaparte 1991), and the short-necked sauropod *Brachytrachelopan* (Rauhut et al. 2005).

As already noted by Lockley (1999a), these are macromorphological A-P polarities that are obviously convergent with the Schadian model for mammals. Moreover, they provide strong support for the idea that saurischians (especially theropods) are closely related to birds, because it is mainly in this

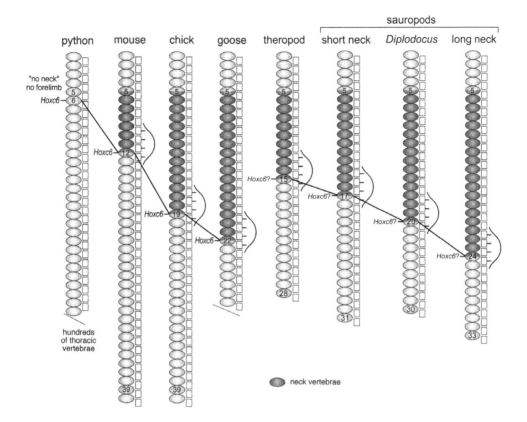

Figure 9.16. The relationship between the cervical-thoracic vertebral boundary in modern snakes, mammals, and birds as a reflection of the anterior expression of *Hoxc6* (modified after Carroll 2005), with inferred arrangement in saurischian dinosaurs (theropods and sauropods), based on vertebral formula, for comparison. Theropod formula (10c + 13d) is constant, but sauropod formula varies between short-neck and long-neck extremes. Numbers at the base of the 4 right columns indicate head, neck, and thorax totals for dinosaurs. Numbers in the 4 leftmost columns are after Carroll (2005).

group, not in ornithischians, that the variation in neck and forelimb development is so pronounced. By contrast, in general among ornithischians, the necks are short (as few as 7–8 cervicals in primitive forms and up to no more than 10 in most other forms, except in the iguanodontids and hadrosaurs, where the number may reach 11–15), but it is in the heads that we see the most enlargement and elaboration, especially in derived forms.

Thus, we have a plausible explanation for the small arms of *T. rex* and the long necks and forelimbs of coelurosaurs and sauropods that is entirely consistent with the broader pattern of *Hox* gene (*Hoxc6*) expression throughout 3 major classes of extant vertebrates: mammals, birds, and reptiles (sensu Carroll 2005). Although it is tempting to credit *Hox* genes with providing the ultimate explanation for such distinctive morphological traits, we have already suggested that *Hoxc6* does not provide a compelling causal explanation for the long-neck versus short neck polarity *within* the theropods.

Evo-devo proponents admit that the actual cause of the switching on and off of homeotic genes like *Hoxc6* is still a complex mystery. The gene has a pivotal relationship in defining the morphological boundary between 3 organs (neck, forelimb, and thorax or torso), but is only one of many dynamic factors that influence the morphology of these organs. Moreover, *Hoxc6* is dependent on other genes to switch it on or off. If we are correct in inferring the role of this gene in influencing differential organ development between theropods and sauropods, and within sauropods, but not within theropods, we can ask whether this gene plays a role in the development of the small arms of *T. rex* (and *Carnotaurus*). For consistency, we can point to the recursive nature of the morphological changes seen within the theropods (and in the birds, where *Hoxc6* does play a role in differentiating the vertebral morphology of major groups) and infer that some similar influence is at work (Fig. 9.16). These arguments favor an understanding of dinosaur morphology in the context of the developmental morphodynamics that affected these organs.

Previous Interpretations of *T. rex* Forelimbs

Viewed from the above perspective of limbs as organs in complex systems, we must try to understand their morphology in relation to adjacent organs. Thus, given the broad perspective demanded by modern biological thinking, the idea that *T. rex* forelimbs served some specific individual purpose may now be questionable. Nevertheless, as pointed out by Carpenter and Smith (2001), the question "What, if anything, did *Tyrannosaurus* do with its puny front legs?" has been asked on many occasions (Gould 1980). Thus, although the question remains why the arms are "outrageously" puny (Fastovsky and Weishampel 2005, p. 283), it is likely the wrong question, at least from the developmental viewpoint. Preferably, we should be thinking about how inherent morphodynamic organization leads to exaggeration of some organs and reduction of others, and how these patterns may be recognized as a staggered or organized sequence of morphologies within clades.

Many theories of *T. rex* arm function have essentially been negative in the sense that they point out that the arms could not function to reach their mouth, or were unlikely to be of much use to push themselves up from a prone position (e.g. Newman 1970; Fastovsky and Weishampel 2005). Positive functional theories include the idea that the arms were strong and useful for clutching prey (Carpenter and Smith 1995, 2001), or that they served some sexual function (Osborn 1906).

Paul (1988, p. 320) stated that "much speculation has been directed towards the use of these forelimbs . . . [but that] . . .this obsession is misplaced. They were not important to their owners, so they should not be important to us." From a functional perspective, we agree, but we hold that from the perspective of the compensation dynamic, the small size is developmentally significant. Paul (1988, p. 320) also noted that "in *Tyrannosaurus* and even more so in *Albertosaurus* the forelimbs became even smaller with time and were on their way to complete loss." This statement suggests some sort of intrinsic biological dynamic in the Tyrannosauridae, although no explicit cause is suggested. (Complete loss of forelimbs does occur in reptiles [snakes] in conjunction with an excessively anterior expression of *Hoxc6*.) It has also been noted in other tetrapods, from Paleozoic lepospondyls to Mesozoic and modern squamates (Caldwell 2003, p. 573), where it is clearly been tied to "developmental genetic models . . . associated with morphogenesis" and not primarily to function. Paul's avoidance of a functional explanation is perhaps symptomatic of the evident difficulties encountered when trying to find one that is compelling. This may account for the fact that even some popular treatments of *T. rex* largely avoid the subject (Horner and Lessem 1993).

An interesting suggestion was made by Fastovsky and Weishampel (2005) that the small arms helped to balance the big head and prevent the animal from tipping forward. This biomechanical suggestion does involve the principle of compensation, but it is distinct from our suggestion that the compensation is due to the inherent (formal) increase in anterior emphasis within the head as another example of a cephalization trend. The idea of biomechanical function is unconvincing because animals are not machines, and to be consistent, the same argument would need to be applied to all other dinosaurs and their organs. Thus, bipedal animals with large forelimbs (including most other theropods) might be interpreted as having developed such limbs in order to compensate for small heads so as to avoid falling backward. Given the reality of anteriorization trends, it seems more parsimonious to simply infer that some species were more anteriorly developed than others, and that the expression of morphological diversity is an inherent property of development within a complex organic system such as a clade— possibly a type of within-clade homeostasis. When we do this, and when we consider how these various degrees or stages of development would be expressed in different organs of the body while obeying the compensation principle, we quickly find that we are looking at the types of morphological variation that we find in nature—in this case, among theropods and saurischians (Fig. 9.14).

Discussion

As noted by Lockley (1999a), the Schadian morphodynamic paradigm allows us to draw some compelling inferences about saurischian behavior and physiology, and to understand their biological organization within the context of the whole clade. For example, small, primitive theropods such as *Coelophysis* typically have short limbs and long, narrow bodies, tails, and feet. Their morphogenesis inherently emphasized the posterior organs; even their teeth are laterally compressed and posteriorly recurved. Footprints and trackways are also inherently narrow, and they sometimes show evidence of the animals sitting back on their posterior organs (metatarsals and ischia). Trackways also suggest that they moved quickly (Farlow 1981) but were typically not gregarious. They probably had large eyes and relatively large brains, perhaps with hind brain emphasis. All these characteristics are convergent with rodents to varying degrees, and although they do not suggest a close biological affinity, they suggest a morphodynamic, and presumably genetic, organization that was convergent with and connected to characteristically small body size. Additionally, their predominantly paedomorphic characteristics probably indicate that they had relatively short life spans.

By contrast, large (predominantly peramorphic) derived theropods like *T. rex* have generated much debate regarding their locomotor capabilities (speed), diet, metabolism, and aggressive, predatory, and/or scavenging behaviors (Paul 1988; Horner and Lessem 1993). Morphologically, they tend to be longer limbed (hind limbs) with shorter, wider torsos and somewhat shorter tails (Bakker and Bir 2004). Their teeth are also much wider (less narrow), and in some large theropods, there is evidence of a wider trackway (Lockley 2001). They also developed larger heads, sometimes with small crests or horns (as in *Carnotaurus*). There is also much evidence from healed and resorbed bone that they had active metabolisms, which may well have permitted them to scavenge old meat. All these characteristics place large forms like *T. rex* at what Schad (1977) calls the metabolic-limb end of the theropod spectrum. However, we can go one step further and say that within the coelurosaurs, it appears that the tyrannosaurids represent the more metabolic exaggeration and the ornithomimids express the greater limb emphasis. Hone et al. (2004) even infer that there may be an ideal mean dinosaur body size (which they specify as 7.8 m in length) above which there is a strong tendency toward Cope's rule. They even use the chaos theory metaphor of a possible "attractor" for such "an apparent stable equilibrium body size" (p. 594). Thus, they not only imply that morphology within dinosaur clades and subclades has plastic and dynamic qualities that express considerably diversity of form during the course of development and evolution, but they also infer that there may be some sort of larger organizational patterning around homeostatic equilibrium, or an attractor. In the context of our discussion, *T. rex*, like most large derived tyrannosaurids, is on the large peramorphic side of the equilibrium point, whereas most small gracile ornithomimids are on the small paedomorphic size with respect to overall size.

A brief perspective on sauropod characteristics provides a sense of the polarity inherent in the morphodynamics of saurischians as a whole. Sauropods were all large and quadrupedal, with long necks. They were predominantly peramorphic as a result of accelerated growth (sensu Erickson et al. 2001, 2004) and likely long lived. They may have been derived heterochronically from prosauropods by retention of juvenile prosauropod proportions during prolonged growth (Reisz et al. 2005). Smaller primitive forms such as the diplodocids had a more posterior and narrow emphasis (long tails, small forelimbs and skulls, and narrow trackways), whereas derived forms like the brachiosaurs and titanosaurs had greater anterior emphasis (short tails, longer necks, somewhat larger skulls, longer forelimbs, and wide trackways; Farlow 1992; Lockley et al. 1994b). All appear to have been gregarious and were vegetarian, with rounded teeth. In comparison with their theropod cousins, the sauropods were larger overall (more peramorphic), with greater metabolic-limb emphasis (vegetarians with erect limbs), thus suggesting convergence with ungulates. (This morphodynamic trend toward ungulate convergence becomes even more pronounced in derived ornithischians; Lockley 1999a.) Sauropods also reveal a long-limb, short-foot design, compared with the relatively short-limb, long-foot design among most theropods, especially the more primitive forms (Fig. 9.5).

The importance of emphasizing these polarities and inherent compensations in the context of heterochronic language is to stimulate interest and observation of these recursive fractal patterns, and to observe dinosaur clades as complex organic systems, like cells or whole individuals, but on larger scales. Again, as scientific language evolves, dynamic terms like *morphological equilibrium point* or *attractor* may replace the more static concepts of *mean* or *average* and allow us to see clades more like dynamic or homeostatic superorganisms, and less like a classification of separate, individually different forms. The evo-devo paradigm supports such holistic morphodynamic thinking, but it essentially proceeds from the bottom up (molecules to cells and embryonic morphologies). The fossil record offers a different type of macromorphological evidence that requires us to work from the top down. This is where the morphodynamic paradigm can be most useful if sufficiently understood. Not only does it help us understand the inherent, formal growth dynamics that lead to differential organ emphasis (e.g., heads, necks, and limbs), but it also gives valuable clues to how these relate to physiology and behavior.

How can we make sense of all these parallel and equal and opposite directional morphodynamic growth trends? There are clear recursive patterns. During ontogeny, many growth trends proceed in the A-P direction, also known as the cephalo-caudal direction (sensu Kingsbury 1924; Verhulst 2003), and by definition, as an organism grows, it gets older. Does this mean that growth in the opposite direction (P-A) somehow represents a juvenilization process? There are many lines of evidence that suggest that during phylogeny, descendants often begin to look more like juveniles representatives of their ancestors. Thus, embryonic apes look human (Verhulst 2003), and many birds resemble juvenile theropods. There has been much debate on this topic (Gould 1977, 2002), which is outside the scope of this chapter to deal with in detail. Nevertheless, the large head is characteristic of early embryological development in most vertebrates, which subsequently have to grow into their bodies, so that their heads get relatively smaller during ontogeny, even though they grow larger in absolute terms. Thus, the long-term evolutionary trends (correlated progressions, sensu Kemp 1999), which have led to an anteriorization or encephalization in many groups, suggest a type of progressive juvenilization (Suchantke 1995). Given the formative importance of early embryonic development in shaping the morphology of adults, it makes sense to consider that the heterochronic shifts that allow increased developmental activity early in ontogeny will allow for greater evolutionary flexibility and novelty. The suggestion that such juvenilization (paedomorphosis) has played an important role in evolution is familiar to students of heterochrony and morphodynamics (Gould 1977; Schad 1977; McKinney and McNamara 1991; McNamara 1997; Verhulst 2003). We conclude that such morphodynamic trends likely influenced dinosaur phylogeny, as they have been inferred to have done with other vertebrate groups. Given that all species display a mixture of peramorphic and paedomorphic characteristics (McNamara 1997), as in the case of the ostrich, which is peramorphically large in terms of its while body but pedomorphic in terms of its juvenile plumage and partial nakedness, it could be that *T. rex* was the coelurosaurian version of this "overgrown baby" phenomenon: i.e., peramorphic in overall size but paedomorphic in the sense that it had a big skull.

We note in the case of tyrannosaurids that recent studies make increasing reference to hypermorphosis (Carr 1999), peramorphosis (Xu et al. 2004), and acceleration (Carr 1999; Erickson et al. 2004). We suggest that this renewed interest in heterochrony is a healthy sign that we are beginning to better understand the developmental dynamics of an extinct group of animals whose biology cannot be studied directly.

Conclusions

Most functional interpretations of *T. rex* forelimbs are simplistic or unconvincing. No one has made a compelling case that the arms were used for mating, clutching, or getting up from a prone position. Likewise, the suggestion that small arms provided some kind of biomechanical balancing mechanism to compensate for a large head is hard to prove. However, we agree that during all vertebrate development, there is good evidence for the so-called balancement of organs. This primary or formal developmental dynamic may lead to the secondary appearance or manifestation of a functional or biomechanical compensation.

Morphodynamic studies of mammals and dinosaurs suggest that we can trace A-P and/or P-A currents of macromorphological growth expressed as relative exaggeration or reduction of organs. Such morphodynamic organization is inherently heterochronic and generally emphasizes posterior development in small (mainly paedomorphic), primitive organisms and anterior emphasis in large (predominantly peramorphic), derived forms. On the basis of such a paradigm, *T. rex* and *Carnotaurus* were the largest and most derived (anteriorized) taxa in their respective clades (coelurosaurs and ceratosaurs), and they can be compared with their more primitive relatives in these same clades where morphological emphasis is centered more in the neck and forelimbs.

We use the compensation principle, which is also inherently heterochronic, to argue that the small head, long neck, and long forelimb versus large head, short neck, and short forelimb polarity repeats at many levels in the taxonomic hierarchy of saurischians (and other clades), and it is thus an example of lawful or ordered patterning in morphodynamic systems linked to P-A encephalization.

The evo-devo paradigm, and especially the pivotal role of *Hoxc6* in defining the cervical-thoracic boundary and forelimb budding, is compelling evidence that we can begin to understand differential theropod and saurischian morphology from these formal, developmental perspectives.

During ontogeny of vertebrates, the whole body and its organs generally grow from anterior to posterior, although growth rates vary from species to species and organ to organ. However, during evolution (phylogeny), there are many trends in the opposite (posterior to anterior) direction. This is seen by some authors as a juvenilization trend.

Recent published work on tyrannosaurids frequently refers to various forms of peramorphosis (acceleration and hypermorphosis) as important developmental dynamics that help us integrate our understanding of this popular and well-studied group. We suggest that these same dynamics recur fractally even within species, but most visibly throughout the Coelurosauria, Theropoda, Saurischia, and Dinosauria, thus allowing us to see these dynamics in a broader context.

Acknowledgments

Thanks to the Black Hills Geological Institute for the invitation to participate in the 100 Years of *Tyrannosaurus rex* Symposium (June 10–11, 2005). Thanks also to Kenneth Carpenter, Denver Museum of Nature & Science, for suggestions regarding references, and his comments on the hypothetical juvenile *Albertosaurus* restoration of Russell (1970). We also thank Wolfgang Schad (Witten-Herdecke University) and Ken McNamara (Western Australia Museum) for helpful discussion of heterochrony and morphodynamics. We also thank Ken McNamara, Kenneth Carpenter, and Peter Larson for their helpful and at times challenging reviews of the chapter in manuscript.

References Cited

Arthur, W. 2006. D'Arcy Thompson and the theory of transformations. *Nature* 7: 401–406.

Bakker, R. T., and Bir, G. 2004. Dinosaur crime science investigations: theropod behavior at Como Bluff, Wyoming and the evolution of birdness. P. 301–342 in Currie, P. J., Koppelhus, E. B., Shugar, M. A., and Wright, J. L. (eds.) *Feathered Dragons*. Indiana Univesity Press, Bloomington.

Bird, R. J. 2004. *Chaos and Life: Complexity and Order in Evolution and Thought*. Columbia University Press, New York.

Bonaparte, J., and Vince, M. 1979. El hallazgo del primer nido de dinosaurios triásicos (Saurischia, Prosauropoda) Triásico superior de Patagonia, Argentina. *Ameghiniana* 16: 173–182.

Caldwell, M. W. 2003. "Without a leg to stand on": the evolution and development of axial elongation and limbnessness in tetrapods. *Canadian Journal of Earth Sciences* 40: 573–588.

Carr, T. D. 1999. Craniofacial ontogeny in Tyrannosauridae (Dinosauria, Coelurosauria). *Journal of Vertebrate Paleontology* 19: 497–520.

Carrano, M. T., and O'Connor, P. M. 2005. Bird's eye view. *Natural History* 114: 42–47.

Carpenter, K., and Smith, M. B. 1995. Osteology and functional morphology of the forelimb of tyrannosaurids as compared with other theropods (Dinosauria). *Journal of Vertebrate Paleontology* 15: 21A.

———. 2001. Forelimb osteology and biomechanics of *Tyrannosaurus rex*. P. 90–116 in Tanke, D. H., and Carpender, K. (eds.) *Mesozoic Vertebrate Life*. Indiana University Press, Bloomington.

Carroll, S. B. 2005. *Endless Forms Most Beautiful: the New Science or Evo-Devo*. Norton, New York.

Chatterjee, S. 1997. *The Rise of Birds*. Johns Hopkins University Press, Baltimore, MD.

Colbert, E. H. 1989. *The Triassic Dinosaur Coelophysis*. Museum of Northern Arizona Bulletin 57.

Conway, M. S. 2003. *Life's Solutions*. Cambridge University Press, Cambridge.

Currie, P. J. 2003. Allometric growth in tyrannosaurids (Dinosauria; Theropoda) from the Upper Cretaceous of North America and Asia. *Canadian Journal of Earth Sciences* 40: 651–665.

de Beer, G. 1940. *Embryos and Ancestors*. Oxford University Press, Oxford.

DeRobertis, E. M., and Sasai, Y. 1996. A common plan for dorso-ventral patterning in Bilateria. *Nature* 380: 37–40.

Duboule, D., and Morata, G. 1994. Colinearily and functional hierarchy among genes of the homeotic complexes. *Trends in Genetics* 10: 358–364.

Erickson, G. M., Curry-Rogers, K., and Yerby, S. A. 2001. Dinosaurian growth patterns and rapid avian growth rates. *Nature* 412: 429–433.

Erickson, G. M., Makovicky, P. J., Currie, P. J., Norell, M. A., Yerby, S. A., and Brochu, C. A. 2004. Gigantism and comparative life-history parameters of tyrannosaurid dinosaurs. *Nature* 430: 772–775.

Farlow, J. O. 1981. Estimates of dinosaur speeds from a new trackway site in Texas. *Nature* 294: 747–748.

———. 1992. Sauropod tracks and trackmakers: integrating the ichnological and skeletal records. *Zubia* 10: 89–138.

Fastovsky, D. E., and Weishampel, D. B. 2005. *The Evolution and Extinction of the Dinosaurs*. 2nd ed. Cambridge University Press, Cambridge.

Foote, M. 1991. Morphologic patterns of diversification: examples from trilobites. *Palaeontology* 34: 461–485.

Gilbert, S. F. 2000. *Developmental Biology*. Sinauer Associates, Sunderland, MA.

Goethe, J. W. 1795. Erster Entwurf einer allgemeinen Einteitung in die vergleichende Anatomie, ausgehend von Osteologie. In *Goethes Naturwissenschaftliche Schriften*. Hrsg. R. Steiner, Dornach.

Gould, S. J. 1977. *Ontogeny and Phylogeny*. Harvard University Press, Cambridge, MA.

———. 1985. Geoffroy and the homeobox. P. 205–218 in Slavkin, H. C. (ed). *Progress in Developmental Biology, Part A*. Alan R Liss, New York.

———. 1980. *The Panda's Thumb*. Norton, New York.

———. 2002. *The Structure of Evolutionary Theory*. Harvard University Press, Cambridge, MA.

Haeckel, E. 1866. *Generelle Morhologie der Organsmen: Allgemeine Grundzüge der organischen Formen-Wissenschaft, mechanisch begründet durch die von Charles Darwin reformirte Descendenz-Theorie*. 2 vols. George Reimer, Berlin.

Hopson, J. A. 1977. Relative brain size and behavior in archosaurian reptiles. *Annual Review of Ecology and Systematics* 8: 429–448.

———. 1979. Paleoneurology. P. 39–146 in Gans, C., Northcutt, R. G., and Ulinski, P. (eds.). *Biology of the Reptilia*. Vol. 17. Academic Press, London.

Horner, J. R., and Lessem, D. 1993. *The Complete T. rex*. Simon & Schuster, New York.

Howe, D. W. E., Keesey, T. M., Pisani, D., and Purvis, A. 2004. Macroevolutionary trends in the Dinosauria: Cope's rule. *Journal of Evolutionary Biology* 18: 587–595.

Hutt, S., Nash, D., Martill, D. M., Barker, M. J., and Newberry, P. 2001. A preliminary account of a new tyrannosauroid from the Wessex Formation (Early Cretaceous) of southern England. *Cretaceous Research* 22: 227–242.

Jerison, H. J. 1969. Brain evolution and dinosaur brains. *American Naturalist* 103: 575–588.

Kaas, J. H. 2000. Why brain size is so important: design problems and solutions as neocortex gets bigger or smaller. *Brain and Mind* 1: 7–23.

Kellner, A. 2003. Pterosaur phylogeny and comments on the evolutionary history of the group. P. 105–138 in Buffetaut, E., and Mazin, J. M. (eds.). *Evolution and Palaeobiology of Pterosaurs*. Geological Society of London Special Publication.

Kemp, T. S. 1999. *Fossils and Evolution*. Oxford University Press, Oxford.

Kingsbury, B. F. 1924. The significance of the so called law of cephalocaudal differential growth. *Anatomical Record* 27: 305–321.

Larsson, H. C. E., Sereno, P. C., and Wilson, J. A. 2000. Forebrain enlargement among non avian theropods. *Journal of Vertebrate Paleontology* 20: 615–618.

LeGuyader, H. 2004. *Geoffrey Saint-Hilaire: Visionary Naturalist*. University of Chicago Press, Chicago.

Lenoir, T. 1987. The eternal laws of form: morphotypes and the condition of existence in Goethe's biological thought. P. 17–28 in Amrine, F., Zucker, F. J., and Wheeler, H. (eds.). *Goethe and the Sciences: A Reappraisal*. Boston Studies in the Philosophy of Science 97.

Lockley, M. G. 1999a. *The Eternal Trail: A Tracker Looks at Evolution.* Perseus Books, Reading, MA.

———. 1999b. Einblicke in die Gestaltebiologie der Dinosaurier anhand ihrer Fahrtenspuren. *Tycho de Brahe–Jahrbuch für Gotheanismus* 134–166.

———. 2001. Trackways—dinosaur locomotion. P. 412–416 in Briggs, D. E. G., and Crowther, P. (eds.). *Paleobiology: A Synthesis.* Blackwell, Oxford.

———. 2004. Beyond feet and footprints: what morphodynamics and heterochrony tell us about the relationships between feet limbs and the whole body. P. 6–8 in Buatois, L. A., and Mangano, G. M. (eds.). *Ichnia 2004, First International Congress on Ichnology.* Trelaw, Patagonia.

———. In press. The morphodynamics of dinosaurs, other archosaurs and their trackways: holistic insights into relationships between feet, limbs and the whole body. In Bromley, R., and Melchor, R. (eds.). *Ichnology at the Crossroads: A Multidimensional Approach to the Science of Organism-Substrate Interactions.* Society of Economic and Paleontologists and Mineralogists Special Publication.

Lockley, M. G., and Jackson, P. In press. Morphodynamic perspectives on convergence between the feet and limbs of sauropods and humans: two cases of hypermorphosis. *Ichnos.*

Lockley, M. G., and Kukihara, R. 2005. A morphodynamic analysis of the Triassic theropod dinosaur *Coelophysis*: dimorphism provides a clue to intrinsic biological organization in saurischian dinosaurs. P. 11–12 in *Tracking Dinosaur Origins: The Triassic/Jurassic Terrestrial Transition,* Abstract Volume. Dixie College, St. George, UT.

Lockley, M.G., dos Santos, V. F., Meyer, C. A., and Hunt, A. P. (eds). 1994b. *Aspects of Sauropod Paleobiology. Gaia: Revista de Geociencias, Museu Nacional de Historia Natural, Lisbon,* 10.

Long, J. A., and McNamara, K. J. 1995. Heterochrony in dinosaur evolution. P. 151–168 in McNamara, K. (ed.). *Evolutionary Change and Heterochrony.* Wiley, Chichester, UK.

———. 1997. Heterochrony: the key to dinosaur evolution. P. 113–123 in Wolberg, D. L., Stump, E., and Rosenberg, G. D. *Dinofest International Proceedings.* Academy of Natural Sciences, Philadelphia.

McKinney, M. L., and McNamara, K. J. 1991. *Heterochrony: The Evolution of Ontogeny.* Plenum Press, New York.

McNamara, K. 1995. Sexual dimorphism: the role of heterochrony. P. 55–89 in McNamara, K. (ed.) *Evolutionary Change and Heterochrony.* Wiley, Chichester, UK.

———. 1997. *Shapes of Time: The Evolution of Growth and Development.* John Hopkins University Press, Baltimore, MD.

Middleton, K. M., and Gatesy, S. M. 2000. Theropod forelimb design and evolution. *Zoological Journal of the Linnean Society* 128: 149–187.

Newman, B. H. 1970. Stance and gait in the flesh-eating dinosaur *Tyrannosaurus. Biological Journal of the Linnean Society* 2: 119–123.

Norell, M. A., Gaffney, E. S., and Dingus, L. 1995. *Discovering Dinosaurs.* A. A. Knopf, New York.

Osborn, H. F. 1906. *Tyrannosaurus,* Upper Cretaceous carnivorous dinosaur. *Bulletin of the American Museum of Natural History* 22: 281–296.

Paul, G. 1988. *Predatory Dinosaurs of the World.* Simon & Schuster, New York.

Pike, A., Versus, L., and Alexander, R. M. 2002. The relationship between limb-segment proportions and joint kinematics for the hind limbs of quadrupedal mammals. *Journal of Zoology* 258: 427–433.

Portmann, A. 1964. *New Paths in Biology.* Harper & Row, New York.

Rauhut, O. W. M., Remes, K., Fechner, R., Cladera, G., and Puerta, P. 2005. Discovery of a short-necked sauropod dinosaur from the Late Jurassic period of Patagonia. *Nature* 435: 670–672.

Reisz, R. R., Scott, D., Sues, H. D., Evans, D. C., and Raath, M. A. 2005. Embryos of an Early Jurassic prosauropod dinosaur and their evolutionary significance. *Nature* 309: 761–764.

Riegner, M. 1985. Horns, spots and stripes: form and pattern in mammals. *Orion Nature Quarterly* 4(4): 22–35.

———. 1998. Horns, hooves, spots and stripes: form and pattern in mammals. P. 117–212 in Seamon, D., and Zajonc, A. (eds.). *Goethe's Way of Science: A Phenomenology of Nature*. State University of New York Press, New York.

Russell, D. A. 1970. Tyrannosaurs from the Late Cretaceous of Western Canada. *National Museum of Natural Science Publications in Paleontology* 1: 1–34.

Russell, D. A., and Seguin, R. 1982. Reconstruction of a small Cretaceous theropod *Stenonychosaurus inequalis* and a hypothetical dinosauroid. *Syllogeus* 37: 1–43.

Saint-Hilaire, G. 1822. Considérations générales sur la vertèbre. *Memoires du Muséum d'Histoire Naturelle* 9: 89–114.

Salgado, L., and Bonaparte, J. F. 1991. Un nuevo sauropodo Dicraeosauridae, *Amargasaurus cazaui* gen. et sp. nov., de la Formacion La Amarga, Neocomiano de la Provincia del Neuquen, Argentina. *Ameghiniana* 28(3–4): 333–346.

Schad, W. 1977. *Man and Mammals: Towards a Biology of Form*. Waldorf Press, New York.

———. 1992. The Heterochronic Role of Evolution in the Vertebrate Classes and the Hominids. Ph.D. diss., University of Witten, Herdecke.

Seamon, D., and Zajonc, A. (eds.). 1998. *Goethe's Way of Science: A Phenomenology of Nature*. State University of New York Press, New York.

Shubin, N. H., and Alberch, P. 1986. A morphogenetic approach to the origin and basic organization of the vertebrate limb. *Evolutionary Biology* 20: 319–387.

Suchantke, A. 1995. The metamorphosis of plants as an expression of juvenilisation in the process of evolution. P. 47–69 in Bockemuhl, J., and Suchantke, A. (eds.). *The Metamorphosis of Plants*. Novalis Education Series, Cape Town, South Africa.

Thom, R. 1975. La théorie des catastrophes et ses applications. *Reflexions sur Nouvelles Approches dans l'étude des Systèmes*. Editions de l'ENSTA, Paris.

Thompson, D'A. W. 1917. *On Growth and Form*. Cambridge University Press, Cambridge.

Verhulst, J. 2003. *Developmental Dynamics in Humans and Other Primates*. Adonis Press, New York.

Wallenfand, M. R., and Seydoux, G. 2000. Polarization of the anterior-posterior axis of *C. elegans* is a microtubule-directed process. *Science* 408: 89–92.

Weishampel, D. B., Dodson, P., and Osmòlska, H. (eds.). 2004. *The Dinosauria*. 2nd ed. University of California Press, Berkeley.

Xu, X., Norell M. A., Kuang, X., Wang, X., Zhao, Q., and Jia, C. 2004. Basal tyrannosauroids from China and evidence for protofeathers in tyrannosauroids. *Nature* 431: 680–684.

Xu X., Clark, J. M., Forster, C. A., Norell, M. A., Erickson, G. M., Eberth, D. A., Jia, C., and Zhao, Q. 2006. A basal tyrannosauroid dinosaur from the Late Jurassic of China. *Nature* 439: 715–718.

LOOKING AGAIN AT THE FORELIMB OF *TYRANNOSAURUS REX*

10

Christine Lipkin and Kenneth Carpenter

Introduction

The large theropod *Tyrannosaurus rex* is the archetype carnivorous dinosaur ever since it was named in 1905 (Osborn 1905). Even its name, "tyrant-lizard king," invoked it as a top predator. Recent challenges to its title as the largest terrestrial carnivore (e.g., Sereno et al. 1996; Calvo and Coria 1998) have not diminished its popularity. Osborn reasoned that *Tyrannosaurus* was a predator on the basis of its teeth. Coprolitic material, some containing fossilized soft tissue, supports a carnivorous diet (e.g., Chin et al. 2003), as does tooth-grooved bones of prey (Erickson and Olson 1996) and evidence of a failed attack on a hadrosaur (Carpenter 1998) and ceratopsian (Happ this volume).

Although Osborn (1905, 1912) was clear that he considered *Tyrannosaurus* a predator, Lambe (1917) dissented, considering tyrannosaurids to be scavengers instead, primarily because of the apparent absence of tooth wear. Tyrannosaurids as obligatory scavengers never gained popularity until recently, when the hypothesis was reintroduced by Horner and Lessem (1993). By and large, though, *Tyrannosaurus* has been considered an active predator, although the mode of attack remains controversial and includes flank bite and run (Paul 1987), opportunistic (Farlow 1994), and neck or snout crushing (Molnar 1998). The method of killing is assumed to have been the jaws, and the bite force has been variously calculated or estimated to have been 13,400 N (Rayfield et al. 2001), 6400–13,400 N (Erickson et al. 1996), or even 183,000–235,000 N (Meers 2002). As Meers has noted, the high forces are consistent with either scavenging or predation.

In arguing against predation, Horner and Lessem (1993) cite the shortness of the forelimbs as being useless for holding prey. In support of this position, Lingham-Soliar (1998) considered head shaking as a means of flesh removal from small prey and direct ripping of flesh from large prey. Carpenter (2002) has shown that without exception, no theropod could extend its forelimb beyond the snout, thus limiting its usefulness as a prey-grasping organ before the mouth was engaged (Fig. 10.1). Does this mean, however, that the forelimbs of *Tyrannosaurus* were as useless as portrayed? Paul (1988, p. 320) considers such the question irrelevant: "the reduced size of the forelimb shows they were not important to their owners, so they should not be important to us." This position, supported by Lockley et al. (this volume), is based on an unsubstantiated assumption ("were not important to their owners"), which is just as useless as the untestable speculations of Lockley et al. (this volume). In point of fact, Carpenter and Smith (2001) concluded that the forelimb was powerful, an interpretation previously made by Brown (1915, p. 271): "front limbs exceedingly small but set for a powerful clutch."

Figure 10.1. *Comparison of maximum forelimb motion in 3 well-known theropods. None of the dinosaurs can reach its manus to its mouth as a result of constraints in the shoulder (see Carpenter 2002). Note that **Tyrannosaurus** has the greatest range of retraction. Not to scale.*

New material has led to our reassessing the forelimb of *Tyrannosaurus*, and a stronger case is made for forelimb use during predation. Some of this new material displays pathologies, which is important because they are a reflection of lifestyle behaviors (Rothschild and Martin 1993). As Paul has noted (this volume), *Tyrannosaurus* must have had a rough, active life.

The materials used in this study are as follows: scapula of BHI 3033, DMNH 2827, and MOR 555; furcula of FMNH PR2081, MOR 980, and TCM 2001.90.1; humerus of BHI 6230, FMNH PR2081, and MOR 555; ulna and radius of FMNH PR2081, MOR 555, and MOR 980; and manus of FMNH PR2081, MOR 555, MOR 980, and BHI 6230.

New Information on the Pectoral Girdle and Forelimb

The pectoral girdle and forelimb of *Tyrannosaurus* have been described by Carpenter and Smith (2001) and by Brochu (2002). Since then, the furcula has been described (Larson and Rigby 2005), the third metacarpal and semilunate carpal have been found (described below), and new information is available for the scapula and coracoid. To date, no ossified sternal plates are known, but these are predicted to resemble those of *Gorgosaurus* as described by Lambe (1917). Brochu (2002) has discussed the possibility of ossified sternal plates in *Tyrannosaurus* but came to no conclusion.

Furcula

The furcula of *Tyrannosaurus*, the presence of which was predicted by Carpenter and Smith (2001), is now known for several specimens (Larson and Rigby 2005). It is broadly U or boomerang shaped (Fig. 10.2), with a roughened sutural scar (acrominal facet) on the epicleidium for a ligamentous attachment to the acromion of the scapula (Fig. 10.3; see also below). In birds, the rami of the furcula have a nearly circular or laterally compressed cross-sectional geometry that allows them to act as a spring, with laterally directed tension added on the downstroke and medial directed recoil on the upstroke (Jenkins et al. 1988; Boggs et al. 1997; Hui 2002). In *Tyrannosaurus*, however, the furcula is clearly designed to resist extreme lateral forces: (1) the rami are anteroposteriorly flattened (cf. Fig. 10.2F, G), thus prohibiting lateromedial springlike action; (2) the rami diverge, thus directing lateral stresses down the shaft (Fig. 10.4A; and (3) the rami deepen distally from the epicleidium and the furcula is deepest (thickest vertically in the anatomical position) near the midline to counter the stresses directed down the rami. In most theropods, the furcula is nearly uniform throughout its length (see Chure and Madsen 1996, fig. 2, 3; Makovicky and Currie 1998, fig. 1; Currie et al. 2005, fig. 16.8; Larson and Rigby 2005, fig. 12.3). Therefore, the great depth of the *Tyrannosaurus* furcula is unusual and is clearly adapted to resist stress.

Three of the 5 known *Tyrannosaurus* furculae are pathologic (Fig. 10.2) and were examined by computed tomographic scanning and X-ray. Two of them are missing portions of a ramus (FMNH PR2081 [Fig. 10.2A, B] and TCM 2001.90.1 [Fig. 10.2E, F]), and 2 show localized swelling of the cortical bone (MOR 980 [Fig. 10.2D] and TCM 2001.90.1 [Fig. 10.2F]).

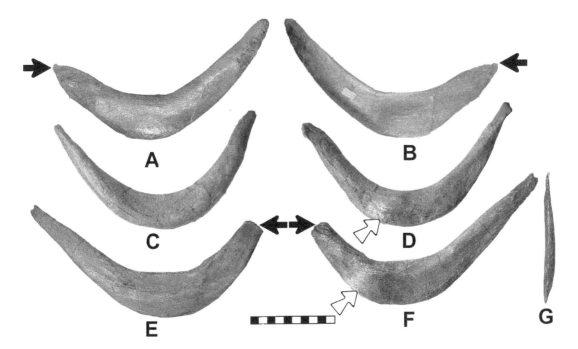

Figure 10.2. Furcula of *Tyrannosaurus rex* include several with pathologies, including fractures (black arrows) and localized exostosis of stress fractures (white arrows). FMNH PR2081 in posterior (A) and anterior (B) views; MOR 980 in posterior (C) and anterior (D) views; TCM 2001.90.1 in posterior (E), anterior (F), and lateral (G) views. Scale in centimeters.

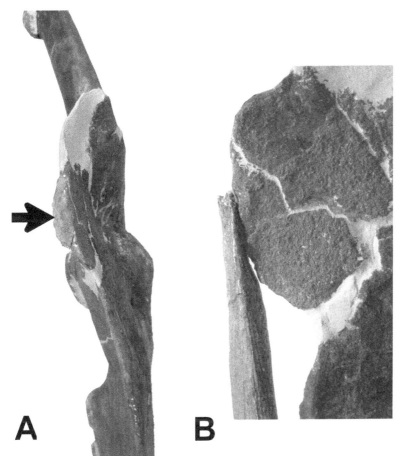

Figure 10.3. Close-up of the epicleidial facet (arrow) on the dorsal edge of the scapula DMNH 2827 (A) and lateral view with the furcula articulated (B).

The missing sections of the rami were broken in life, with subsequent remodeling of the fracture surface. Surprisingly, none show a pseudoarthrosis joint at the site of the break, thus indicating significant displacement of the broken portion, possibly due to the contraction of the M. supracoracoideus brevis, which probably inserted along the ramus.

The amount of force needed to break a ramus was calculated from the largest furcula, TCM 2001.90.1. The fracture is 2.5 cm wide; however, the bone is remodeled on the anterior and posterior sides so as to exaggerate the original thickness of the bone. Therefore, the anterior-posterior thickness is approximated from the undamaged ramus, where it is has the same width, which gives a thickness of 1.35 cm. The ramus can be modeled as an ellipse; therefore the area of the break is given by the following:

$$A_r B_r, \quad (1)$$

or 2.6 cm², where A_r is the radius of the ramus width and B_r is the radius of the thickness. Given that the shear strength of living cortical bone is conservatively ~10^2 N/cm² (Currey 2002), about 26,000 N (i.e., 2652 kg of force) was required to break the furcula. As seen by computed tomography, trabecular bone occupies a small portion of the normal ramus, so it was not considered in the calculations because it would have reduced the fracturing force only slightly.

Two of the pathologic furculae also show a characteristic bony callus from a healed stress fracture (Rothschild 1988) located on the posterior side (Fig. 10.2D, F; Fig. 10.5). Stress fractures occur as a result of repetitive loading on bone, which leads to mechanical failure and microfracturing (Resnick 2002; Rothschild and Martin 1993; Rothschild 1988). Rothschild and Tanke (2005) and Rothschild and Molnar (this volume) note a high incidence of stress fractures in the manual elements of tyrannosaurids. They also note, "Active resistance of prey is required to overstress the manus," which would also essentially include the rest of the forelimb and pectoral girdle. Because a stress fracture is a partial fracture, the amount of force must be less than the maximum required to completely break the bone. Again, the furcula of TCM 2001.90.1 was used to calculate the force. The normal portion of the furcula corresponding to the pathological portion measures 4.5 cm dorsoventrally and 1.7 cm anteroposteriorly. Assuming the regions of the stress fracture were approximately the same dimensions, then the force must have been less than 60,080 N (6128 kg of pressure), resulting from equation 1.

Scapula

The scapula or *T. rex* was mostly described by Carpenter and Smith (2001). New information is available based on DMNH 2827. The acromion is a thin plate that is often damaged or lost in other specimens (e.g., MOR 555). Fortunately, this region is preserved in DMNH 2827 and shows a small facet for the epicleidium near the scapulocoracoid suture (Fig. 10.3). This epicleidial facet measures 4.8 cm by 1.1 cm. Placement of the furcula connecting the epicleidial facets of the left and right scapula show how close

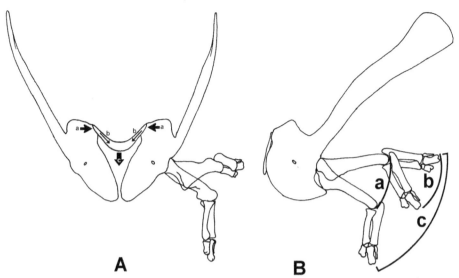

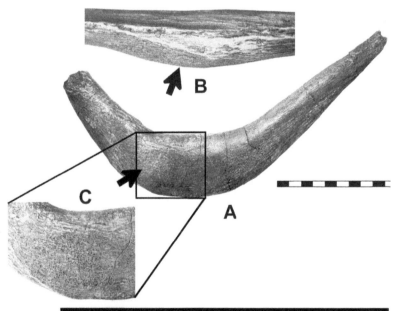

Figure 10.4. Reconstruction of the pectoral girdle and forelimb of **Tyrannosaurus** showing (A) the distribution of force from the scapula (a arrows), through the furcula (b arrows), which results in cumulative force (c arrow) at the middle of the furcula. Resisting force c explains why the furcula is deepest at the midline, which is unlike any other theropod furcula. The range of motion for the forelimb segments (B), angle represented by arc a = 40°, arc b = 60°, arc c = 67°.

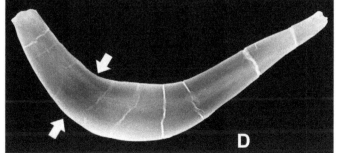

Figure 10.5. Evidence for stress fracture in the furcula of TCM 2001.90.1 is the prominent callus (A), seen clearly in dorsal view (B) and in close-up showing periosteal reactive bone (C). The region is X-ray opaque because of the greater deposit of bone (D, between arrows). Scale for A in centimeters.

Looking Again at the Forelimb   171

together the coracoids really were in life (Fig. 10.4A, Fig. 10.6), a position supported by a nearly uncrushed *Tyrannosaurus* chest region found in situ that is currently under study (Lipkin and Sereno 2004). A similar close placement is also known in hadrosaurs (e.g., Osborn 1912), suggesting that coracoids were closely placed in all dinosaurs (including sauropods), as has been discussed elsewhere (Carpenter 2002; Carpenter et al. 1994).

DMNH 2827 also shows an unusual pathology of the glenoid, which is partially collapsed as a result of ventroposterior rotation of the coracoid. Although the glenoid was partially damaged during preparation (the bone in the region was crumbly), it is clear that the 2 bones were initially damaged before co-ossification because the 2 bones are now firmly fused by remodeled bone (Fig. 10.7). The rotation is less near the acromion and greater near the glenoid, suggesting that great rotational forces were applied to the coracoid in a posteroventral direction, thereby partially collapsing the glenoid. Although the damage may have resulted from a fall onto the chest, the direction of rotation also corresponds to the vector for the M. coracobrachialis brevis ventralis (although the terminology is retained for "dorsal" versus

Figure 10.6. *Position of the furcula relative to the scapula-coracoids as seen in a mounted skeleton of **Tyrannosaurus** (cast of BHI 3033). Human (Neal L. Larson) for scale.*

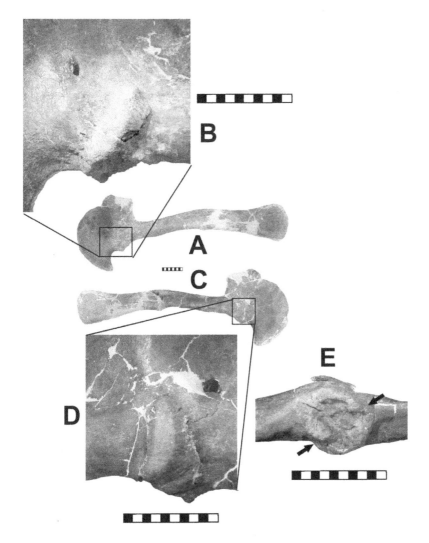

Figure 10.7. *Partial collapse of the glenoid in DMNH 2827 as seen in lateral view (A), with close-up (B); in medial view (C), with close-up (D); and ventral view showing the telescoping that occurred between the arrows. Sclerotic bone overhangs the lateral surface. The amount of deformation decreases dorsally to about the level of the coracoid foremen and indicates a posteroventral rotation of the coracoid due to great stress.*

Looking Again at the Forelimb   173

Figure 10.8. Comparison of normal right humerus of MOR 690 and its pathological left in anterior (A, D), lateral (B, E), and posterior (C, F) views. Note spur at dart in (E), perisoteal reactive bone opposite arrow in (F), with close-up in (G). Region between darts in (F) are shown in lateral view in (H) and in close-up in (I). See text for discussion. Scale in centimeters.

"ventral" muscles, we are aware that the more vertical position of the humerus in *Tyrannosaurus* indicates a need for modified terminology). It is therefore possible that the damage occurred when the individual was young and the forelimbs were pulling struggling prey toward the chest. Unfortunately, so little of the skeleton was recovered (see N. L. Larson this volume) that the extent and location of damage to other bones is unknown (e.g., right scapula-coracoid, humerus, gastralia). The distribution of pathologies elsewhere on the skeleton might resolve between the 2 possibilities.

Humerus

Several additional humeri are now known, including more with pathologies that reflect behavior. Overall, these specimens resemble those described by Carpenter and Smith (2001), differing only in minor detail. Two of the new specimens are of gracile morphs (e.g., MOR 980 and BHI 6230), which are probably male (P. Larson this volume). The robust morph (e.g.,

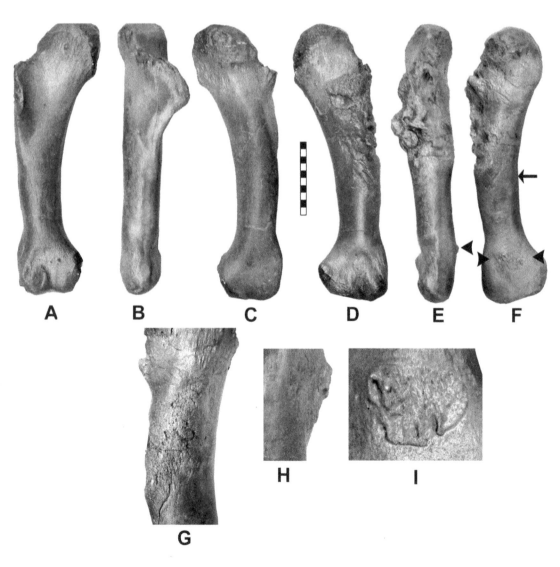

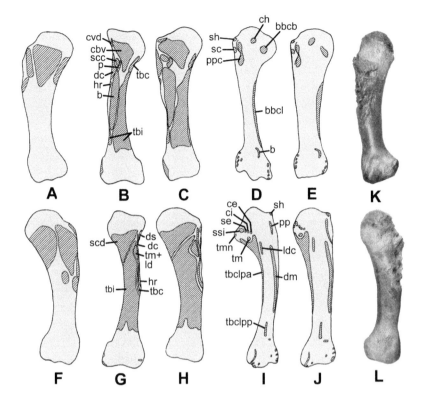

Figure 10.9. Muscle maps for humerus in *Tyrannosaurus*, *Alligator*, and *Gallus*. Top row is anterior, bottom row is posterior. Muscle map based on scars (A, F) *Tyrannosaurus*. Map for *Alligator* (B, G) and predicted for *Tyrannosaurus* (C, D) based on deformation of *Alligator* humerus. Map for *Gallus* (D, I) and predicted for *Tyrannosaurus* (E, J) based on deformation of *Gallus* humerus. Note that predicted scars for deformed *Alligator* (C, H) are a better match for the scars of *Tyrannosaurus* (A, F). This prediction is also supported by the pattern of avulsion seen in a *Tyrannosaurus* humerus (K, L). See Figure 10.13 and text for further explanation.

FMNH PR2081) is broader proximally than the gracile morph (cf. Carpenter and Smith 2001, fig. 9.4, with Fig. 10.8).

As noted by Carpenter and Smith (2001), the humerus of FMNH PR2081 shows several pathologies, as does the new left humerus of MOR 980 (e.g., Fig. 10.8). This latter humerus shows an extensive juxtacortical pathology located proximally on the anterolateral, lateral, and posterolateral sides of the diaphysis, which has all but obliterated the deltopectoral crest (Fig. 10.8D, E). The pathology shows extreme disruption of the cortical bone and irregular patches of sclerotic bone that suggest uneven loss and regrowth of the periosteum. The resulting periostitis implies extensive inflammation of the region. The region of the pathology corresponds with the crocodile humerus to the insertions of the M. pectoralis, the common insertion for the M. supracoracoideus complex, M. deltoideus clavicularis, common insertion for the M. teres major and M. latisimus dorsi, and origins of the M. brachialis, M. humeroradialis, and part of the triceps brevis cranialis (cf. Figs. 10.9B with 10.9K, G with L). There is also a juxtacortical lesion on the posterior side that, in the crocodile, corresponds with the origin for the proximal part of the M. triceps brevis intermedius (Fig. 10.8F, G; compare 9G with L). More distally, there is a small spur on the posterior side as well (Fig. 10.8E, F, H, I). These 2 lesions blend with the diaphysis proximally but have sharp, overhanging margins distally that may be due to osteoblastic response to the extensor motion of the overlying M. triceps group.

The large pathology on the proximal end is irregular or knobby. Superficially, it suggests a malignant neoplasm or extensive osteomyelitis, but as Donnelly et al. (1999) have noted, site location is important for differential diagnosis. The location and extent of the pathology is most probably due to avulsion of muscles in the vicinity of the deltopectoral crest, and the posterior pathologies due to periostitis from trauma to the periosteum associated with the event that caused the avulsion. The aggressive appearance of the proximal pathology, caused by bone resorption resulting in a lytic appearance of bone, is characteristic of a healing avulsion and can mimic the appearance of osteomyelitis or skeletal Ewing sarcoma (Stevens et al. 1999). An avulsion is a failure of bone at a tendinous or aponeurotic insertion of muscle (Tehranzadeh 1987; El-Khoury et al. 1997). Avulsions can either be acute (the result of extreme, abnormal muscle contractions) or chronic (the result of repeated microtrauma or overuse) (Stevens et al. 1999), where reoccurrence of the injury is more frequent than the ability of the tissue to repair itself (El-Khoury et al. 1997). The presence of a stress fracture in the furcula associated with this specimen suggests that repetitive overuse of the forelimb may have been a major factor leading to the avulsion. A secondary factor may have been a violent stress on the arm because so many different muscles were apparently affected. Such stresses would be generated in the sudden pull of the arm toward the chest at the same time prey was struggling in the opposite direction. It may not be incidental that MOR 980 is a young adult because Tehranzadeh (1987) noted that the incident of avulsion is highest in younger human individuals. Regardless of the cause, these pathologies support the hypothesis of forelimb use in *Tyrannosaurus*.

Manus

The incomplete manus of *Tyrannosaurus* was described by Carpenter and Smith (2001). New specimens, MOR 980 and BHI 6230, provide new information, including a carpal and metacarpal III (Figs. 10.10, 10.11). The carpal (Fig. 10.10A–F) is distal carpal I ("semilunate") and shows a facet on the proximal surface for the missing radiale. This carpal shows that the damaged, incomplete one described by Carpenter and Smith (2001) is in fact a distal carpal I. Ventrally, the new carpal is faceted and fits snugly between the proximal ends of metacarpals I and II (Fig. 10.11C), rather than across them as in *Deinonychus* (Carpenter 2002). Metacarpal III is slender, posteriorly curving, and tapering, and lacks a distal condyle, so it thus had no phalanges. MOR 980 is from a larger individual than BHI 6230, and the differences between them are probably ontogenetic. These metacarpals suggest that as the individual grew, the metacarpal became more robust and the proximal end formed a broader attachment to metacarpal II.

All of the new information on the forelimb was used to create a new view of the pectoral girdle and forelimb of *Tyrannosaurus* (Fig. 10.12) and was used in following biomechanical study.

Reconstructing Forelimb Musculature

Mathematical models for scavengers not withstanding (e.g., Ruxton and Houston 2003), many of the same criteria and assumptions for scavenging theropods also hold for predatory theropods as well (see Holtz this volume). In reality, there are too many unknown variables (e.g., population densities,

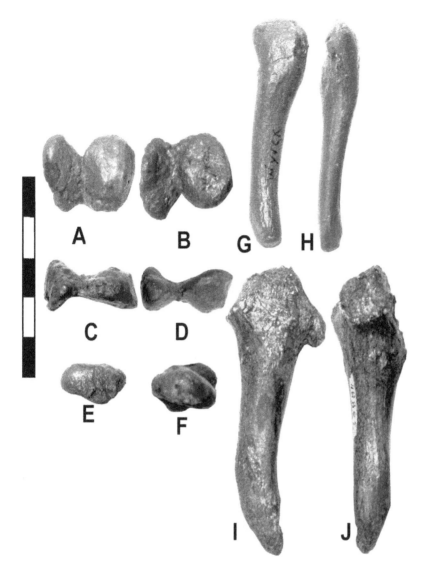

Figure 10.10. *Distal carpal (BHI 6230) of **Tyrannosaurus** in multiple views: proximal or dorsal (A); distal or ventral (B); anterior (C); posterior (D); extensor side (E); palmar side (F). Metacarpal III of BHI 6230 in lateral (G) and extensor side (H). Metacarpal III of MOR 690 in lateral (I) and extensor side (J). Scale in centimeters.*

energentics, locomotion capabilities of both predator and prey) to ever have testable results. With so many assumptions built on assumptions, a house of cards results. Our approach is to minimize assumptions (including minimizing untested "common sense," referred to by Ruxton and Houston 2003) and to rely on evidence and testable models.

Although we have attempted to minimize our assumptions, some are required as a result of the nature of fossilized remains. We assume the following: (1) The power output of muscles of extinct tetrapods is comparable to that of extant tetrapods—that is, muscles were not weaker or stronger than they are today. (2) Scars on fossilized bone that are clearly not pathological (e.g., lack of obvious remodeling; see Rothschild and Martin 1993) indicate insertion or origin for muscles or for ligaments (but see below). (3) Homologous origin-insertion scars, as determined by extant phylogenetic bracketing, identify each muscle (but see below). (4) The extent of the joint surface can be determined from the smooth surfaces of the joints, which are separated

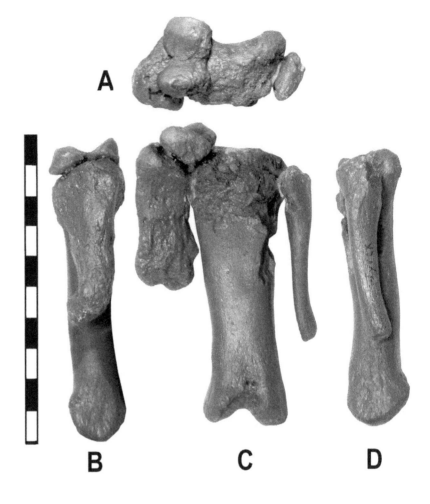

Figure 10.11. *Distal carpal and metacarpals in articulation (BHI 6230). Proximal (A), digit I side (B), extensor side (C), digit III side (D). Scale in centimeters.*

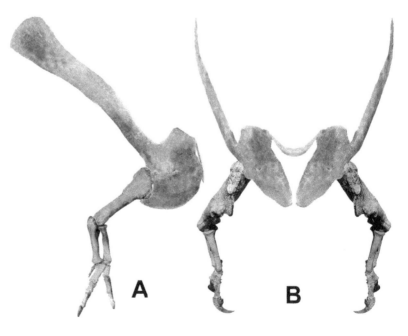

Figure 10.12. *Articulated pectoral girdle and forelimb of **Tyrannosaurus** in lateral (A) and anterior (B) views.*

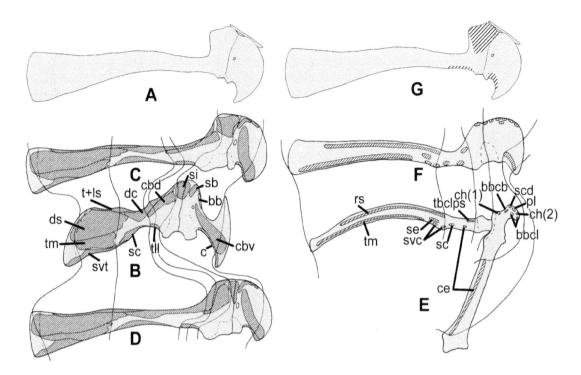

Figure 10.13. Deformation of a crocodilian and avian scapula and coracoid (SC) to approximate that of *Tyrannosaurus*. Wavy vertical lines show direction and degree of morphing. This method allows for the prediction of the position and shape of various muscles on the *Tyrannosaurus* SC (see text). (A) *Tyrannosaurus* scapula-coracoid used as the end point to morphing of the (B) crocodilian and (E) avian pectoral girdles. Note that 2 possible scenarios (C, D) occur in the morphing of the crocodilian SC depending on what portions of the crocodilian scapula is morphed into the acromion process of the *Tyrannosaurus* scapula. Location of actual muscle scars (G). Abbreviations: bb, M. biceps

from the diaphysis by a rim or abrupt textural transition, and denote the area capped by joint cartilage. (5) The movement of the joints was less than the area covered by the cartilage cap (based on dissections of birds) (Carpenter 2002). With these basic assumptions, we reanalyze the forelimb of *Tyrannosaurus* below.

A testable method of predicting musculature patterns for the forelimb elements of *Tyrannosaurus* is presented that uses extant phylogenetic bracketing, which is based on comparable elements of the crocodilian and bird. Previously, Carpenter and Smith (2001) had presented muscle maps for the forelimb of *Tyrannosaurus*, the results of which were partially criticized by Brochu (2002). The forelimb map was admittedly influenced too greatly by the phylogenetic placement of *Tyrannosaurus* as closer to modern birds than to modern crocodilians (e.g., Brochu 2002). It was believed that the muscle patterns should reflect this phylogenetic closeness, thus resulting in a loss of objectivity from the start (the loss of objectivity in theropod studies was subsequently criticized by Carpenter 2002, pp. 72–73).

The scapula and coracoid of *Tyrannosaurus* (and dinosaurs in general) are the most difficult elements on which to map muscles because few muscle scars are present and the shape of the coracoid differs; these issues hamper the application of phylogenetic bracketing. Nevertheless, the scapula and coracoid are crucial in muscular reconstructions because the muscle origin patterns affect the muscle patterns for the rest of the forelimb. Although muscle scars remain the chief means for mapping muscles, supplemental information is needed in the case of the scapula and coracoid. This supplemental information, as a testable hypothesis, is obtained by deforming the scapulocoracoids of both a crocodile (*Alligator*) and bird (*Gallus*) to approxi-

(continued) brachii; c-M. costocoracoideus; cbd, M. coracobrachialis brevis dorsalis; cbv, M. coracobrachialis brevis ventralis; ce, M. coracobrachialis externus; ch, M. coracohumeralis; dc, M. deltoideus clavicularis; ds, M. deltoideus scapularis; l, M. levator scapulae; rs, M. rhomboideus superficialis; sb, M. supracoracoideus brevis; sc, M. scapulohumeralis cranialis; scd, scapulohumeralis caudalis; se, M. subscapularis externus; si, M. supracoracoideus intermedius; svc, M. subscapularis ventralis cranialis; svt, M. serratus ventralis thoracis; t, M. trapezius; tbclps, M. triceps brachii; caput longus pars scapularis; tll, M. triceps longus lateralis; tm, M. terres major. Figure (B) and terminology adapted from Meers (2003); figure (E) and terminology adapted from Yasuda (2002). Although the terminology is retained for "dorsal" versus "ventral" muscles (e.g., m.c.b. ventralis), the more vertical position of the humerus in **Tyrannosaurus** indicates a need for modified terminology.

mate that of *Tyrannosaurus* (Fig. 10.13). The technique does not involve morphing of one scapulocoracoid to another because the scapulocoracoid of *Tyrannosaurus* is not used as one end point. Although the technique uses a Cartesian grid, it does not attempt to explain homologous points of 2 forms in the manner used by D'Arcy Thompson (1961). Instead, the technique attempts to predict the muscle origin and insertion patterns of the scapulocoracoid of *Tyrannosaurus* as it would be if the scapulocoracoid of the crocodile or bird were deformed into that of *Tyrannosaurus*. The results can then be used to determine which of the 2 muscle patterns, crocodilian or avian, is most like that seen on the bones of *Tyrannosaurus*.

The technique begins by scanning scapulocoracoid outlines on which the muscles have been mapped (Fig. 10.13B, E). Minimum thickness of the scapular neck was selected to standardize the scapulocoracoids of the *Alligator*, *Gallus*, and *Tyrannosaurus* (Figs. 10.13A, B, E) because of the peculiar shape of the avian coracoid precluded scapula-coracoid length as the standard. The Mesh Warp feature of Corel PhotoPaint 7 was used to deform the images using a 10 by 10 grid. By manually moving each intersect of the gridlines (node), a small area of the scapulocoracoid surrounding each node could be deformed. The deformation was smooth, meaning that no sharp angles and lines resulted, thus approximating changes in a biological structure. Nodes were moved until the outline of the scapulocoracoid closely approximated the that of *Tyrannosaurus* (cf. Fig. 10.13A and 10.13C, D, F). Because moving the nodes also moved the contents of each grid, the result is a prediction of what the resultant muscle pattern would be like. As used, Mesh Warp is not mathematically as rigorous as the thin plate spline of Bookstein (1991) because measuring the change in landmark position is irrelevant. Two versions are presented for the deformed crocodile scapulocoracoid, with results differing in the acromion. Figure 10.13D assumes the dorsal prominence just anterior to the scapular neck of the crocodile is homologous to the posterodorsal corner of the acromion in *Tyrannosaurus*, whereas Figure 10.13C does not. This is tested below.

The muscle patterns of the deformed crocodile and avian scapulocoracoids were then compared against the few muscle scars on the scapulocoracoid of *Tyrannosaurus* (DMNH 2827). As can be seen in Figure 10.13G, several muscle scars on the scapulocoracoid of *Tyrannosaurus* seem to be homologous with the origins for the M. costocoracoideus, M. triceps longus lateralis, and M. supracoracoideus intermedius on the deformed crocodilian scapulocoracoid, than with any muscle origin on the scapulocoracoid of the bird. We may infer, then, that the other muscles, for which scars are not evident on the scapulocoracoid, were more homologous with those of the crocodile than of the bird. The large depression or fossa on the acromion of *Tyrannosaurus* seems to better match the pattern for the M. supracoracoideus intermedius in Figure 10.13C than for the multiple muscles in this region, as seen in Figure 10.13D. This suggests that the dorsal prominence of the crocodile is not homologous to the posterodorsal corner of the acromion in *Tyrannosaurus*.

Some independent support for the shoulder of *Tyrannosaurus* having the crocodilian muscle pattern rather than the avian pattern is seen in the

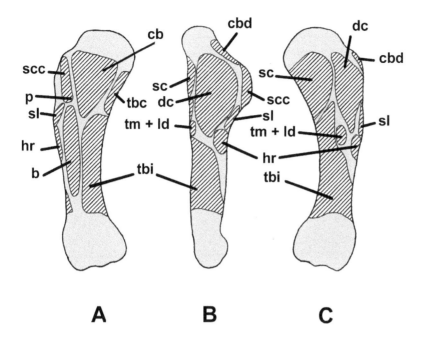

Figure 10.14. Resultant muscle map for the humerus of *Tyrannosaurus* based on actual muscle scars and those inferred from Figure 10.13. Some differences between actual and predicted include relative sizes of scars (e.g., m. deltoideus clavicularis), as well as position (e.g., M. terres major + M. latissimus dorsi). Where muscle scars are ambiguous, the prediction was used as a guide constrained by unambiguous scars (e.g., M. triceps brevis). Abbreviations: b, M. brachialis; cb, M. coracobrachialis brevis; cbd, M. coracobrachialis brevis dorsalis; dc, M. deltoideus clavicularis; hr, M. humeroradialis; p, M. pectoralis; sb, M. supracoracoideus brevis; sc, M. scapulohumeralis cranialis; sc, scapulohumeralis caudalis; scc, supracoracoideus complex; sl, M. supracoracoideus longus; tbi, M. triceps brevis intermedius (+ cranialis?); tm, M. terres major. Terminology adapted from Meers (2003).

scapular blade. In a random sample of various bird skeletons (DMNH avian collection, including *Aechmophorus*, *Buteo*, *Cygnus*, *Corvus*, *Aquila*, *Gymnogyps*, *Gallus*, and *Struthio*), a faint longitudinal trough is present in the distal half of the scapular blade. Interpreted another way, there is thickening of the scapular surface along the origins of the M. terres major and M. rhomboideus superficialis; the trough is the unthickened bone between these origins. In contrast, the crocodile has a single longitudinal thickening or ridge on the scapular surface that corresponds roughly to the common margin of the M. terres major and M. deltoideus scapularis. The scapula of *Tyrannosaurus* also has a similar ridge, not the trough seen in birds.

The deformation method outlined above was applied to the humerus (Fig. 10.9B–E, G–J) and tested against the muscle scars (Fig. 10.9A, F). Overall, the pattern most closely resembles that of the crocodile, with notable exceptions. Aside from differences in relative proportions of some muscles between the actual and the predicted (cf. Fig. 10.9A and C, F and H), there are also some positional differences. For example, the common insertion for the M. terres major and M. latissimus dorsi is lower on the diaphysis and more centrally located in *Tyrannosaurus* (cf. Fig. 10.9F, G, and H). Furthermore, there seems to be a distinct scar on all *Tyrannosaurus* humeri, suggesting that the M. supracoracoideus longus had a separate insertion slightly below the peak of the deltopectoral crest (Fig. 10.10)—either that or the pectoralis inserted more lateral to the deltopectoral crest than medial, but that seems highly unlikely. As noted above, the pathology on MOR 980 better matches the muscle pattern of the crocodile than the bird. With this information, it is possible to reconstruct the muscles of the pectoral girdle and arm of *Tyrannosaurus* (Fig. 10.15).

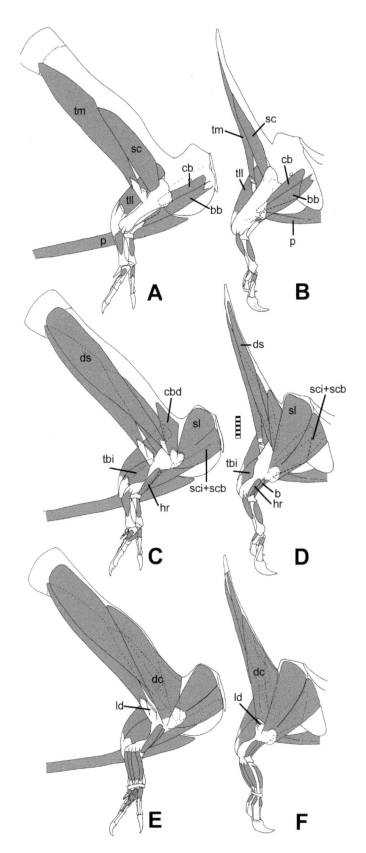

Figure 10.15. Reconstruction of forelimb and pectoral girdle musculature in *Tyrannosaurus* based on results of Figures 10.13 and 10.14. Deep muscles in lateral (A) and anterior (B) views; intermediate muscles in lateral (C) and anterior (D) views; surficial muscles in lateral (E) and anterior (F) views. Scale in centimeters. Abbreviations: b, M. brachialis; bb, M. biceps brachii; cb, M. coracobrachialis brevis; cbd, M. coracobrachialis brevis dorsalis; dc, M. deltoideus clavicularis; ds, M. deltoideus scapularis; hr, M. humeroradialis; ld, tendon for M. latissimus dorsi; p, M. pectoralis; sb, M. supracoracoideus brevis; sc, M. scapulohumeralis cranialis; scd, M. scapulohumeralis caudalis; sci, M. supracoracoideus intermedius; sl, M. supracoracoideus longus; tbi, M. triceps brevis intermedius; tll, M. triceps longus lateralis; tm, M. terres major. Terminology adapted from Meers (2003).

Biomechanical Analysis of the Forelimb in *Tyrannosaurus rex*

In this section, we do a reanalysis of Carpenter and Smith (2001) and an extension of the biomechanical properties of the forelimb in *Tyrannosaurus rex* based mostly on FMNH PR2081. In order to work out the forces acting on the forelimb, we start by modeling the forelimb as a third-class lever (Fig. 10.16). In a third-class lever, the effort force (M. biceps muscle) is applied between the fulcrum (elbow joint) and the resistance force. The elbow joint is a hinge where the humerus, ulna, and radius articulate. Of all of the muscles coordinating and controlling the movement of the elbow, the M. biceps is the most powerful flexor of the elbow joint (Özkaya and Nordin 1999). Our model assumes that the M. biceps is the major flexor and that the line of action (the tension) at the biceps is vertical.

Anatomical measurements were used to derive the motive force arm (MFA) and the resistive force arm (RFA) (Table 10.1). For the ulna, the MFA was measured from the sigmoid notch to the midscar of the insertion point for the M. biceps (motive force, MF), and the RFA was derived from measuring the ulna from the sigmoid notch to the distal end (Fig. 10.16). To obtain the MFA of the radius, we measured from the radial head to the midscar of the insertion point for the M. biceps and from the radial head to the distal end to determine the RFA.

A tendon tensile strength of 100 MPa, the global mean across all species, was used to estimate the tendon tensile strength in *Tyrannosaurus* (Nigg and Herzog 1999). The safety factors in the values of bird tendons range from 1.19 to 4.10 (Van Snik et al. 1994; Alexander 1981). A safety factor of 3 will be used in this study.

The normal working range (NWR) is one-third the safety factor (Carpenter and Smith 2001). Although the size, shape, and the biomechanical behavior of each tendon differs, the basic structure of tendons and their mechanical properties are similar (Jòzsa and Kannus 1997). The size of the cross-sectional area of a tendon is directly related to the size of the load that can be carried before failure (Butler et al. 1978).

The surface area of the scar for the insertion of the M. biceps is 122.11 mm^2 on the radius and 192 mm^2 on the ulna. The conversion for the tendon strength, expressed as MPa, is 1 MPa = 1,000,000 Pa, with 1 Pa = 1 N/m^2. Therefore, 1 MPa = 1 N/mm^2. The maximum working range (MWR) and the NWR are calculated from the tensile strength of the tendon. The formula for estimated tendon tensile strength is as follows:

tendon tensile strength/area^2 × surface area of the scar for insertion of the M. biceps = estimated tendon tensile strength   (1)

Tendon tensile strength for the radius is

100 N/mm^2 × 122.11 mm^2 = 12,211 N

where MWR is 12,211 N/3 = 4070 N, and NWR is 4070 N/3 = 1357 N

Tendon tensile strength for the ulna is:

100 N/mm^2 × 192 mm^2 = 19,200 N

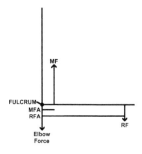

Figure 10.16. Free-body diagram (simplified model) of the forelimb of FMNH PR2081. Abbreviations: MF, motive force; MFA, motive force arm; RF, resistive force; RFA, resistive force arm.

Table 10.1. *Power Analysis Measurements*

Abbreviations.—MFA, motive force arm; RFA, resistive force arm.

Measurement	Radius	Ulna
MFA	15.2 mm (0.0152 m)	45.8 mm (0.0458 m)
RFA	166.2 mm (0.166 m)	186.6 mm (0.187 m)
MANUS	177.6 mm (0.178 m)	177.6 mm (0.178 m)
RFA including manus	343.8 mm (0.344 m)	364.2 mm (0.364 m)

where MWR is 19,200 N/3 = 6400 N, and NWR is 6400 N/3 = 2133 N.

The values for the MWR and NWR represent the estimated strength of the tendon at the insertion of the M. biceps and are used as the MF in the analysis of the power of the *Tyrannosaurus* forelimbs. The following equations are used to estimate the amount of force the arm of *Tyrannosaurus* can resist (resistive force, or RF):

$$MF \times MFA = T \quad (2)$$

$$RF \times RFA = T \quad (3)$$

Measurement of the manus (177.6 mm) was taken from a cast of FMNH PR 2081, from the proximal end of the wrist to the proximal end of the claws. It was then added to the RFA (166.2 mm).

MWR for the radius of *T. rex* is as follows:

$$4,070 \text{ N} \times 0.0152 \text{ m} = 61.86 \text{ Nm}$$
$$RF \times 0.3438 \text{ m} = 61.86 \text{ Nm}$$
$$RF = 179.93 \text{ N (or 18.36 kg)}$$

NWR for the radius of *T. rex* is as follows:

$$1,357 \text{ N} \times 0.0152 \text{ m} = 20.63 \text{ Nm}$$
$$RF \times 0.3438 \text{ m} = 20.63 \text{ Nm}$$
$$RF = 60.01 \text{ N (or 6.12 kg)}$$

MWR for the ulna of *T. rex* is as follows:

$$6,400 \text{ N} \times 0.0458 \text{ m} = 293.12 \text{ Nm}$$
$$RF \times 0.3642 \text{ m} = 293.12 \text{ Nm}$$
$$F = 804.83 \text{ N (or 82.13 kg)}$$

NWR for the ulna of *T. rex* is as follows:

$$2,133 \text{ N} \times 0.0458 \text{ m} = 97.69 \text{ Nm}$$
$$RF \times 0.3642 \text{ m} = 97.69 \text{ Nm}$$
$$RF = 268.23 \text{ N (or 27.37 kg)}$$

Adding the resistive forces of the radius and ulna results in 984.76 N (100.49 kg or 221.10 pounds) for the MWR (no safety factor) and 328.24 N

(33.49 kg or 73.70 pounds) for the NWR (with safety factor). The conversion factor for kilograms to newtons is 1 kg = 9.8 N. These results are summarized in Table 10.2.

An average strength of 5 kg/cm² per cross-sectional area of muscle was used to determine the cross-sectional area of the M. biceps in *Tyrannosaurus* (Carpenter and Smith 2001). The NWR of the tendon tensile strength for the radius and ulna were added together to get the MF: 1357 N + 2133 N = 4490 N (356.12 kg). The formula used to determine the cross-sectional area of muscle is MF (kg)/strength (kg × cm^{-2}) = cross-sectional area (cm²). Thus, the estimated cross section of the *Tyrannosaurus* M. biceps is 356.12 kg/5 kg × cm^{-2} = 71.224 cm². This translates into a diameter of 9.52 cm. Of course the M. biceps is not the only arm protractor. In fact, by using half the estimated cross-sectional area of the upper arm (based on a diameter of 25 cm), the amount of force generated is estimated to have been around 1150 kg, or 11,270 N. Of this, the biceps generated about 40%, and thus was a major muscle.

Mechanical Analysis of the Forelimb in *Tyrannosaurus rex*

A small lever arm requires greater muscle tension to balance a load. Therefore, while resisting prey or holding prey, it is disadvantageous to have a muscle attachment close to the elbow joint. The advantage to having the muscle attachment close to elbow joint is that it will have a larger range of motion of the elbow flexion-extension, and therefore the hand can move faster toward the upper arm or shoulder (Özkaya and Nordin 1999).

The mechanical advantage is the amount of force a given effort can produce. It can be expressed as a ratio of the resistive force to the MF, or as a ratio of the MFA to the RFA (Kreighbaum and Barthels 1985). Both of the equations produce the same result.

The *Tyrannosaurus* forelimb is found to have a mechanical advantage of the 0.09 (RFA measurement including the hand) and 0.18 (RFA measurement excluding the hand). The mechanical advantage of a human forearm is 0.07 (RFA measurement including the hand) and 0.13 (RFA measurement excluding the hand).

Next we evaluate the force at the elbow joint. The sum of the MFs (NWR + MWR) at the radius (138.3 kg) and the ulna (217.5 kg) minus the RF at the manus (33.5 kg) must equal the force at the elbow for a static configuration. Therefore, *Tyrannosaurus* has a force of 138.3 + 217.5 − 33.5 = 322.3 kg at the elbow joint for the NWR. This compares with a force of about 128.25 kg at the elbow joint of an average adult male human for the NWR.

Table 10.2. *Power Analysis Summary*

	RF (kg)		
Range	Radius	Ulna	Combined
MWR	18.36	82.13	100.49
NWR	6.12	27.37	33.49

Abbreviations.—RF, resistive force; MWR, maximum working range; NWR, normal working range.

Acceleration

From the torque of the forearm, the force that could be applied at the manus and the resultant force at the elbow joint were determined. We now estimate the acceleration that could be generated at the claws using the moment of inertia. The fleshed-out version of the arm of *Tyrannosaurus* (Figs. 10.15–10.17) was converted to a closely packed series of elliptical cylinders. The cross sections of each elliptical cylinder were determined from Figure 10.15. Assuming a density of 1000 kg/m^2 for tissue, the data from these cylinders result in a mass of 1.8 kg for the forearm plus manus. The integral of the density times the perpendicular distance to the pivot point results in a moment of inertia of 0.06 kg m^2. Because the NWR torque of the forearm and hand to be 118.3 Nm, the angular acceleration (torque divided by the moment of inertia) is 1983 s^{-2}. The angular acceleration can be converted to a linear acceleration by multiplying it by the distance (0.354 m) from the pivot point to the claws, which results in a linear acceleration of 702 ms^{-2}. This is likely an overestimate because the skin and the claw sheath are not factored in. Also, this only gives the initial acceleration. The force from muscles is known to rapidly reduce at high speeds (Hill 1938).

Conclusions

As we have shown, the forearm of *Tyrannosaurus* was capable of resisting large forces and moving at high accelerations. These results strengthen the hypothesis that the forelimbs were used during predation. However, because of the small size of the forelimb relative to the body size, it is unlikely that the *Tyrannosaurus* would use the manus for striking prey, as discussed in Carpenter (2002). Rather, the forelimbs may have been used to cling to prey. Our results of finding large forces at the elbow joint and possible signs of injury at the furcula further support this hypothesis.

Finally, in contrast to the belief of Lockley et al. (this volume) that "no useful function is plausible" to explain the forelimb of *Tyrannosaurus*, our results support the previous assertion that the forelimb played a functional role in predation. By implication, the short forelimbs of other tyrannosau-

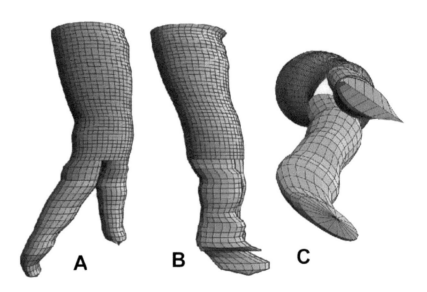

Figure 10.17. *A 3-D representation of the forearm and manus in the* **Tyrannosaurus rex** *FMNH PR2081. In (A) lateral, (B) anterior, and (C) reaching views.*

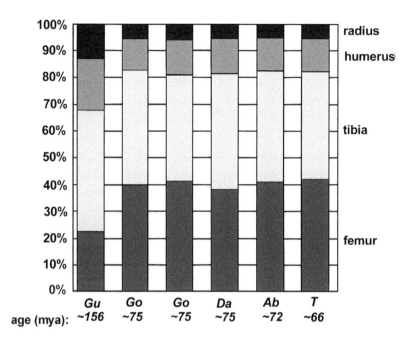

Figure 10.18. Comparison of forelimb length to hind limb length shows that a progressive reduction in forelimb length does not occur in the Tyrannosauridae. Abbreviations: Gu, *Guanlong* (basal tyrannosauroid); Go, *Gorgosaurus*; Da, *Daspletosaurus*; Ab, *Albertosaurus*; T, *Tyrannosaurus*.

rids had a similar function. In support of this, we note that a progressive reduction in the forelimb does not occur in the Tyrannosauridae (Fig. 10.18), contrary to Paul (1988) and Lockley et al. (this volume). In point of fact, once the shortened forelimb of the tyrannosaurids was established, it remained proportionally stable relative to hindlimb length.

Acknowledgments

We are honored to contribute to this volume, and we thank Neal Larson and Peter Larson for the organization of the symposium that preceded it. We are indebted to Bill Simpson and Peter Makovicky for access to FMNH PR 2081, and C.L. would like to thank Paul Sereno for access to the cast of FMNH PR 2081, and Michael Benton, Paul Sereno, Gordon, Jürgen Kriwet, Simon Braddy, Lorrie McWhinney, and Don Henderson for helpful discussions. K.C. thanks John Daggett, Bill Simpson, and Peter Larson for loans of specimens or casts, and Matt Smith for our previous joint work on *T. rex* forelimb analysis. CT scans are by Steven White, Kaiser Permanente, Denver.

References Cited

Alexander, R. McN. 1981. Factors of safety in the structure of animals. *Science Progress* 67: 109–130.

Boggs, D. F., Jenkins, F. A., and Dial, K. P. 1997. The effects of the wingbeat cycle on respiration in black-billed magpies (*Pica pica*). *Journal of Experimental Biology* 200: 1403–1412.

Bookstein, F. L. 1991. Morphometric tools for landmark data. *Geometry and Biology*. Cambridge University Press, Cambridge.

Brochu, C. A. 2002. *Osteology of Tyrannosaurus rex: Insights from a Nearly Complete Skeleton and High-Resolution Computed Tomographic Analysis of the Skull*. Journal of Vertebrate Paleontology Memoir 7.

Brown, B. 1915. *Tyrannosaurus*, the largest flesh-eating animal that ever lived. *Museum Journal* 15: 271–279.

Butler, D. L., Grood, E. S., Noyes, F. R., and Zernicke, R. F. 1978. Biomechanics of ligaments and tendons. *Exercise Sports Science Review* 6: 125–181.

Calvo, J. O., and Coria, R. 1998. New specimen of *Giganotosaurus carolinii* (Coria & Salgado, 1995), supports it as the largest theropod ever found. P. 117–122 in Pèrez-Moreno, B. P., Holtz, T. J., Sanz, J. L., and Moratalla, J. (eds.). *Aspects of Theropod Paleobiology. Gaia: Revista de Geociencias, Museu Nacional de Historia Natural, Lisbon*, 15.

Carpenter, K. 1998. Evidence of predatory behavior by carnivorous dinosaurs. P. 135–144 in Pèrez-Moreno, B. P., Holtz, T. J., Sanz, J. L., and Moratalla, J. (eds.). *Aspects of Theropod Paleobiology. Gaia: Revista de Geociencias, Museu Nacional de Historia Natural, Lisbon*, 15.

———. 2002. Forelimb biomechanics of nonavian theropod dinosaurs in predation. *Senckenbergiana Lethaea* 82: 59–76.

Carpenter, K., Madsen, J., and Lewis, A. 1994. Mounting of fossil vertebrate skeletons. P. 285–322 in Leiggi, P., and May, P. (eds.). *Vertebrate Paleontological Techniques*. Cambridge University Press, New York.

Carpenter, K., and Smith, M. 2001. Forelimb osteology and biomechanics of *Tyrannosaurus rex*. P. 90–116 in Tanke, D., and Carpenter, K. (eds.). *Mesozoic Vertebrate Life*. Indiana University Press, Bloomington.

Chin, K., Eberth, D. A., Schweitzer, M. H., Rando, T. A., Sloboda, W. J., and Horner, J. R. 2003. Remarkable preservation of undigested muscle tissue within a Late Cretaceous tyranosaurid coprolite from Alberta, Canada. *Palaios* 18: 286–294.

Chure, D. J., and J. H. Madsen, Jr. 1996. On the presence of furculae in some non-maniraptoran theropods. *Journal of Vertebrate Paleontology* 16: 573–577.

Currey, J. D. 2002. *Bones: Structure and Mechanics*. Princeton University Press, Princeton, NJ.

Currie, P. J., Trexler, D., Koppelhus, E. B., Wicks, K., and Murphy, N. 2005. An unusual multi-individual bonebed in the Two Medicine Formation (Late Cretaceous, Campanian) of Montana (USA). P. 313–324, in Carpenter, K. (ed.). *The Carnivorous Dinosaurs*. Indiana University Press, Bloomington.

Donnelly, L. F., Helms, C. A., and Bisset, G. S. 1999. Chronic avulsive injury of the deltoid insertion in adolescents: imaging findings in three cases. *Radiology* 211: 233.

El-Khoury, G. Y., Daniel, W. W., and Kathol, M. H. 1997. Acute and chronic avulsive injuries. *Radiological Clincs of North America* 35: 747–766.

Erickson, G. M., and Olson, K. H. 1996. Bite marks attributable to *Tyrannosaurus rex*: preliminary description and implications. *Journal of Vertebrate Paleontology* 16: 175–178.

Erickson, G. M., Van Kirk, S. D., Su, J., Levenston, M. E., Caler, W. E., and Carter, D. R. 1996. Bite-force estimation for *Tyrannosaurus rex* from tooth-marked bones. *Nature* 382: 706–708.

Farlow, J. O. 1994. Speculations about the carrion-locating ability of tyrannosaurs. *Historical Biology* 7: 159–165.

Hill, A. V. 1938. The heat of shorting and dynamic constants of muscle. *Proceedings of the Royal Society of London* 141: 314–320.

Horner, J. R., and Lessem, D. 1993. *The Complete T. rex*. Simon & Schuster, New York.

Hui, C. A. 2002. Avian furcula morphology may indicate relationships of flight requirements among birds. *Journal of Morphology* 251: 284–293.

Jenkins, F. A., Dial, K. P., and Goslow, G. E. 1988. A cineradiographic analysis of bird flight: the wishbone in starlings is a spring. *Science* 241: 1495–1498.

Jàzsa, L. G., and Kannus, P. 1997. *Human Tendons: Anatomy, Physiology, and Pathology.* Human Kinetics, Champaign, IL.

Kreighbaum, E., and Barthels, K. M. 1985. *Biomechanics: A Qualitative Approach for Studying Human Movement.* Burgess Publishing, Minneapolis, MN.

Lambe, L. M. 1917. *The Cretaceous Theropodous Dinosaur Gorgosaurus.* Geological Survey of Canada Memoir 100.

Larson, P., and Rigby, J. K. 2005. Furcula of *Tyrannosaurus rex.* P. 247–255 in Carpenter, K. (ed.). *The Carnivorous Dinosaurs.* Indiana University Press, Bloomington.

Lingham-Soliar, T. 1998. Guess who's coming to dinner: a portrait of *Tyrannosaurus* as a predator. *Geology Today* 14: 16–20.

Lipkin, C., and Sereno, P. C. 2004. The furcula in *Tyrannosaurus rex. Journal of Vertebrate Paleontology* 24(Suppl. to 3): 83A.

Makovicky, P., and Currie, P. J. 1998. The presence of a furcula in tyrannosaurid theropods, and its phylogenetic and functional implications. *Journal of Vertebrate Paleontology* 18: 143–149.

Meers, M. B. 2002. Maximum bite force and prey size of *Tyrannosaurus rex* and their relationships to the inference of feeding behavior. *Historical Biology* 16: 1–12.

———. Crocodylian forelimb musculature and its relevance to Archosauria. *Anatomical Record* Part A, 274A: 891–916

Molnar, R. E. 1998. Mechanical factors in the design of the skull of *Tyrannosaurus rex* (Osborn, 1905). P. 193–218 in Pérez-Moreno, B. P., Holtz, T. J., Sanz, J. L., and Moratalla, J. (eds.). *Aspects of Theropod Paleobiology. Gaia: Revista de Geociencias, Museu Nacional de Historia Natural, Lisbon*, 15.

Nigg, B. N., and Herzog, W. 1999. *Biomechanics of the Musculo-Skeletal System.* 2nd ed. J. Wiley, New York.

Osborn, H. F. 1905. *Tyrannosaurus* and other Cretaceous carnivorous dinosaurs. *Bulletin of the American Museum of Natural History* 21: 259–265.

———. 1912. Integument of the iguanodont dinosaur *Trachodon. Memoirs, American Museum of Natural History* 1: 33–54.

Özkaya, N., and Nordin, M. 1999. *Fundamentals of Biomechanics: Equilibrium, Motion, and Deformation.* Springer, New York.

Paul, G. S. 1987. Predation in the meat eating dinosaurs. P. 173–178 in Currie, P., and Koster, E. (eds.). *Fourth Symposium on Mesozoic Terrestrial Ecosystems, Short Papers.* Occasional Papers of the Tyrrell Museum of Palaeontology 3.

———. 1988. *Predatory Dinosaurs of the World.* Simon & Schuster. New York.

Rayfield, E. J., Norman, D. B., Horner, C. C., Horner, J. R., Smith, P. M., Thomason, J. J., and Upchurch, P. 2001. Cranial design and function in a large theropod dinosaur. *Nature* 409: 1033–1037.

Resnick, D. 2002. *Diagnosis of Bone and Joint Disorders.* W. B. Saunders, Philadelphia.

Rothschild, B. M. 1988. Stress fracture in a ceratopsian phalanx. *Journal of Paleontology* 62: 302–304.

Rothschild, B. M., and Martin, L. D. 1993. *Paleopathology: Disease in the Fossil Record.* CRC Press, Boca Raton, FL.

Rothschild, B. M., and Tanke, D. H. 2005. Theropod paleopathology. P. 351–365 in Carpenter, K. (ed.). *The Carnivorous Dinosaurs.* Indiana University Press, Bloomington.

Ruxton, G. D., and Houston, D. C. 2003. Could *Tyrannosaurus rex* have been a scavenger rather than a predator? An energetics approach. *Proceedings: Biological Sciences* 270: 731–733.

Sereno, P. C., Dutheil, D. B., Iarochene, M., Larsson, H. C. E., Lyon, G. H., Magwene, P. M., Sidor, C. A., Varricchio, D. J., and Wilson, J. A. (1996). Predatory dinosaurs from the Sahara and Late Cretaceous faunal differentiation. *Science* 272: 986–991.

Stevens, M. A., El-Khoury, G. Y., Kathol, M. H., Brandser, E. A., and Chow, S. 1999. Imaging features of avulsion injuries. *Radiographics* 19: 655–672.

Tehranzadeh, J. 1987. The spectrum of avulsion and avulsion-like injuries of the musculoskeletal system. *Radiographics* 7: 945–974.

Thompson, D'Arcy W. 1961. *On Growth and Form*. Cambridge University Press, Cambridge.

Van Snik, G., Olmos, M., Casinos, A., and Planell, J. A. 1994. Stresses in leg tendons of birds. *Netherlands Journal of Zoology* 44: 1–14.

Yasuda, M. 2002. The Anatomy of *Gallus*. Tokyo, University of Tokyo Press.

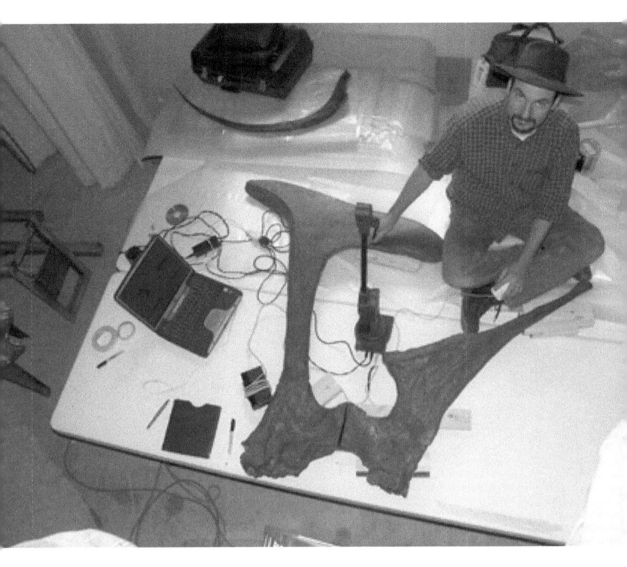

Figure 11.1. *Ray Wilhite using an Immersion Microscribe digitizer on casts of elements of the pelvic girdle of **Tyrannosaurus rex** specimen BHI 3033 (Stan). Photo courtesy Ray Wilhite and Virtual Surfaces Inc.*

REX, SIT: DIGITAL MODELING OF *TYRANNOSAURUS REX* AT REST

11

Kent A. Stevens, Peter Larson, Eric D. Wills, and Art Anderson

Introduction

The great theropod *Tyrannosaurus rex* is usually depicted in an active, bipedal pose, perhaps in pursuit of prey or facing off an opponent. Some artists, e.g., Lawrence Lambe (1917), Gregory S. Paul (1988), John Sibbick (Norman 1991, p. 72), and Michael Skrepnick (Currie et al. 2004), have provided views of these animals in other, less active postures, including lying prone or squatting. Presumably the animal would rest with a substantial portion of its body mass supported by the prominent pubic boot. Trace fossils of small crouching theropods show both tarsal and pubic-ischiatic impressions (e.g., Gierliński et al. 2005). In the great theropods, descending from a standing pose to a rest position was presumably a straightforward matter of squatting, a process considerably less involved than the complex sequencing of folding movements that some modern large quadrupeds, such as camelids (Gauthier-Pilters and Daag 1981) and bovids, use to lower their mass to the ground. *Tyrannosaurus rex* might simply have settled vertically in one continuous flexion movement involving the hip, knee, and ankles.

It is in rising from a prone or squatting rest position that some concern for the mechanics of the tyrannosaurid frame might present itself. How could the center of mass (COM) be controlled so that the animal was stable while rising? Was there sufficient mechanical advantage in the major extensor muscles to provide a direct ascent that retraces the trajectory followed in descending to the ground? Were the forelimbs useful in stabilizing the body and in providing thrust during the initial stages of the ascent? To address some of these questions, a fully articulated digital model of *Tyrannosaurus rex* was created where limb movements are delimited by anatomically based estimates of achievable range of motion, and the position of the instantaneous COM of the animal can be visualized in order to judge balance and stability.

Extant bipeds that might serve as analogs for studying the sitting and standing movements of *T. rex*, include members of the Macropodoidea, notably the large red kangaroo (*Macropus rufus*) and a variety of birds, particularly the large ratites such as the emu (*Dromaius novaehollandiae*) and the ostrich (*Struthio camelus*). As animal mass increases, muscular strategies cannot be expected to scale indefinitely (Alexander 1989); the effortless rise of a small passerine from rest to a bipedal stance might require multiple, more deliberate stages of limb extension in a biped of several orders greater weight. The biomechanical principles governing the choice of strategy, particularly as regards scaling with body mass, are not well understood. Motion studies have concentrated on capturing relatively steady-state locomotion (e.g., Muybridge 1899; Jenkins et al. 1988), not the transient body movements associated with sitting or standing.

To examine the potential movements that take the animal from standing to sitting, and vice versa, it is important to begin with an estimation of the typical stand and sit postures. Movements that smoothly transition between these extremes can then be proposed and analyzed. In their analysis, it is important to understand how the COM translates during the movement. Longitudinal (caudal-cranial) pitching movements in particular would produce instability that would have to be corrected at risk of injury to the great theropod. It is also important to examine range of motion issues throughout the sit-stand movements and the mechanical leverage of large muscle groups for providing the necessary movements.

Proposals have been offered for how *T. rex* could sit down on its pubic boot, then rise by first using the forearms as props to help anchor the front of the body while the rear legs were straightened. The upper body would then be tilted back to regain an upright standing posture (Newman 1970). This idea is but one of the potential uses proposed for the forelimbs (Osborn 1906; Horner and Lessem 1993; Carpenter and Smith 2001; Carpenter 2002).

In the following, an articulated, 3-dimensional digital reconstruction is used to explore alternative hypotheses regarding the sit-stand movements of this dinosaur. The process of descending and then ascending is amenable to quantitative modeling, taking into consideration the distribution of mass in the animal and the flexibility of those joints involved in the movements, particularly the ankle, knee, and hip within the hind limb, and the potential role of the forelimbs in the process of rising. QuickTime video showing the action is available Video is available online at https://www.iupress.indiana.edu/media/tyrannosaurusrex/

Creating an Articulated Digital Model

DinoMorph software (Stevens 2002) provides a framework with which to create and pose a digital model of *Tyrannosaurus rex*. The software can accept 3-dimensional data representing bone morphology (e.g., from computed tomographic [CT] scan or hand digitization), as well as more schematic and simplified representations. In this study, the *Tyrannosaurus rex* specimen BHI 3033 (Stan) at the Black Hills Museum of Natural History was used as the source for the digital model. The articulation of the appendicular skeleton and the morphology of the pelvic and pectoral girdles were of particular importance, so they were specifically for this study (Fig. 11.1). Digitization data of the head was provided from an earlier CT scan made by Virtual Surfaces Inc. and the Black Hills Institute. The remainder of the axial skeleton was modeled schematically, with centra, neural spines, lateral processes, chevrons, and ribs in a dimensionally accurate but simplified form (Fig. 11.2).

The next step was to estimate the relative placement of each bone within the overall skeletal framework. Along the presacral axial skeleton, the intervertebral separations and overall curvature were determined from

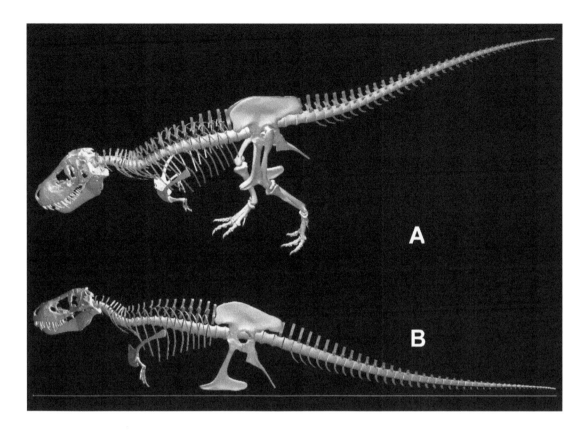

measurements and photographs in lateral view. Likewise, the rib cage was formed by painstakingly adjusting each digitally represented dorsal rib to match the curvature, dimensions, and placement of its counterpart in reference photographs that were underlaid within DinoMorph as background images (Fig. 11.3). To refine the 3-dimensional skeletal model, the trunk was successively viewed in anterior, dorsal, and lateral orientations, and for each view, the curvature and placement of the ribs were adjusted so that the digital ribs superimposed precisely over their photographic counterparts. The pelvic girdles, complete with furcula, were then placed on the rib cage as they are mounted on Stan (Larson and Rigby 2005).

Next, those DinoMorph parameters governing the position and orientation of all appendicular joints were adjusted to create a neutral standing pose, the starting point for this study. Then, for the major appendicular joints important to this study, a range of motion was determined on the basis of an estimate of the thickness and extent of the intervening cartilage in modern avians and direct manipulation of the casts (Kenneth Carpenter and Yoshio Ito, personal communication June 2005). Direct manipulation assisted in determining, for example, the axis of rotation of the femur head within the acetabulum, and in the forelimb the orientation of the fully extended forelimb with respect to the pectoral girdles.

In analyzing potential sit-stand strategies of a theropod dinosaur weighing several metric tonnes, it is important to track the trajectory undertaken by the COM during hypothesized movement. The COM is computed in DinoMorph by assigning both a volume and a density to each

Figure 11.2. (A) Dino-Morph model of ***Tyrannosaurus rex*** specimen BHI 3033 (Stan). The appendicular skeleton and head were digitized whereas the axial skeleton was represented in schematic form, with important dimensions (e.g., centrum length, neural spine height, and intervertebral separations) dimensionally accurate. (B) The axial skeleton was laid out with reference to measurements taken from the mount and photographs (see text). Scale bar indicates an overall length of 11.2 m.

Figure 11.3. *(A) Screen image showing the reconstruction of the trunk superimposed on a reference photograph of an assembly of casts of the Stan specimen. Background image courtesy Black Hills Institute of Geologic Research. (B) DinoMorph model shown with addition of digitized pectoral girdles (including furcula), forelimbs, and pelvic girdle, all based on Stan.*

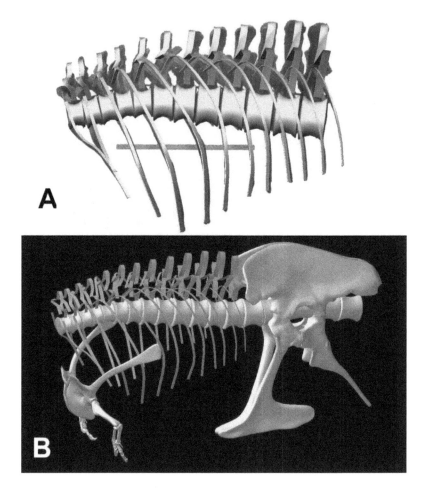

Figure 11.4 (opposite). *(A) Visualization of the distribution of body mass, based on a parametric fit to segments of body cross section of the axial and appendicular skeleton. (B) Computed COM visualized just anterior pubis, and above pes, as required for static bipedal balance.*

discrete segment of the skeleton, such as each interval of the axial skeleton associated with an individual vertebra. Fitted conical and elliptical cylinders are used as a first-order approximation to the body cross-sectional area, governed by adjustable parameters (Fig. 11.4). The density (i.e., specific gravity) associated with each segment was adjusted to roughly reflect cranial and axial pneumaticity, air sacs, and lungs. The COM was computed by summating the gravitational moments associated with each segment throughout the skeleton. By assigning densities of 0.8–1.0 to presacral regions and 1.0 to segments of the appendicular skeleton and caudal vertebral series, the overall COM was located just anterior to the pubic shaft (see Fig. 11.4B), consistent with estimates by Henderson (1999) and Hutchinson and Garcia (2002). Small but potentially significant shifts in the instantaneous COM during movements could be detected visually as the movement unfolded. For this study, a lithe reconstruction of the cross sections of soft tissue associated with each segment of the skeleton was chosen to corresponding to recent computations by one of us (P. L.). The resulting overall mass for BHI 3033 (Stan) was estimated as ~4400 kg., or about 80% of the mass estimated for the more robust specimen FMNH PR2081 (Sue) by using the same techniques (Stevens et al., in preparation). Although it was possible to estimate as little as 3800 kg for the same skeletal

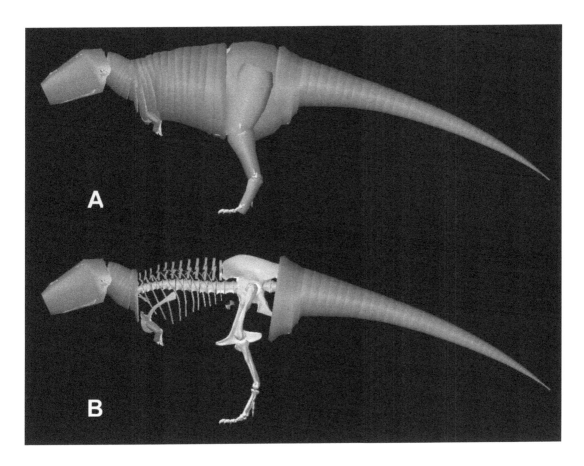

structure by progressively reducing body bulk, particularly in the pelvic region, for this study, the position of the COM was of greater importance than the magnitude.

Reconstructing a Sitting Movement

It is reasonable to assume that *Tyrannosaurus rex*, and other theropods with distally expanded pubic boots, lowered itself until the majority of its mass bore down on the pubis. Upon ground contact, the orientation of pelvic girdle would have shifted slightly so that the elongate ventral surface of the pubis laid generally parallel to the horizontal ground. The elongate pubic shaft of *T. rex* places the ventral surface of the pubic boot just below the knee, permitting simultaneous ground contact at the knees and along the pubic boot (Figs. 11.5–6). The pubis thus likely provided a stable means to offload the great majority of the animal's weight, limiting pressure on the respiratory system, and to permit repositioning of the hind limbs without requiring a shifting of weight. The limit of hip flexion (femoral protraction) is difficult to estimate because it was governed by soft tissues, but it likely was sufficient to permit achieving the protraction shown in Figure 11.5A so that the tarsus could lie flat on the ground. The limits of knee and ankle flexion are more obvious in the osteology. It is noteworthy that in a full

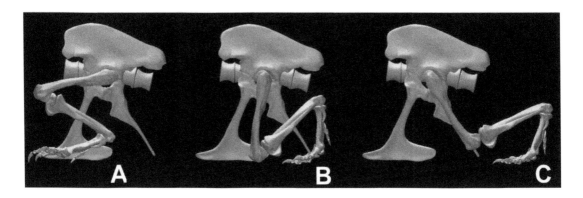

Figure 11.5. *DinoMorph* model of *T. rex* Stan demonstrating that the pubic shaft is sufficiently long that, with the animal's weight resting on the pubic boot, the hind limb is free to assume a broad range of positions from (A) crouch, to (B) kneel to (C) moderate extension of the hip and knee, to full leg extension when stretched behind the hips (not shown). Note that in these 3 images the tarsus is near the limit of flexion; in (A) and (B), the knee is fully flexed. Observe that the knee just clears the ground as it swings through hip flexion-extension, permitting repositioning of either hind limb while continuously resting on the pubic boot. Joint flexibility estimated in collaboration with Yoshio Ito and Kenneth Carpenter.

squat (Fig. 11.5A), which brings the pubic boot in contact with the ground, the knee and ankle are nearly fully flexed.

To visualize the descent from a neutral standing pose to a squat, Dino-Morph was used to interpolate between these 2 extremes of pose. By constraining all joints to movements within their respective ranges of motion, it was determined that a descent movement could be performed that did not induce any significant longitudinal shift in the position of the COM, i.e., it was possible to move the COM in a purely vertical direction, until contact was made between the pubic boot and the ground. The COM, however, then tends to shifts posteriorly during the final settling of weight from the hind feet onto the pubic boot, and it must return anteriorly through some trajectory in the process of rising and regaining bipedal balance.

Reconstructing Alternative Standing Movements

In rising, *Tyrannosaurus rex* has to cope with lifting the COM by approximately 1.4 m vertically starting with the pubic boot in ground contact and ending in the neutral standing pose, as seen in Figure 11.4. This could be achieved in principle by pure muscular exertion of the large extensor muscle groups of the hind limb, particularly the M. caudofemoralis longus, the largest contributor to femoral retraction, and secondarily the knee and tarsus extensors. The M. caudofemoralis longus, however, is simultaneously in significant stretch (roughly 115% of that while standing), and the moment (or lever) arm is greatly foreshortened (Fig. 11.7). Depending on the particular position of the femur in this deep squat, the moment arm may be less than 40% of that provided when standing.

Tyrannosaurus rex, while resting on the pubis, could freely retract one or both femora (Figs. 11.5 and 11.6) and hence vary the stretch on the M. caudofemoralis. Optimal mechanical advantage occurs when the femur is roughly vertical (i.e., associated with the thrust phase in locomotion). Although the femoral position providing greatest muscle moment would correspond to roughly that in Figure 11.5B, the placement of the pes in Figure 11.5A would appear better suited for elevating bipedally because the hind feet are then under the COM. If *T. rex* were not to slowly rise vertically into a stationary standing position, but instead accelerate diagonally from the squat in Figure 11.5A, then, provided the hind limbs direct the ground reaction force diagonally through the COM, no net pitching moment would be created as the animal

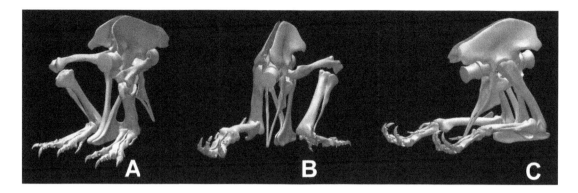

rose. Much as a sprinter begins a race accelerating and rising gradually out of the blocks, it is not inconceivable that *T. rex* could have accelerated diagonally upward from sitting into forward locomotion. Although it is more likely an option for light young tyrannosaurids, it remains a matter of quantitative modeling to estimate whether that was achievable by an adult.

If the femoral retractor muscles were not in an advantageous state for lifting *Tyrannosaurus rex* vertically out of a squat into a balanced standing position, what were the alternatives? One suggestion (Phillip Manning, personal communication June 2005) borrows from modern analogues. In ratites in particular, the M. gastrocnemius comes into play: the Achilles tendon stores energy when in a state of stretch, which is trapped when the animal's weight bears down on the tarsus while sitting. By leaning forward onto the its knees, the tendon is released, and the hind limb receives a passive boost. Whether recovery of stored mechanical energy would scale to be of significant value in helping boost *Tyrannosaurus rex* from sitting to standing would require quantitative study. One further concern, beyond the matter of scaling to be effective on a 4000-kg animal, is whether the stored energy would dissipate during the period of rest as the Achilles tendon would stretch.

Another approach is to enlist the forelimbs, as suggested by Newman (1970), as a potential use for these appendages. When sitting, they are close to the ground and are brought into contact by a slight tipping of the pelvis about the prepubis (Fig. 11.8). They could have been instrumental in rising back into a standing position. As shown in Figure 11.9, although the forelimb range of

Figure 11.6. With the body mass supported by the pubic boot, the hind limbs appear to have been able to shift from (A) a sitting position (with hip flexed) to (B) kneeling on one knee or (C) both knees, without having to lift the body weight off of the pubic boot. Although the axislike insertion of the femur head within acetabulum suggests little femoral abduction was possible, there was likely sufficient flexibility to provide lateral stability. Moreover, the posterolateral angulation of the acetabular axis caused the knees to splay with femoral protracted, again aiding stability against lateral tipping in addition to clearing the rib cage as necessary in locomotion.

Figure 11.7. In ascending from repose, the M. caudofemoralis longus is in stretch (~115%) and the moment arm is greatly foreshortened compared with its neutral state when standing, thus providing poor mechanical advantage.

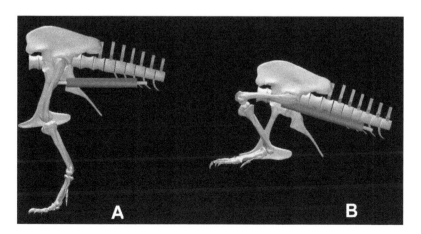

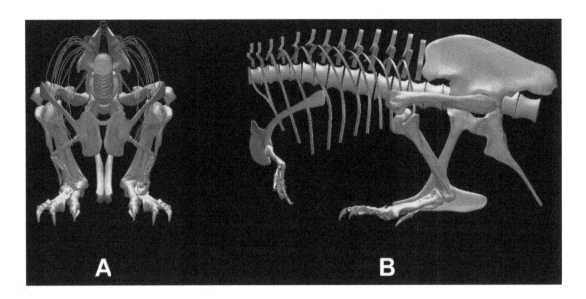

Figure 11.8. When resting on the pubis, the forelimbs are near ground level. They are brought closer to ground level as a consequence of initiating a standing movement from this resting pose. A modest tipping of the body, by pivoting about the curved anteroventral surface of the prepubis, would have shifted the overall COM anteriorly and returned the point of balance to between the hindfeet. In the process, the forelimbs would have been available to assist in stabilizing, if not actively contributing toward raising, the body, by pushing against the ground.

Figure 11.9. (A) Reconstruction of the pectoral girdles and forelimbs based on CT data (except for the radiale and distal carpal, which were reconstructed within Dino-Morph). Three superim-

motion is curiously limited (Carpenter and Smith 2001), when each forelimb is extended laterally, with elbow straight and manus extended as well, the arrangement resembles a jack stand (or a pair of bicycle kick stands).

The stout forelimbs, fully extended and acting as struts anchored into the ground by strong manual unguals (which are also well oriented for this anchoring task), are well placed for stabilizing the anterior of this giant theropod in preparation for rising. As the animal shifts its weight, ground reaction forces would have been directed nearly perpendicularly into the cup shape of the glenoid fossa; the compressive load would then distribute along the scapulocoracoid over a span of ribs. If the stout forelimbs were indeed involved in stabilizing the body, it is noteworthy that the ground reaction forces would have communicated directly to the vicinity of the acromion process of the coracoid, and therefore place the furcula, which is directly aligned with this force vector, under significant bending stress. As noted by others (Larson 2001; Larson and Rigby 2005; Lipkin and Carpenter this volume; Rothschild and Molnar this volume), the furcula is frequently found with evidence of healed stress fractures and breaks.

The sprint start discussed earlier would have been assisted by braced and stabilizing the anterior portion of the body by holding the forelimbs strutlike. Indeed, the resemblance of the initial pose to that of a human sprinter is striking (Fig. 11.10).

Alternatively, with the forelimbs serving to anchor the animal, the posterior musculature of the hind limbs could come into play more gradually to elevate the COM even though it was located ahead of the hind feet. With the animal's weight positioned fractionally between anchored forelimbs and extending hind limbs, the COM could be elevated with the additional mechanical advantage of a second-class lever. The animal would remain in a stable quadrupedal stance during this initial stage of elevation, and progressively, as the femora and knee come out of the deep crouch, the mechanical advantage of the large femoral retractors and knee extensors would have increased. If not intending a sprint start, but merely wishing to regain a stand-

posed poses are assumed symmetrically by left and right forelimbs. With elbows and manus extended, the forelimbs can act as a jack stand to stabilize the body during ascent, but the line of action of the ground reaction force would have placed the furcula under significant bending stress, consistent with commonly observed healed fractures. Forelimb range of motion estimated in collaboration with Kenneth Carpenter.

ing posture, the great theropod would likely have (1) tipped forward slightly, pivoting about the prepubis, until (2) the forelimbs were in ground contact and helping to anchor the giant, then (3) it would have first raised its rump, much as large herbivores do today, then (4) either step into forward movement or ascended symmetrically into a standing posture.

Unlike quadrupeds such as bovids, the disparity between forelimb and hind limb length in *Tyrannosaurus rex* limited the extent to which it could ascend rump first while maintaining a purchase on the ground with the forelimbs. Before achieving full extension of the hind limb, the animal would have had to break contact between its forelimbs and the ground, and either take a step with one hind limb in order to regain its balance, or remain symmetrically posed on 2 hind limbs, and by means of momentum, body movements, and strength, bring the COM back between the hind feet. Perhaps the furcula injuries reflect mishaps that occurred while attempting to regain its balance, particularly when lame as a result of other injuries, or they may reflect the amount of stress imposed on the shoulder girdle during these maneuvers. In the event of a misstep or other failure to achieve balance between the hindfeet, 4 or more metric tonnes falling on the forelimbs could have precipitated such fractures.

Conclusion

The great bulk of an adult *Tyrannosaurus rex* was capable of being gracefully lowered until it settled its weight on the elongate pubic boot, freeing the animal to adjust its legs much as a sports spectator would use a portable one-legged stool. When it came to rising again to a bipedal stance, the options, particularly for a small tyrannosaurid, would be a sprint start with or without assistance from the forelimbs, or a more gradual elevation using the hind limbs during which the forelimbs played an essential role. The latter was energetically more efficient and might have been preferable for the adult. The forelimbs were literally pivotal in this operation, and mishaps might have resulted in transmission of enormous compressive forces on the pectoral girdles and the delicate furcula that spanned the acromion processes. Although it was perhaps ungainly for the tyrant king to rise rump first, its ascent was likely more elegant than that of modern bovids rising from repose.

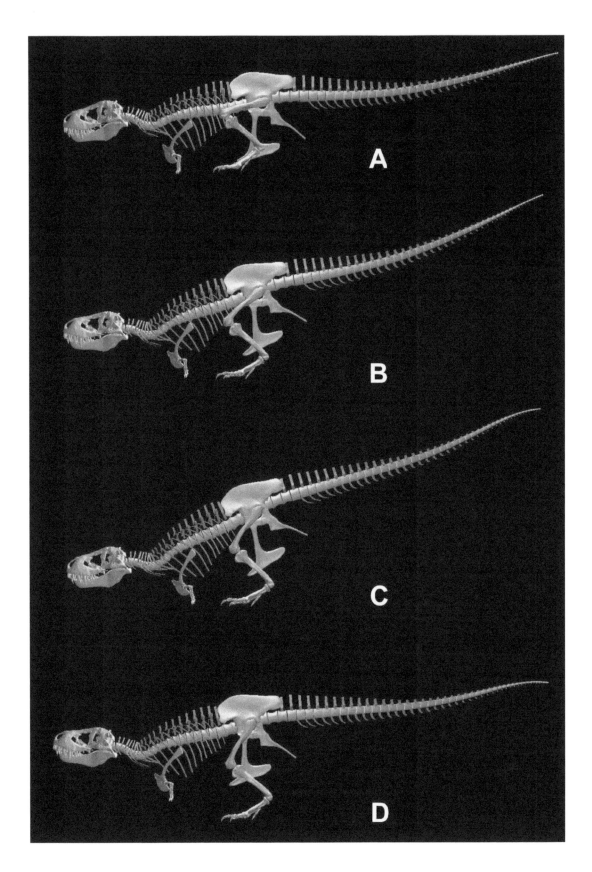

Acknowledgments

We are grateful to Ray Wilhite for digitization of *T. rex* bones used in this study, and to the Black Hills Institute of Geologic Research for hosting the 100 Years of *Tyrannosaurus rex* Symposium and for providing access to specimens. Thanks also to Phillip Manning, Yoshio Ito, and Kenneth Carpenter for helpful suggestions regarding range of motion and movements.

References Cited

Alexander, R. McN. 1989. *Dynamics of Dinosaurs and Other Extinct Giants.* Columbia University Press, New York.

Carpenter, K. 2002. Forelimb biomechanics of nonavian theropod dinosaurs in predation. *Senckenbergiana Lethaea* 82: 59–76.

Carpenter, K., and Smith, M. 2001. Forelimb osteology and biomechanics of *Tyrannosaurus rex*. P. 90–116 in Tanke, D., and Carpenter, K. (eds.). *Mesozoic Vertebrate Life*. Indiana University Press, Indiana.

Currie, P. J., Koppelhus, E. B., Shugar, M. A., and Wright, J. L. 2004. *Feathered Dragons*. Indiana University Press, Bloomington.

Gauthier-Pilters, H., and Daag, A. I. 1981. *The Camel: Its Ecology, Behavior and Relationship to Man*. University of Chicago Press, Chicago.

Gierliński, G., Lockley, M., and Milner, A. R.C. 2005. Traces of early Jurassic crouching dinosaurs. P. 4 in *Tracking Dinosaur Origins: The Triassic/Jurassic Terrestrial Transition Abstract Volume*. Dixie State College, St. George, UT.

Henderson, D. M. 1999. Estimating the masses and centers of mass of extinct animals by 3-D mathematical slicing. *Paleobiology* 25: 88–106.

Horner, J. R., and Lessem, D. 1993. *The Complete T. rex*. Simon & Schuster, New York.

Hutchinson, J. R., and Garcia, M. 2002. *Tyrannosaurus* was not a fast runner. *Nature* 415: 1018–1021.

Jenkins, F. A., Dial, K. P., and Goslow, G. E. 1988. A cineradiographic analysis of bird flight: the wishbone is a spring. *Science* 241: 1495–1498.

Lambe, L. M. 1917. *The Cretaceous Theropodous Dinosaur Gorgosaurus*. Canada Department of Mines, Geologic Survey of Canada, Memoir 100.

Larson, P. L. 2001. Paleopathologies in *Tyrannosaurus rex* (in Japanese). *Dino Press* 5: 26–35.

Larson, P. L., and Donnan, K. 2002. *Rex Appeal: The Amazing Story of Sue, the Dinosaur that Changed Science, the Law and My Life*. Invisible Cities Press, Montpelier, VT.

Larson, P. L., and Rigby, K., Jr. 2005. The furcula of *Tyrannosaurus rex*. P. 247–255 in Carpenter, K. (ed.). *Carnivorous Dinosaurs*. Indiana University Press, Bloomington.

Muybridge, E. 1899. *Animals in Motion*. London: Chapman & Hall. Dover reprint, 1957.

Newman, B. H. 1970. Stance and gait in the flesh-eating dinosaur *Tyrannosaurus*. *Biological Journal of the Linnean Society* 2: 119–123.

Norman, D. 1991. *Dinosaur!* Prentice Hall General Reference, New York.

Osborn, H. F. 1906. *Tyrannosaurus*, Upper Cretaceous carnivorous dinosaur. *Bulletin of the American Museum of Natural History* 22: 281–296.

Paul, G. S. 1988. *Predatory Dinosaurs of the World*. New York Academy of Sciences, New York.

Stevens, K. A. 2002. DinoMorph: parametric modeling of skeletal structures. *Senckenbergiana Lethaea* 82(1): 23–34.

Figure 11.10. *Elevation of the posterior while anchoring the anterior body by the forelimbs, creating a pose much like a sprint start. The mechanical advantage of a second-class lever is provided during extension of the hind limbs in raising the COM. With sufficient elevation achieved, the animal could push back and regain bipedal balance, and complete its ascent to a standing pose.*

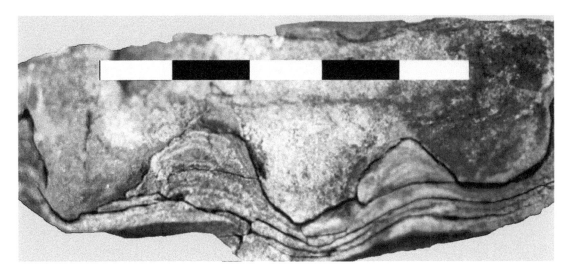

Figure 12.1. *Cross section through digits of a tridactyl Middle Jurassic theropod track from the Saltwick Formation, Whitby, UK. Scale bar = 5 cm.*

T. REX SPEED TRAP

Phillip L. Manning

12

Introduction

The underlying assumption of many track interpretations is that what is preserved represents a surface trace. Track geometry and morphology (e.g., track length [FL] and track width [FW], digit length, number of digits, interdigital angles) are based on what is visible, often recorded as a 2-dimensional (2D) feature. The pages of this volume record information on complex 3-dimensional (3D) bones, but fossil bones are defined by obvious boundaries. Tracks are not. The tracks of dinosaurs are truly the icebergs of the ichnological world, with the majority of the structure expressed below the track surface horizon. The fossil record rarely offers more than a 2D slice of this complex multitiered structure (Fig. 12.1). Although this surface trace provides an opportunity to map the track (another source of error when defining which part of a surface to measure), it is the relationship to underlying transmitted track surfaces that is the key to unlocking this 3D puzzle. A fossil track is more than a simple 2D outline; they are complex 3D structures that have volume that can be visualized as 3D failure envelopes (Margetts et al. in press).

Laboratory track simulations can provide some insight to the complex subsurface sediment deformation associated with track formation (Allen 1989, 1997; Gatesy et al. 1999, 2005; Manning 2004). The recovery of subsurface layers provides insight to track morphology relative to the true surface track. When compared with fossil tracks and the tracks of extant avian theropods (birds), a more complete understanding of tracks and their formation can be undertaken.

The tracks and trackways of theropod dinosaurs are common in the Mesozoic and can provide useful data on the locomotor ability of large predators. However, speed estimates calculated from the trackways of large theropods can potentially under- or overestimate the speed (u) at which an animal was traveling, if the FL (used to calculate hip height) is misinterpreted.

I examine track maker anatomy and gait, fossil and laboratory-simulated tracks, the complexities of track formation, track preservation, and what understanding can be gained from this dynamic source of information.

Historical Tracks

The study of vertebrate tracks and traces, vertebrate paleoichnology, has largely concentrated on describing a trace, often with little or no interpretation of the formation or preservation of that trace. It is surprising how such an important and revealing area of trace fossil interpretation is so sparsely represented in past literature. However, seek and you shall find!

Figure 12.2. *Edward Hitchcock. Courtesy of the Pratt Museum, Amherst College, Massachusetts.*

William Buckland was the first to undertake laboratory experiments to help interpret the origin of a fossil trackway (Sarjeant 1974; Pemberton and Gingris 2003). Buckland's "culinary" approach to deciphering the identity of the track maker was simple. He persuaded a crocodile and then a tortoise to walk across a soft piecrust (presumably made of dough) and then over wet sand and soft clay. The resulting impressions left Buckland in no doubt that a tortoise had left the fossil trackway (Duncan 1831), yielding the first of many ichnospecies, *Testudo duncani* (Sarjeant 1990; Pemberton and Gingris 2003).

Edward Hitchcock (Fig. 12.2) should be considered the father of dinosaur ichnology, even though he thought his tracks were traces of ancient giant birds. However, it is quite ironic that in the 21st century, *Hoxd12* and *Hoxd13* genes (Vargas and Fallon 2005) have proved him partially right.

Hitchcock made many smart observations on the preservation of tracks from the Lower Jurassic rocks of the Connecticut Valley, USA. The most acute of his observations was that he recognized transmitted tracks and illustrated a stacked sequence of tracks (Hitchcock 1858). These repeated echoes of the surface trace in successive subsurface layers were found commonly among the Connecticut Valley tracks. He referred to these layers, often physically bound together like stony books, as his "fossil volumes" (Hitchcock 1858). These volumes were recognized as a dynamic record of movement, as Hitchcock remarked on the anterior travel of transmitted tracks relative to the surface track. It would take over a hundred years before ichnologists would begin to recognize the significance of what Hitchcock (1858) had correctly interpreted.

Triassic tracks from Milford, NJ, led Baird (1957) to conclude that tracks and trackways were not just a simple record of anatomy, but one that belays the movement of the foot on, and in some cases through, sediment. Baird saw track formation as a dynamic interaction that was in turn controlled by substrate type.

Allen (1989) used the mechanical indenter theory (Calladine 1969; Hill 1971; Johnson et al. 1982) to interpret fossil tracks, supported by complementary scaled laboratory experiments. He used a circular indent on laminated plasticine and sectioned the resultant tracks to display the distribution of subsurface track features. He noted that the undertraces in his experiments were a substantially less perfect record of the shape and size of the face of the indenter, and concluded that general indenter theory, complemented by laboratory experiments, could provide insight to the formation and preservation of vertebrate tracks and trackways.

The most comprehensive study on the mechanics of the formation, preservation, and the distribution of vertebrate tracks was also undertaken by Allen (1997). The study looked in detail at subfossil mammalian tracks from Flandrian deposits in the Severn Estuary, Southwest Britain. Allen (1997) suggested that understanding the mechanics of track making and the taphonomy of traces has continued to lag behind the descriptions of anatomical aspects and the distribution of tracks, despite its relevance in taxonomy and ecological interpretations. The 2 approaches used in interpreting the mechanics of track formation can be divided into those that use live animals walking over prepared substrates (McKee 1947; Farlow 1989; Padian and Olsen 1984, 1989) and the application of indenter theory as first undertaken by Allen (1989).

Allen (1997) continued the experimental laboratory approach to a theoretical mechanical model as a means of interpreting and understanding vertebrate track preservation and morphology. He suggested that the model of mechanical theory offered a number of insights into the likely character of animal tracks in the field. He found that the use of an indented plastic material in laboratory tests was supported by results that qualitatively reproduced all the essential features of real tracks.

The distribution and preservation of subfossil tracks in the Severn Estuary are presented in outcrop in several different modes as undertraces of varying degrees of detail, as overtraces from a range of levels in the shafts, as emptied shafts, and as tracks. Allen (1997) noted that only a small proportion of tracks in the area were capable of yielding unchallengeable taxonomic information about the animals that made them. The indenter model, coupled with laboratory and field experiments, provided a robust means to test the information recoverable from the fossil track record; however, it was suggested that further study was required to permit more solid taxonomic, ecological, and environmental inferences about extinct species (Allen 1997).

Reviews on the history of vertebrate track ichnology can be found in Sarjeant (1974) and Thulborn (1990). Although the use of laboratory studies to interpret the formation and preservation of dinosaur tracks has been limited, the multidisciplinary approach Padian and Olsen (1984) suggested, which uses anatomy, kinematics, and the nature of the substrate to interpret tracks, should have been embraced by the ichnological world. It was not, until very recently.

Excellent progress is now being made with computer-aided ichnology. Stephen Gatesy's work brings ichnology firmly into the 21st century (Gatesy 2001, 2003; Gatesy et al. 1999, 2005). He recognized the complex surface relief of Triassic tracks from East Greenland could replay the movements of the limbs that created the tracks (Gatesy et al. 1999). We will return to Gatesy's splendid work later in this chapter.

While I was happily chasing emu, generating experimental tracks and hunting vertebrate traces in the field, Jesper Milán was doing likewise in Denmark. His research (Milán 2003; Milán et al. 2004) clearly demonstrates the value of experimental work when interpreting fossil tracks.

Although laboratory tracks are useful, it must not be forgotten that their function is to aid the interpretation of fossil ones. It is a little surprising, but given that vertebrate ichnology has close to 2 centuries of history, tyrannosaur tracks are conspicuous by their rarity.

T. rex Tracks

Tyrannosaurus tracks are rare in the fossil record. Many tracks initially identified as being made by *T. rex* have subsequently been identified as belonging to their prey: hadrosaurs! Tracks that were allegedly made by *T. rex* from the Mesa Verde Group of Carbon County, UT, were given the suitable ichnospecies *Tyrannosauropus* (Haubold 1971). However, subsequent research (Lockley and Hunt 1995) has shown this ichnospecies could not have been made by *T. rex*. The Mesa Verde Group is too old (Campanian) to have the tracks of *T. rex*, and most workers agree that the large tridactyl tracks found belonged to a hadrosaur. However, the ichnospecies *Tyrannosauropus* stands—an unfortunate assignment for a hadrosaur track.

The tracks of *T. rex* (Fig. 12.3.) have been described from the late Cretaceous Laramie (Colorado) and Raton Formations (Northern Mexico), named *Tyrannosauripus pillmorei* (Lockley and Hunt 1994). The Raton Formation track displays several features that might be expected for one made by *T. rex*—length (0.85 m long) that exceeds width and a clear hallux impression—and to date, *T. rex* is the only large theropod that fits this track in the late Cretaceous (Lockley and Hunt 1994). The existence of a *T. rex*

 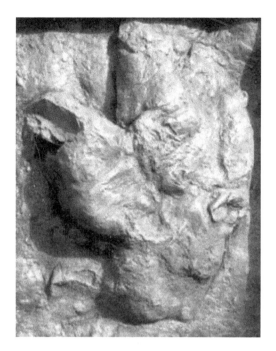

track is important for ichnologists, but unfortunately, it is very much in the singular. A trackway of a *T. rex* had not been identified until recently.

The Black Hills Institute of Geologic Research has recognized a series of theropod trackways in the Lance Formation (Maastrichtian, Upper Cretaceous) of Wyoming. Some of the trackways have been interpreted as being made by a tyrannosaur (Peter Larson, personal communication). The large (0.83 m FL) tridactyl tracks have low relief, almost certainly represent transmitted features, and are preserved in fine- to medium-grained sandstone (Fig. 12.4A). They are associated (on the same bedding plane) with smaller oviraptorid-type tracks (Fig. 12.4B) and large hadrosaurine tracks (Fig. 12.4C), displaying concentric shear failure (Manning 1999, 2004).

The Lance Formation tracks, although important, do not have any of the animals breaking into runs, merely walking across the substrate. These trackways will be the focus of future research and will certainly add more

Figure 12.3. ***Tyrannosauripus pillmorei*** *(Lockley and Hunt 1994). Footprint (left) and cast (right). Scale bar = 50 cm.*

Figure 12.4. *(A)* ***Tyrannosaurus rex*** *track? (B) Oviraptorid dinosaur track. (C) Hadrosaur track displaying concentric shear failure. Lance Formation, Wyoming. Scale bar = 5 cm.*

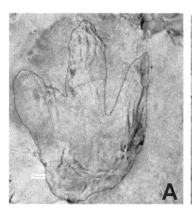 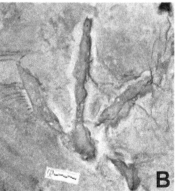 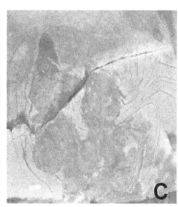

Figure 12.5. *Mongolian theropod trackway. Trackway on left with latex cast of trackway on right. Courtesy of Yoshiki Koda.*

to our knowledge of this dinosaur assemblage. Although tracks of medium to large theropods attaining significant speeds are rare, they are known (Farlow 1981; Viera and Torres 1995).

An additional series of theropod trackways recently found in Mongolia support that some medium-large theropods were capable of running at moderate to high speeds (Fig. 12.5). The fossil trackways of the theropod (hip height [h] 1.23–1.71 m) from Inner Mongolia potentially shows an animal moving at 7.15–10.6 m/s (25.7–38 km/h)—relatively fast for such a large animal! Is it possible that *T. rex* might also have achieved such speeds?

The following sections will begin to unpick a complicated story of anatomy, limb movement (kinematics), and substrates, a story that holds many clues to unraveling some of the secrets locked within tracks and trackways of theropods and other bipedal dinosaurs.

Anatomy of Locomotion

The running ability of *T. rex* has had a checkered history that is based on body fossil evidence, from a sprinting, cursorial predator (Paul 1998; Sellers and Manning, in preparation) to a graviportal scavenger (Horner and Lessem 1993). The running ability of any organism is intimately tied to the anatomy of the animal. The relationship between anatomy and locomotion was recognized as far back as the fifth century B.C., by Aristotle (Padian 1995). The form of vertebrate skeletons has a direct influence on limb function because bones provides the anchor points for the tendons, musculature, and ligaments that drive locomotion and delineates the degree of movement. The body fossil record preserves information on the skeletal anatomy, size, and inferred gait (based on joint articulation and geometry) of some dinosaurs. This has enabled workers to generate functional models and to reconstruct the locomotion for some dinosaurs (Romer 1923, 1927, 1956; Walker 1977; Tarsitano 1983: Alexander 1996; Johnson and Ostrom 1995; Farlow et al. 2000; Jones et al. 2000; Biewener 2002; Hutchinson and Garcia 2002; Carrano and Hutchinson 2002; Hutchinson 2004).

T. rex, along with all theropods, were obligatory bipeds. The majority of theropods walked on 3 toes (tridactyl), but some retained vestigial digits I and V. The phalangeal formulae remain remarkably conservative for the group and its descendants, the birds (I prefer the term *avian theropods*). Like many other dinosaurs, theropods stood, walked, ran, and jumped on their toes (digitigrade), like reptilian prima ballerinas, with bite! The metatarsi varied considerably in form, ultimately fusing in the derived avian condition (tarsometatarsus).

The theropod skeleton shares many features among tetrapods, such as the common possession of specific bones in the manus (hand) and pes (foot). Some of these features provide useful homologous characters when tracing theropod evolution, as well as the associated evolution of the locomotor trends in more distant groups (Parrish 1986). Analogous characters shared by distinct groups with separate evolutionary lineages can also provide useful comparative information on the form and function of independently evolved characters (Gatesy and Biewener 1991).

The primary components of the tetrapod skeleton concerned with locomotion are the limbs and their attachments. However, before we discuss the limbs, it is worth considering storage of energy within an organism in relation to locomotion. Muscular energy is of prime importance contributing to locomotion, but the vertebral column and the flexibility of bones in almost the whole skeleton are an important passive component (Alexander et al. 1985). The vertebral column of mammals is quite different from that of reptiles, but given the similarities in gait and posture to dinosaurs, it seems quite likely that their vertebral column (as in all tetrapods) played an important role in locomotion. The tendons that braced the backs of many dinosaurs would have played a crucial role in storing energy for locomotion, as they do today for mammals, from shrews to elephants (Sellers and Manning, in preparation 2006). It has been suggested that back tendons account for about 70% of the total strain energy stored in the vertebral column, muscles contribute about 10%, and the elasticity of the vertebrae themselves accounts for about 20% (McGowan 1999). Put simply, bones provide a framework for muscles to deliver their work, but to fully understand the energy inputs to locomotion, a whole-organism approach is ideal. The physical and mechanical properties of bone, muscle, tendon, ligament, keratin, and other biomaterials all have a role to play and must be accounted for when unraveling the locomotory abilities of extinct animals.

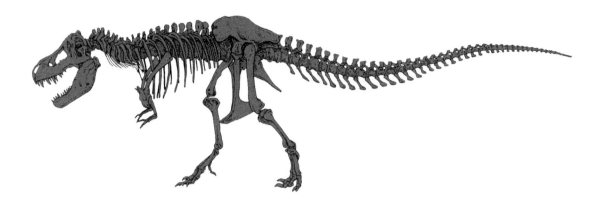

Figure 12.6. *Tyrannosaurus* Stan. Stan shows a typical skeletal layout for many large theropod dinosaurs.

Theropod skeletons display typical form expected for a terrestrial cursorial tetrapod (albeit a biped), but the posture of the limbs is not typical for reptiles, by the degree of adduction of the hind limb, allowing an erect, or parasagittal, posture and resultant gait (Fig. 12.6). The erect posture, in which the plane of the legs is perpendicular to the plane of the torso, has the limbs slung under the body in a typically avian or mammalian posture.

Tracking Extant Avian Theropods

Theropod descendants, the birds, form a diverse derived group, but the anatomy of their feet and limbs remains relatively conservative and similar to their ancestors. However, the loss of the dinosaurian tail and the development of flight resulted in major reorganization of the pelvic girdle (Hutchinson 2001). The many outcomes of this reorganization include an anterior shift in the center of mass and altered limb kinematics to compensate for the evolutionary changes (Gatesy 1990, 1991, 1995).

Birds have adopted the derived knee-based retraction mechanism for locomotion (Gatesy 1990, 1991, 1995; Gatesy and Biewener 1991), making them poor analogues for calculating the position, articulation, and relative limb angles of theropod dinosaurs. However, humans have adopted a hip-based retraction mechanism, functionally comparable with most large theropods (Gatesy 1990, 1991, 1995; Gatesy and Biewener 1991). The limb angle positions, relative to the ground, for a human vary from 73°–> to 74° at heel-down phase to 118°–> to 124° at toe-off phase of the step cycle. The limb posture for the hip-based retraction mechanism of humans has been used to simulate the limb posture angles in laboratory-simulated dinosaur tracks (Manning 1999, 2004). Fossil tracks can display evidence of a 3-phase distorted force bulb, inferring a heel-down, roll forward, and toe-off phase during track formation, supporting a hip-based retractor mechanism for some theropods. Triassic theropod tracks from Greenland (Gatesy et al. 1999) also demonstrate that track morphology can elucidate the movement of the step cycle that created them.

Theropods: A Moving Experience

The step cycle of a human can be evaluated by our brains, given that we are quite familiar with the gait and movements adopted by our own species.

Dinosaur gait and movement are problematic, given they are predominantly based on fragmentary fossil evidence. However, movement is a function of the organism's limb anatomy, joint articulation, development and distribution of musculature, and geometry.

Constraining the limb kinematics for a particular dinosaur is fraught with difficulties, not least of which is the fact that no living relative quite walks the walk of a dinosaur (Manning et al. 2006). It might be possible to infer some of the locomotor ability, gait, and so on by applying the principle of extant phylogentic bracketing (EPB) (Witmer 1995). Might it then be possible to bracket the potential range of locomotor styles and abilities of dinosaurs? Given that dinosaurs fall in the EPB of extant birds and crocodilians, this technique infers the locomotor styles and abilities, like anatomy, physiology, and biology, might also be constrained by the extant group. Reconstructions of the pelvic and hind limb musculature of *T. rex* (Carrano and Hutchinson 2002) have proven how useful the EPB is when unraveling complex soft tissue relationships. Combining EPB with observations of theropod anatomy (Carrano and Hutchinson 2002) and track data (Gatesy et al. 1999; Manning 1999, 2004) indicates that it is possible to reconstruct gait patterns and movement. Once gait pattern and limb movement are constrained, it is possible to begin the process of resurrecting a theropod step cycle and the processes that result in track formation.

The part of an animal's step cycle where the foot is in contact with the ground is known as the support phase (Gatesy and Biewener 1991). It is during the support phase of locomotion that a track is formed. The initial angle at which the foot makes contact with the ground alters, increasing as the animal's body moves over the foot to the toe-off phase of the step cycle. The foot-down and toe-off phase of the step cycle have been studied by several workers (Clark and Alexander 1975; McMahon 1984; Thulborn and Wade 1989; Padian and Olsen 1989; Gatesy 1990; Gatesy and Biewener 1991) for different species of animal, including humans and ratite avian theropods. The knee flexion in birds functions similarly to hip extension in humans, providing the principal angular joint displacement by which the body is moved forward (Cracraft 1971; Jacobson and Hollyday 1982).

The angle of the action of force acting on a foot at the heel-down phase of a limb cycle appears remarkably similar among living bipedal animals (Gatesy and Biewener 1991), with variation between 56°–> and 73° to the ground, depending on the type of animal and the speed at which it was traveling. The angle to which the limb rotates forward before the toe-off phase of the step cycle also shows little variation between several living bipedal animals, between 106°–> to 138° to the ground. Variation in the step cycle limb position angles was greatest in birds, ranging from 56°–> to 63° at heel-down to 106°–> to 138° at toe-off phase of the step cycle. The relative position of an animal's center of mass to the angle of the action of force acting on the foot results in variation in the distribution of pressure over the sole of the foot during a step cycle. Can this variation in pressure possibly be one of the keys to allow the kinematic information stored within a track to be unlocked?

A Critical Eye for Walking

Experimental work on trackways, coupled with considerations of limb kinematics and substrate conditions, permits the most robust inferences about track maker's and fossil footprint data (Padian and Olsen 1984, 1989; Manning 1999, 2004). It is logical that similar trackways indicate analogous kinematics in many large theropods (Padian and Olsen 1989). However, Gatesy (1995) questioned the resolution at which details of limb-segment orientation, kinematics, muscular anatomy, and neuromuscular control could be addressed by means of Padian and Olsen's (1989) technique. He suggested that footprints could not be equally informative about all locomotor categories, even if trackways have helped confirm that birds retained the obligatory, digitigrade bipedalism and highly adducted limb posture of their theropod ancestors. Gatesy (1995) concluded that it was quite possible that such bipeds could make almost identical footprints even if they differed in several locomotor categories. He maintained that trackways could not provide enough detail to discriminate between hip-based (primitive theropod) and knee-based (avian theropod) limb-retraction mechanisms.

The track morphology of the largest living ground birds (ratites), such as emu (Padian and Olsen 1989; Manning 1999; Milán 2003; Milán et al. 2004) and ostriches (Farlow 1989), allow comparison of track morphology generated by either avian knee-based or hip-based (primitive theropod and human) retraction mechanisms by using laboratory-simulated and fossil tracks (Manning 1999, 2004). Observations on the distribution of pressure across ratite feet when walking, inferred from track morphology (Thulborn and Wade 1989; Farlow 1989; Manning 1999, 2004), indicate the pressure distribution across the foot of a knee-based retraction mechanism differs from a track generated by a hip-based retractor mechanism. The distinct heel-down phase in laboratory-simulated tracks and fossil tracks is almost absent from emu and ostrich tracks (Farlow 1989; Manning 1999). The heel-down phase is replaced by what can be described as a contact phase. In the contact phase, an avian theropod tests the ground to assist motor control for the completed step cycle (Gatesy and Biewener 1991). This enables an animal to accommodate for substratum heterogeneity during locomotion in a natural environment (Clark 1988; Gatesy and Biewener 1991), without committing its whole mass over the foot, as with a heel-down phase of a hip-based retractor mechanism. The knee-based retraction mechanism appears to combine the heel-down phase and rotation phase of the step cycle (contact phase), with the greatest force exerted at the distal end of digits at the push-off phase of the step cycle. The variation in the distribution of pressure across the foot is essentially a by-product of the relative position of the animal's center of mass during a step cycle, as the body moves over each foot and step, respectively. This suggests, contrary to Gatesy (1995), that it is possible to differentiate between hip- and knee-based retractor systems from track geometry, given that track relief (surface and subsurface) is a function of the distribution of pressure.

The center of mass of a theropod dinosaur is also reflected in the distribution of the animal's weight on its 2 limbs. If the center of mass is directly over the limbs of a biped, each limb will support an equal amount of weight (when the animal stands still). The posture of a theropod dinosaur had to account for the relative position of the center of mass to be stable, or, when walking, to be dynamically unstable (see Stevens et al. this volume). The difference in the position of the center of mass would certainly have had an effect on all theropod locomotion, including sitting, standing, walking, running, and jumping.

The distribution of pressure exerted across the sole of a foot can vary with subtle directional changes in the load applied during a step cycle (Manning 1999). Laboratory track simulations and force-plate (optical pedobaragraph) experiments can yield useful information on the subsurface morphology of tracks and the distribution of pressures across the sole of a foot (Manning 1999, 2004). The varying degree to which a digit or digits were transmitted to deeper successive layers correlates with the distribution and magnitude of pressure acting on the sole of a foot. Experimental pressure plate systems have been used to track the variation in the center of pressure during a step cycle, making it possible to correlate variation in load with the resultant distribution of pressures over time through a step cycle (Manning 1999, 2004).

The implication of being able to infer the kinematics of a step cycle from a fossil track, coupled with the size of the animal and speed at which it was traveling, could yield important information on the locomotion of all dinosaurs. The 3D subsurface track record of the relative magnitude and distribution of pressure across a foot can assist in assigning theropod tracks to a primitive (hip-based) or derived (knee-based) locomotor system. This could provide useful data on evolutionary trends in theropod locomotion in the fossil track record, thus supporting the evidence from the body fossil record (Gatesy 1990, 1991, 1995). It may now be possible to differentiate from Late Cretaceous avian theropod and bipedal ornithopod tracks on the basis of subsurface deformation by the presence or absence of a heel-down phase during locomotion.

A Middle Jurassic theropod track from the Scalby Formation, Yorkshire, UK, provides an example of a 3-phase track (Thulborn and Wade 1989) (Fig. 12.7), with features typically expected for a hip-based limb retraction mechanism (Manning 2004). The cross section along the median line of digit III shows a region of downwarped sediment in the heel area (A), which delineates the deformation caused at the heel-down phase of the step cycle (Fig. 12.7). The bridge of the foot (B) of the section of digit III delineates the second forward rotation phase of the step cycle (Fig. 12.7). The third and most distinct point of the step cycle, the toe-off phase (C), is clearly delineated by a severely downwarped area of sediment laminae, coupled with liquefaction failure (Fig. 12.7). The track displays all 3 phases of a step cycle (Thulborn and Wade 1989) that would be expected for a hip-based retractor mechanism, typical of a Middle Jurassic theropod dinosaur (Manning 1999, 2004).

To test this hypothesis further would require the sectioning of many complete fossil tracks from the Jurassic and Cretaceous, to compare and contrast the distribution of pressure across the foot in relation to subsurface features, which will (curators willing) be the subject of future work.

Figure 12.7. Cross section along medial line of digit III from a Middle Jurassic theropod track, Scalby Formation, Burniston, UK. Areas A to C represent phases of the step cycle (see text). Scale bar = 10 cm.

Firm Grounding to Build On

Previous work on dinosaur trackways has tended to concentrate on identifying the animal that may have produced the track, and the speed, gait, and size of the animal. The way in which sediments behave before, during, and after a track is formed and the subsequent processes that may further modify, enhance, or disguise a track has been much neglected. The poor fossil record for *T. rex* tracks and trackways suggests that either more tracks must be found, or an alternative to fossil tracks be generated to study what potential traces might look like.

A combination of laboratory-controlled track simulations, coupled with field observations, can provide a more complete understanding of how tracks are formed and preserved (Padian and Olsen 1989; Allen 1989, 1997; Manning 1999, 2004; Milàn 2003; Milàn et al. 2004). The morphology of surface and subsurface tracks must be studied with a view to treating tracks as dynamic records of movement and not just static traces.

Did the Earth Move for You?

A footfall of a dinosaur mechanically makes dense the sediment beneath its foot, sometimes resulting in failure, preserved as tracks. For many years, ichnologists have named track species (ichnospecies) at an alarming rate using qualitative characters that vary greatly from one footfall to the next. Few ichnospecies are supported by laboratory-controlled experiments to quantify and qualify interpretations of the fossil tracks. Resultant ichnospecies are 2D shadows of what was a more complex 3D track structure.

The mechanical properties of substrate clearly influence the response and resultant features associated with a footfall. To understand the formation and preservation of dinosaur tracks, it is essential to understand the mechanics of soils. The word *soil* has different meanings to workers from various disciplines. The definition I use here is an engineering one, defined as any loose sedimentary deposit, such as gravel, sand, clay, or a mixture of these materials (Smith 1981).

The size and distribution of particles that form a soil, combined with the air or water occupying voids between the solid particles, affects the mechanical properties of that soil, as does its porosity and permeability. For example, silty fine- to medium-grained sands have little resistance to shearing when dry, but when moisture content increases, so too does shear strength (Manning 1999, 2004). An increase in moisture content effectively increases the bulk density of a sediment as water replaces the air contained in the voids between soil particles. The denser a soil becomes, the greater its shear strength (Karafiath and Nowatzki 1978). However, if the moisture content increases beyond the soil's critical saturation point (critical hydraulic gradient), where the soil particles no longer come into contact as a result of pore-water pressure, the soil fails.

Boussinesq (1883) solved the problem of predicting the distribution of stress at any point in a homogeneous, elastic, isotropic medium as the result of a point load applied at the surface. Boussinesq's elastic analysis is represented by the following equation:

$$\sigma_v = \frac{3P}{2\pi z^2}\left[\frac{1}{1+\left(\frac{r}{z}\right)^2}\right]^{5/2} \quad (1)$$

Boussinesq's equation shows a point load (P), with σ_v the vertical stress at point depth z below the load at a horizontal distance r from the line of action.

Manning (1999, 2004) applied Boussinesq's equation to help provide insight to the distribution of pressure within a volume of sediment, yielding track features. The maximum zone of deformation (MZD) marks the zone of influence (or failure envelope) in track formation, and Boussinesq's equation provides the theoretical distribution of actual pressure in a volume of sediment. However, Boussinesq's theory relates to the distribution of a static load at depth. A track is the result of a dynamic load; therefore, the application of Boussinesq's theory has its limitations. Manning (2004), who used a reconstructed foot to indent constructed sediments, showed that the use of laboratory-generated tracks can yield important information on dynamic sediment failure.

Making Tracks in the Laboratory

Laboratory-simulated tracks can provide a quantitative approach to interpreting the morphology and spatial distribution of failure in a volume of sediment, and the associated sediment conditions and properties at the time of track formation (Allen 1997; Gatesy et al. 1999; Manning 1999, 2004; Milán 2003; Milán et al. 2004). They might also help confirm or disprove track and track maker relationships, given the benefits of a closed system and knowing the experimental parameters.

A quantitative test for fossil dinosaur tracks is a difficult benchmark to achieve, given there are many variables to account for, including moisture content at the time of track formation, weight of the dinosaur, true morphology of the dinosaur's foot, and the exact gait of dinosaur at the time of track formation. Although these variables can be measured in controlled laboratory track simulations, a method to predict these variables from a fossil track has yet to be devised. This means that the track data produced from laboratory simulations can only be used as a qualitative guide to the conditions prevailing, and the foot morphology of the track maker, at the time of track formation.

A simple experimental setup can provide a huge amount of data on the formation, preservation, and subsequent morphology of tracks. Man-

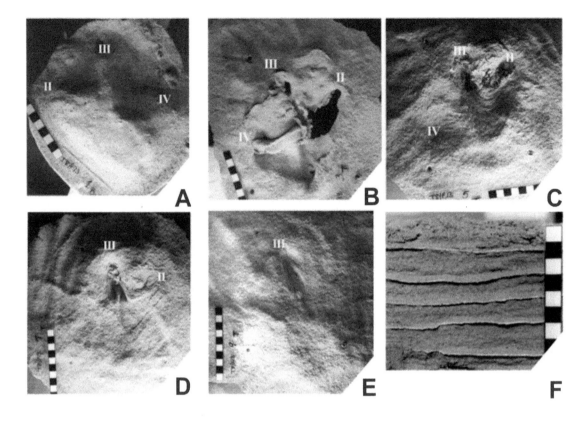

Figure 12.8. Track generated in a saturated (31.2% moisture content), fine-grained sand. (A) Overprint cast recording surface track feature. (B) Track at −1.3 cm. (C) Track at −3.5 cm. (D) Track at −4.9 cm. (E) Track at −6.3 cm. (F) Side view of constructed sediment. All scale bars in centimeters.

ning (1999, 2004) developed a series of experiments to characterize and reconstruct a number of sediments to indent with a standardized, reconstructed tridactyl theropod foot. The resultant tracks could be dissected and surface layers compared to any subsurface deformation (transmitted tracks) (Fig. 12.8).

From the data recovered from the experiments, it was possible to plot (with Petrel software) tracks in 3D showing the spatial relationship of surface and subsurface deformation (Fig. 12.9). The MZD on each track surface (Fig. 12.9A) and the relative position of digits (Fig. 12.9B) were recorded and plotted (see Manning 2004). With Petrel or other suitable visualization packages, it was possible to clearly show the spatial relationship of the track layers within a volume of sediment for the first time.

The experimental tracks indicated that the magnitude of the MZD and related sedimentary features is in proportion to the load applied to an isotropic sediment (Manning 1999, 2004). The MZD marks the maximum extent of deformation within a volume of sediment. A cross section of such deformation produces a distorted onion-shaped force bulb of influence. The deformation of the force bulb is a function of the dynamic nature of track formation, providing evidence of the direction of load applied to a sediment.

The distorted force bulb can be undistorted by plotting the cross section of a laboratory-simulated or fossil track on a normalized axis (Manning 1999). The resultant cross section through the force bulb resembles failure typical of a static load, in effect conforming the Boussinesq (1883) indenter theory.

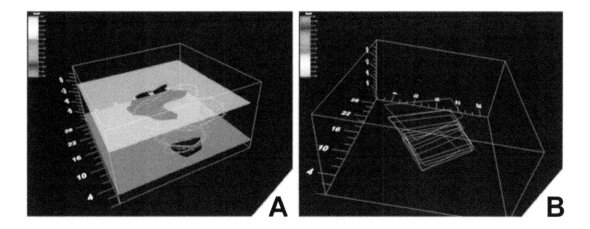

Laboratory-generated tracks can show the potential for errors in the interpretation of fossil track morphology and geometry. This is most evident when using trackways to calculate the speed at which the track maker was moving at the time of track formation, when using crucial parameters derived from track geometry.

Tracks as a Measure of Speed

Alexander (1976) devised a method to estimate an animal's speed from its tracks that used the known parameters of stride length (λ), and estimated parameters such as hip height (h). His formula could then be applied to fossil trackways. Alexander used a nondimensional parameter, the Froude number, to allow meaningful comparisons between animals of varying size by means of physical similarity theory (Duncan 1953). Physical similarity theory predicts that the movements of animals of geometrically similar form, but of different sizes, will be geometrically similar only when they move with the same Froude number (i.e., when the squares of their speeds are proportional to their linear dimensions). Geometrically similar movements require equal values of λ/h—that is, stride length must also be proportional to their linear dimension.

Alexander (1976) used stride length (λ) as the distance between corresponding points on successive tracks of the same foot. His equation can be written so that speed (u) can be estimated from the known values of λ and h can be measured accurately from tracks. Hip height (h) was estimated from FL, multiplying FL by 4 to give h.

$$u \omega \backslash^{0.25} g 0.5 \backslash l \lambda^{1.67} \backslash h^{-1.17} \backslash \qquad (2)$$

However, he recognized that in many bipedal dinosaurs, h did not conform to the rule of 4FL; some were 3.6FL to 4.3FL (Alexander 1976). Even with the potential pitfalls of Alexander's (1976) method, it has been adopted by many (Tucker and Burchette 1977; Thulborn and Wade 1979; Farlow 1981; Thulborn 1981, 1982; Haubold 1984; Lockley et al. 1996). The assumption that a dinosaurs foot has a constant relationship to h is possibly wrong for 3 reasons. First, the h/FL ratio is unlikely to account for

Figure 12.9. *Petrel plots of experimental tracks clearly show the 3D surfaces that combine to create a track. (A) MZD plots. (B) Digit plots showing spatial relationship of digit position with depth.*

variation in geometry between dinosaur taxa (Thulborn 1990). Second, the h/FL ratio would have changed during growth, a function of allometry (Thulborn 1990). Third and finally, the h/FL assumes that what is measured from the track represents the animal's actual foot length, something I will discuss further shortly.

Alexander (1976) did indicate that if h were overestimated by 10%, u would be underestimated by 11%. However, to create even larger discrepancies is not difficult. For example, a fossil FL of 0.10 m would give a hip height (h) of approximately 0.4 m. However, if the true foot length were 0.07 m, h would be calculated as 0.28 m, meaning the speed (u) from a trackway might be underestimated by more than 25%. Such a discrepancy can easily arise from how a track is measured or by being a transmitted feature misinterpreted as a surface trace.

Laboratory track simulations (Manning 1999, 2004) show that the length and width of a track can vary within an individual track, with the original surface track feature often less well defined than those transmitted. Overprint (Thulborn 1990) features can also alter track parameters in comparison to the original surface on which the animal left its trace. Given that dinosaur FL, and in some cases FW, are key parameters used to calculate hip height in calculations of speeds from trackways, here lies a problem (Alexander 1976; Lockley et al. 1983; Sanz et al. 1985; Thulborn 1989). What is the true track (foot) length?

The variation in FL is a function of the relative position of a track layer to the force bulb (MZD/failure envelope) transmitted within a volume of sediment. The deformation of sediment beneath a foot initially expands then diminishes with depth, and with it the track geometry and morphology alters with depth.

Edward Hitchcock (1858), as discussed earlier, made an enormous contribution to the early science of vertebrate ichnology by recognizing transmitted tracks. His "fossil volumes" (Hitchcock 1858) still provide useful data when comparing laboratory-simulated tracks with fossil tracks. The experimental tracks show clearly how deformed tracks are transmitted and distorted, as were many of Hitchcock's tracks (Fig. 12.10).

FL decreases from a–c, increases from c–d, and then decreases again from d–f (Fig. 12.10). The red line (Fig. 12.10) follows this trend and gives clues to that illusive parameter, true foot length, given FL is a function of true foot length. Tracks a–c represent the entrance and exit traces of the foot as it sunk into to the sediment forming the surface track. Track c, although the smallest, is closet to the original sediment-foot interface (the floor of the footprint) and might represent most closely the original foot length of the track maker. Tracks d and f are transmitted tracks, conforming to the behavior expected for a force bulb, increasing and then decreasing in size beneath impact of the foot (Fig. 12.10). Given that the surface track (track a) feature is nearly 33% larger than the potential true foot length (track c), there is significant room for error, depending on which track horizon was used to calculate foot length, hip height, and speed of the track maker.

The critical parameter of FL needs defining to be useful to ichnologists. FL was defined as the distance between the most anterior point and the most

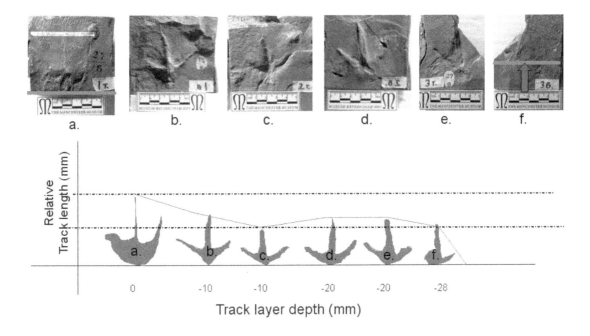

Figure 12.10. *Ornithopus gracilior* transmitted track ("fossil volume"). Tracks a–b, c–d, and e–f represent the top and bottom surfaces, respectively, of 3 layers that combine to create a single track volume. The plotted track outlines (a–f) show variation in FL with depth. The gray line (tracks a and f) marks the posteriormost point of the track, with the arrow (track f) indicating the degree of anterior travel of the track feature with increasing depth.

posterior point of the track, measured parallel to the long axis of the track (Leonardi 1987; Thulborn and Wade 1984). The existing definition unfortunately has the greatest variation in its parameters with depth because the MZD represents the maximum extent of a track. However, if theropod FL were defined as the length of digit III (Manning 1999, 2004), a closer approximation can be made of the true length of digit III, and in turn, a calculation of the total FL can be made. However, this relies on an individual recording FL from surface tracks that display a clearly definable middle digit (digit III) or clear skin impressions (Currie et al. 1990), or can be defined by the relative position of the track's force bulb (as discussed earlier).

Laboratory track simulations have demonstrated that FL variation is as much a function of sediment as it is of the track maker's foot morphology and size (Manning 1999, 2004). For example, the Table 12.1 charts the results from 11 laboratory-generated tracks (Manning 1999). The actual length of foot (template) is a known quantity, as are the sediment characteristics and condition at the time of track formation. The same amount of force was applied to the foot on each track run in the same step cycle. The percentage variation from actual foot (template) length was calculated for the maximum and minimum track size recovered from surface and subsurface tracks.

The variation in track parameters with depth was dependent on the moisture content and sediment used. Track 11 (Table 12.2) and Track 10 (Table 12.3) were generated by the same experimental setup and sediment, with the only variation being moisture content. The fine-grained sand used for both experiments was indented dry (moisture content 1%) and saturated (moisture content 30%).

Track 11 (saturated sediment) showed a large variation in FL from surface layer 1 to basal layer 11 (Table 12.2), from 17.5 to 9.0 cm, respectively, a variation of some 194.4%. The length (FL) of track 11 was then used to calculate h. According to Alexander (1976), h was 36–65.2 cm,

Track Simulation	MZD FL			Digit III		
	Maximum (cm)	Minimum (cm)	Variation Compared with Actual (%)	Maximum (cm)	Minimum (cm)	Variation Compared with Actual (%)
1	19	12.3	152–98.4	16.5	9.5	137–78.9
2	19	4.7	152–37.6	15.5	10.2	128.7–84.7
3	20	7.4	160–59.2	13.5	9.7	112.1–80.5
4	15.4	12.4	131.2–99.2	14.8	12.5	122.8–103.8
5	16.4	10	131.2–80	13.5	10	112.1–83
6	17.8	4.5	132.4–36	15.2	12.5	126.2–103.8
7	18.5	14.8	148–118.4	15.3	13.4	127–111.2
8	24.7	14.5	197.6–116	18.2	11	151.1–91.3
9	23.6	6.3	188.8–50.4	16.4	9.2	136.12–76.4
10	21.5	7.5	172–60	16.3	10.2	135.3–84.7
11	17.6	9	140.8–72	14.3	11.9	118.7–98.8

Table 12.1. *Variation in Maximum Zone of Deformation (MZD) Track Length (FL) and Digit III Compared with True Track Length (Length of Foot Template) from Individual Track Simulations*

Note.—The actual length of the foot template was 12.5 cm. The actual length of digit III was 12.1 cm.

Thulborn (1990) gives h as 40.5–99.97 cm, and Lockley et al. (1983) give h as 45–105 cm—a potential variation of 291.6%, depending on which track layer was used. This potential variation is clearly not restricted to laboratory-generated tracks but is present in fossil tracks (Fig. 12.10).

Track 10 (dry sediment) also exhibited a large variation in FL from surface layer 1 to basal layer 11 (Table 12.3), from 21.5 to 7.5 cm, respectively, a variation of some 286.6%. The length (FL) of track 10 was then used to calculate h. According to Alexander (1976), h ranged 30–81.6 cm, Thulborn (1990) gives h as 33.7–122.5 cm, and Lockley et al. (1983) give h as 37.5–129 cm—a potential variation of 430%, which depends on which track layer and h estimate method are used.

If a fossil trackway exposed relatively deep, transmitted tracks, equivalent to layer 11 of track simulation 10 (Table 12.3), and the FL was used to calculate h and in turn the speed at which the animal was traveling (hypothetical stride length of 2 m), then the following results are obtained:

1. Using lower limit estimation of hip height (Alexander 1976) of 30 cm, Equation 2 gives a speed of 10.29 m/s.
2. Using the upper limit estimation of hip height (Lockley et al. 1983) of 45 cm, Equation 2 gives a speed of 6.4 m/s.

However, if the fossil trackway exposed the equivalent of layer 5 of track simulation 10 (Table 12.3) and the FL was used to calculate h and in turn the speed at which the animal was traveling, the hypothetical stride length of 2 m now reduces to 1.86 m as a result of the increased foot length encroaching into the stride length (Fig. 12.11) would give the following results:

Track 11, Dry Fine-Grained Sand (E)			h in Relation to FL		
Layer	Depth (cm)	FL (cm)	Alexander (1976) h (cm) = FL4	Thulborn (1990) h (cm) = FL4.5–5.7	Lockley et al. (1983) h (cm) = FL5–6
1	0	16.3	65.2	73.35–92.91	81.5–97.8
3	−1.6	14.3	57.2	64.35–81.51	71.5–85.8
5	−3.2	17.5	70	78.75–99.97	87.5–105
7	−4.9	12.7	50.8	57.15–72.39	63.5–76.2
9	−6.3	10.6	42.4	47.7–60.42	53–63.6
11	−8	9	36	40.5–51.3	45–54

1. Using lower limit estimation of hip height (Alexander 1976) of 86 cm, Equation 2 gives a speed of 2.65 m/s.
2. Using the upper limit estimation of hip height (Lockley et al. 1983) of 129 cm, Equation 2 gives a speed of 1.65 m/s.

A fossil trackway, with tracks similar to those transmitted in track simulation 10 (Fig. 12.11), creates the potential difference in speed from varying track layer depths from 1.65 m/s to 10.29 m/s, depending on which layer and hip height values are applied.

By comparing the MZD foot length data from laboratory track simulations (Manning 1999, 2004), there is significant variation in track geometry between surface and transmitted tracks. Track simulation 10 has a variation of up to 286.6% and simulation 11 a variation of up to 194.4%. The only parameter altered between the 2 was moisture content; sediment, foot morphology, force, etc., remained constant. The high moisture content (31.2%) of track simulation 11 indicates that the MZD is reduced with higher moisture contents; effectively, the sediment's bulk density is higher as a result of the increased moisture filling the pore spaces, increasing the shear strength. It is clear that a controlling factor in the formation of track features is the re-

Table 12.2. *Variation in Hip Height (h) Due to Depth and Method Chosen for Estimating Hip Height (h) for Laboratory Track Simulation 11*

Abbreviation.—FL, foot length.

Table 12.3. *Variation in Hip Height (h) Due to Depth and Method Chosen for Estimating h for Laboratory Track Simulation 10*

Abbreviation.—FL, foot length.

Track 10, Dry Fine-Grained Sand (E)			h in Relation to FL		
Layer	Depth (cm)	FL (cm)	Alexander (1976) h (cm) = FL4	Thulborn (1990) h (cm) = FL4.5–5.7	Lockley et al. (1983) h (cm) = FL5–6
1	0	20.4	81.6	91.8–116.3	102–122.4
3	−1.7	20.1	80.4	90.45–114.6	100.5–120.6
5	−3.7	21.5	86	96.75–122.5	107.5–129
7	−5.3	16.5	66	74.25–94	82.5–99
9	−6.87	11.5	46	51.7–65.5	57.5–69
11	−8.43	7.5	30	33.7–42.7	37.5–45

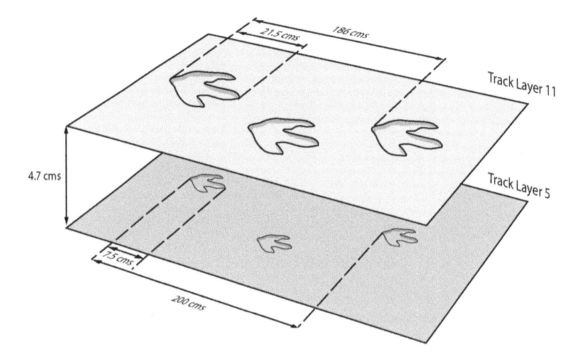

Figure 12.11. Cross section of trackway showing variation of FL and stride length with depth. Data from track simulation 10 (Table 12.3).

lationship between moisture and density that prevails in a volume of sediment at the time of indentation.

FW can also be used to calculate hip height (Thulborn and Wade 1984; Lockley et al. 1986; Thulborn 1990). However, FW also varies within individual tracks in relation to the relative position to the true track surface. Analysis of dinosaur tracks also indicated FL was more variable than FW (Thulborn and Wade 1984), and that FL was the least reliable indicator of foot size.

FL and FW data (Manning 1999) agreed with that of Thulborn and Wade (1984), thus supporting smaller variation in FW when compared with FL (Table 12.4). It is clear that FW rather than FL could be used when calculating the speed of a dinosaur on the basis of measured parameters from trackways. The use of FW can be further justified if a clear relationship between FW and hip height can be established.

Thulborn and Wade (1984) proposed an index of footprint size (SI), which they calculated by using the FW and FL of a fossil track (expressed in the same units of measurement):

$$SI = (FL \times FW)^{0.5}$$

Thulborn and Wade (1984) applied their index of footprint size to 57 fossil trackways (*Wintonopus*) from the Middle Cretaceous, Winton Formation, Queensland, Australia. They concluded that the footprint size index (SI), based on the sample the sample of 57 *Wintonopus* trackways, was the most reliable guide to estimating the size of track maker.

The data from laboratory-simulated tracks provides the unusual situation where both the track size from each layer of a single track volume and the actual foot length (template length) of the track maker were known

Track Simulation	MZD TL			MZD TW		
	Maximum (cm)	Minimum (cm)	Variation Compared with Actual FL (%)	Maximum (cm)	Minimum (cm)	Variation Compared with Actual FL (%)
1	19	12.3	152–98.4	19.4	16	157.1–129.6
2	19	4.7	152–37.6	16	12.3	129.6–99.6
3	20	7.4	160–59.2	17	5.5	137.7–44.6
4	15.4	12.4	131.2–99.2	15.9	14	128.8–113.4
5	16.4	10	131.2–80	19.7	8.6	159.6–69.7
6	17.8	4.5	132.4–36	16.6	3.2	134.5–25.9
7	18.5	14.8	148–118.4	16.3	13.6	132–110.2
8	24.7	14.5	197.6–116	18.3	14	148.2–113.4
9	23.6	6.3	188.8–50.4	18.4	6	149–48.6
10	21.5	7.5	172–60	19.2	12	155.5–97.2
11	17.6	9	140.8–72	20.2	10.5	163.6–85.1

Table 12.4. *Variation in Maximum Zone of Deformation (MZD) Track Length (TL) and Track Width (TW) Compared with the True Track Length within Individual Track Simulations*

Note.—The actual foot length template T8 is 12.5 cm. The actual foot width template T8 is 12.4 cm.

quantities (Manning 1999). This makes it possible to test whether the index of footprint size (SI) more closely reflects the foot geometry of the track maker—in this case, a prosthetic theropod dinosaur foot.

The index of footprint size (SI) was applied to both the maximum (SI.1) and minimum (SI.2) MZD FL and MZD FW data from Manning (1999) (Table 12.4). The SI was calculated using the following equations:

$$SI.1 = (\text{maximum MZD FL} \times \text{maximum MZD FW})^{0.5}$$
$$SI.2 = (\text{minimum MZD FL} \times \text{minimum MZD FW})^{0.5}$$

The SI results for the maximum (SI.1) and minimum (SI.2) were not as close to actual SI foot size as predicted by Thulborn and Wade (1984) (Table 12.5). However, if the SI.1 and SI.2 values were treated as maximum and minimum values from a trackway and fed back through the SI formulae, a different picture emerged. SI.1 and SI.2 in effect provide an index of footprint size (SI) for an individual track (SI.3):

$$SI.3 = (SI.1 \times SI.2)^{0.5}$$

The estimated value of SI.3 provided a closer estimate to the original track maker's foot size (Table 12.5). Because the true foot length and foot width were known for the template that were used in the laboratory simulations, it was also possible to calculate the true index of footprint size SI (true SI), for comparison with SI.1, SI.2, and SI.3 (Table 12.5). SI.3 represented the footprint size index for all track layers from an individual track and could be used to estimate hip height from a multitiered fossil or laboratory-simulated trackway (if all track layers were recoverable).

Table 12.5. *Index of Footprint Size for Laboratory-Simulated Tracks Based on Data from Table 12.4*

Abbreviations.—FL, track length; FW, track width.
* FL = 12.5 cm; FW = 12.4 cm.
† True SI of T8 = (12.5 × 12.4)0.5 = 12.45 cm.
‡ True SI of T8 = (12.5 × 12.4)0.5 = 12.45 cm.
§ True SI of T8 = (12.5 × 12.4)0.5 = 12.45 cm.

Template*	Track Simulation (cm)		
	SI.1†	SI.2‡	SI.3§
1	19.2	14.03	16.41
2	17.45	7.6	11.52
3	18.44	6.38	10.85
4	15.65	13.18	14.36
5	19.97	9.27	12.9
6	17.19	3.79	8.07
7	17.37	14.19	15.7
8	21.26	14.25	17.41
9	20.84	6.15	11.32
10	20.32	9.49	13.9
11	18.86	9.72	13.54

The index of footprint size for the MZD FL (SI.1) showed a percentage variation from the true foot length of 125.2%–170.1%, compared with 36%–197.6% variation in FL. The index of footprint size for the minimum MZD FL (SI.2) showed a percentage variation from the true foot length of 30.6%–114.9%, compared with 25.8%–162.9% variation in FL. However, the revised method for index of footprint size SI.3, using the values for SI.1 and SI.2, gave a percentage variation of only 64.8%–139.8% compared with the true foot length.

The application of the index of footprint size (Thulborn and Wade 1984) to the data from this study suggests that it provides an estimate for the percentage variation in track size in a sequence of tracks. If the trackway is a series of transmitted track features, with tracks represented by the maximum size of the MZD, the h generated will be too high and the speed calculated from the trackway too low. However, if the tracks are from a horizon that represents the minimum development of the MZD, it is possible that h will be underestimated, and the speed calculated would be too high. By combining the laboratory-simulated track data from SI.1 and SI.2 to calculate SI.3 (Table 12.5), a closer estimate of the track size can be made, allowing nearer estimates of h. The FL generated using SI.1 (Table 12.5) gave an average of 18.6 cm, and for SI.2, the average FL was 9.8 cm; however, SI.3 gave an average FL of 13.27 cm, the closest to the true SI FL (12.45 cm). The index of footprint size (SI) is only accurate if the tracks from which the measurements are taken closely resemble in size and proportion the original track maker's foot size parameters. However, by using the revised technique of SI.3 (Table 12.5), it may be possible to estimate h from transmitted tracks and subsequently calculate the speed from a trackway more accurately.

The speed at which a dinosaur was traveling is an important variable to assess if the potential of the 3D preservation of tracks is to be fully utilized. The speed at which an animal travels directly affects the time a foot remains on the ground (duty factor) and the intensity and distribution of the load transmitted in that given time. Thulborn (1990) suggested it was impossible to calculate the duty factor for a dinosaur directly from its trackway; however, a 3D approach to track subsurface deformation could possibly alter this. The anterior displacement of track features with increasing depth (Fig. 12.11) is a result of the physical properties of sediment and the dynamic load encountered during track formation. The distribution of pressure (load) over the sole of a foot correlates with the resultant subsurface track relief. If the track and associated features allow an estimate of the conditions prevailing at the time of track formation, the subsurface deformation could be coupled with the calculated speed of a trackway to enable an estimate of the duty factor. The subsurface relief (contours) of the fossil track could provide a means to reconstruct the dynamic pressure distribution over the sole of a dinosaur's foot. This, coupled with a known speed, might provide insight into the amount of time a dinosaur's foot remained on the ground (duty factor) during locomotion.

Summary

It is clear that the use of FL in calculations of speed from a trackway (Alexander 1976) should account for transmitted track features because they can potentially vary speed estimates by 10-fold. The use of dinosaur tracks in comparative multivariate studies should be restricted to surface track features for comparison with other surface track features. The inclusion of transmitted tracks in such studies invalidates any taxonomic or osteological relationships inferred as a result of the disparity between surface and subsurface track morphology. Any multivariate study based on morphological variability in tracks and trackways can only be viable if the 3D variability of track morphology is understood. Future multivariate studies must approach the task of understanding the 3D components of a track before valid comparison can be made with other tracks within a 3D framework.

This chapter has opened a possible ichnotaxonomic can of worms. The shifting sands of time have disguised so much of the process of track formation and preservation that potentially very little of a track makers foot morphology might be faithfully locked in stone (Manning 2004). Vertebrate ichnotaxa should reflect the morphological differences resulting from behavior, not the affinity of an alleged track maker or artifacts of track formation and preservation (Manning 2004). What is clear is that the interpretation of fossil tracks requires the application of more robust quantitative methodologies. I hope that if an illusive trackway of a trotting or running *T. rex* is found in the future, its documentation and interpretation will not fall into the potential traps discussed in this chapter. Who knows—the ichnospecies could even be named after a hadrosaur!

Acknowledgments

I thank Peter Larson, Neal Larson, and Bob Farrar of the Black Hills Institute of Geologic Research (BHIGR) for organizing the 100 Years of *Tyrannosaurus rex* Symposium in Hill City (2005). The BHIGR personnel were perfect hosts and provided access to their wonderful collections. A special thanks to Chris Ott, who was the instigator of the whole *rex* symposium. Many thanks to Whitey Hagadorn (Amherst College) for access to the Hitchcock fossil track collection (Pratt Museum). Also many thanks to Amherst College for their generous grant from the General Eastman Fund to assist in my research trip to the Pratt Museum. I thank the University of Sheffield for a Home/EC Bursary that made the research possible, and also Mike Romano and Martin Whyte. Many thanks to Emma Schachner for permission to use the line drawing of *T. rex* and to Richard Hartley for redrawing Figure 12.11. Thanks to Emma Finch for squeezing and manipulating data into Petrel. Many thanks to my wife and daughters, who permit me the time to undertake this research. Finally, a special thanks to Marion Zenker (BHIGR), who kept chasing this chapter and finally found a strategic place to insert a rocket for me to get it finished.

References Cited

Alexander, R. M. 1976. Estimates of speeds in dinosaurs. *Nature* 261: 129–130.

———. 1996. *Tyrannosaurus* on the run. *Nature* 379: 121.

Alexander, R. M., Dimery, N. J., and Kerr, R. F. 1985. Elastic structures in the back and their role in galloping in some mammals. *Journal of Zoology* 183: 125–146.

Allen, J. R. L. 1989. Short paper: fossil vertebrate tracks and indenter mechanics. *Journal of the Geological Society, London* 146: 600–602.

———. 1997. Subfossil mammalian tracks (Flandrian) in the Severn Estuary, S. W. Britain: mechanics of formation, preservation and distribution. *Philosophical Transactions of the Royal Society, London B* 352: 381–518.

Baird, D. T. 1957. Triassic reptile faunales from Milford, New Jersey. *Bulletin of the Museum of Comparative Zoology* 117: 449–520.

Biewener, A. A. 2002. Biomechanics: walking with tyrannosaurs. *Nature* 415: 971–973.

Boussinesq, J. 1883. *Application des potentials à l'étude de l'équilibre et du mouvement des solides élastiques*. Gauthier-Villars, Paris.

Carrano, M. T., and Hutchinson, J. R. 2002. Pelvic musculature of *Tyrannosaurus rex* (Dinosauria: Theropoda). *Journal of Morphology* 253: 207–228.

Calladine, C. R. 1969. *Engineering Plasticity*. Pergamon Press, London.

Clark, B. D. 1988. Mechanics of the Hindlimb of Bobwhite Quail Running and Landing on Substrates of Unpredictable Stiffness. Ph.D. thesis, University of Chicago.

Clark, J., and Alexander, R. M. 1975. Mechanics of running by quail (*Coturnix*). *Journal of Zoology, London* 176: 87–113.

Cracraft, J. 1971. The functional morphology of the hindlimb of the domestic pigeon, *Columbia livia*. *Bulletin of the American Museum of Natural History* 144: 175–268.

Currie, P. J., Nadon, G., and Lockley, M. G. 1990. Dinosaur footprints with skin impressions from the Cretaceous of Alberta and Colorado. *Canadian Journal of Earth Sciences* 28: 102–115.

Duncan, H. 1831. An account of the tracks and footprints of animals found impressed on sandstone in the quarry of Corncockle Muir in Dumfries-shire. *Transactions of the Royal Society Edinburgh* 11: 194–209.

Duncan, W. J. 1953. *Physical Similarity and Dimensional Analysis*. Arnold, London.

Farlow, J. O. 1981. Estimates of dinosaur speeds from a new trackway site in Texas. *Nature* 294: 747–748.

———. 1989. Ostrich footprints and trackways: implications for dinosaur ichnology. P. 243–248 in Gillette, D. D., and Lockley, M. G. (eds.). *Dinosaur Tracks and Traces.* Cambridge University Press, Cambridge.

Farlow, J. O., Gatesy, S. M., Holtz, T. R., Hutchinson, J. R., and Robinson, J. M. 2000. Theropod locomotion. *American Zoology* 40: 640–663.

Gatesy, S. M. 1990. Caudofemoral musculature and the evolution of theropod locomotion. *Palaeobiology* 16: 170–186.

———. 1991. Hind limb scaling in birds and other theropods: implications for terrestrial locomotion. *Journal of Morphology* 209: 83–96.

———. 1995. Functional evolution of the hind limb and tail from basal theropods to birds. P. 219–234 in Thomason, J. J. (ed.). *Functional Vertebrate Morphology in Vertebrate Paleontology.* Cambridge University Press, Cambridge.

———. 2001. Skin impressions of Triassic theropods as records of foot movement. *Bulletin of the Museum of Comparative Zoology* 156: 137–149.

———. 2003. Direct and indirect track features: what sediment did a dinosaur touch? *Ichnos* 10: 91–98.

Gatesy, S. M., and Biewener, A. A. 1991. Bipedal locomotion: effects of speed, size and limb posture in birds and humans. *Journal of Zoology, London* 224: 127–147.

Gatesy, S. M., Middleton, K. M., Jenkins, F. A., and Shubin, N. H. 1999. Three-dimensional preservation of foot movements in Triassic theropod dinosaurs. *Nature* 399: 141–144.

Gatesy, S. M., Shubin, N. H., Neil, H., and Jenkins, A. 2005. Anaglyph stereo imaging of dinosaur track morphology and microtopography. *Paleontologia Electronica* 8(1): 10p.

Haubold, H. H. 1971. Ichnia amphibiorum et reptiliorum fossilium. In Kuhn, O. (ed.). *Handbuch der Paläoherpetologie, Part 18.* Gustav Fisher Verlag, Stuttgart.

———. 1984. *Saurierfährten.* 2nd ed. A. Ziemsen Velag, Wittenberg Lutherstadt.

Hill, R. 1971. *The Mathematical Theory of Plasticity.* Clarendon Press, Oxford.

Hitchcock, E. 1858. *Ichnology of New England. A Report on the Sandstone of the Connecticut Valley, Especially Its Fossil Footmarks.* Wm. White, Boston, MA.

Horner, J. R., and Lessem, D. 1993. *The Complete T. rex.* Souvenir Press, New York.

Hutchinson, J. R. 2001. The evolution of pelvic osteology and soft tissues on the line to extant birds (Neornithes). *Zoological Journal of the Linnaean Society* 131: 123–168.

———. 2004. Biomechanical modelling and sensitivity analysis of bipedal running ability. II. Extinct taxa. *Journal of Morphology* 262: 441–461.

Hutchinson, J. R., and Garcia, M. 2002. *Tyrannosaurus* was not a fast runner. *Nature* 415: 1018–1021.

Jacobson, R. D., and Hollyday, M. 1982. A behavioural and electromyographic study of walking in the chick. *Journal of Neurophysiology* 48: 238–256.

Johnson, R. E., and Ostrom, J. H. 1995. The forelimb of *Torosaurus* and an analysis of the posture and gait of ceratopsian dinosaurs. P. 205–218 in Thomason, J. J. (ed.). *Functional Vertebrate Morphology in Vertebrate Paleontology.* Cambridge University Press, Cambridge.

Johnson, W., Sowerby, R., and Venter, R. D. 1982. *Plane-Strain Slip-Line Fields for Metal Deformation Processes.* Pergamon Press, Oxford.

Jones, T. D., Farlow, J. O., Ruben, J. A., Henderson, D. M., and Hillenius, W. J. 2000. Cursoriality in bipedal archosaurs. *Nature* 406: 716–718.

Karafiath, L. L., and Nowatzki, E. A. 1978. *Soil Mechanics for Off-road Vehicle Engineering*. Clausthal, Aedermannsdorf, Switzerland.

Leonardi, G. 1987. *Glossary and Manual of Tatrapod Footprint Ichnology*. Conselho Nacional de Desenvolvimento Cientifico e Technologico, Brazil.

Lockley, M. G., and Hunt, A. P. 1994. A track of the giant theropod dinosaur *Tyrannosaurus* from close to the Cretaceous/Tertiary Boundary, northern New Mexico. *Ichnos* 3: 213–218.

———. 1995. *Dinosaur Tracks and Other Fossil Footprints of the Western United States*. Columbia University Press, New York.

Lockley, M. G., King, M., Howe, S., and Sharp, T. 1996. Dinosaur tracks and other archosaurs footprints from the Triassic of South Wales. *Ichnos* 5: 23–41.

Lockley, M. G., Young, B. H., and Carpenter, K. 1983. Hadrosaur locomotion and herding behaviour: evidence from footprints in the Mesaverde Formation, Grand Mesa coal field, Colorado. *Mountain Geologist* 20: 5–14.

Manning, P. L. 1999. Dinosaur Track Formation, Preservation and Interpretation: Fossil and Laboratory Simulated Track Studies. Ph.D. thesis, University of Sheffield, England, UK.

———. 2004. A new approach to the analysis and interpretation of tracks: examples from the Dinosauria. P. 93–123 in McIlroy, D. (ed.). *The Application of Ichnology to Palaeoenvironmental and Stratigraphic Analysis*. Geological Society, London, Special Publication 228.

Manning, P. L., Payne, D., Pennicott, J., and Barrett, P. 2006. Dinosaur killer claws or climbing crampons? *Royal Society Biology Letters* 2(1): 110–112.

Margetts, L., Smith, I. M., Leng, J., and Manning, P. L. In press. Parallel three-dimensional finite element analysis of dinosaur trackway formation. *Numerical Methods in Geotechnical Engineering*. Graz University of Technology.

McGowan, C. 1999. *A Practical Guide to Vertebrate Mechanics*. Cambridge University Press, Cambridge.

McKee, E. D. 1947. Experiments on the development of tracks in fine cross-bedded sand. *Journal of Sedimentary Petrology* 17: 23–28.

McMahon, T. A. 1984. *Muscles, Reflexes, and Locomotion*. Princeton University Press, Princeton, NJ.

Milán, J. 2003. Experimental Ichnology—Experiments with Track and Undertrack Formation Using Emu Tracks in Sediments of Different Consistencies, with Comparisons to Fossil Dinosaur Tracks. M.Sc. thesis, University of Copenhagen.

Milán, J., Clemmensen, L. B., and Bonde, N. 2004. Vertical sections through dinosaur tracks (Late Triassic lake deposits, East Greenland)—undertracks and other subsurface deformation structures revealed. *Lethaia* 37: 285–296.

Padian, K. 1995. Form versus function: the evolution of a dialectic. P. 264–277 in Thomason, J. J. (ed.). *Functional Vertebrate Morphology in Vertebrate Paleontology*. Cambridge University Press, Cambridge.

Padian, K., and Olsen, P. E. 1984. Footprints of the Komodo Monitor and the trackways of fossil reptiles. *Copeia* 3: 662–671.

———. 1989. Ratite footprints and the stance and gait of Mesozoic theropods. P. 51–56 in Gillette, D. D., and Lockley, M. G. (eds.). *Dinosaur Tracks and Traces*. Cambridge University Press, Cambridge.

Parrish, M. J. 1986. Locomotor adaptations in the hindlimb and pelvis of the Thecodontia. *Hunteria* 1(2): 1–35.

Paul, G. S. 1998. Limb design, function and running performance in ostrich-mimics and tyrannosaurs. P. 257–270 in Pérez-Moreno, B. P., Holtz, T. J., Sanz, J. L., and Moratalla, J. (eds.). *Aspects of Theropod Paleobiology. Gaia: Revista de Geociencias, Museu Nacional de Historia Natural, Lisbon*, 15.

Pemberton, S. G., and Gingris, M. K. 2003. The Reverend Henry Duncan (1774–1846) and the discovery of the first fossil footprints. *Ichnos* 10: 69–75.

Romer, A. S. 1923. The pelvic musculature of saurischian dinosaurs. *Bulletin of the American Museum of Natural History* 48: 605–617.

———. 1927. The pelvic musculature of ornithischian dinosaurs. *Acta Zoologica* 8: 225–275.

———. 1956. *Osteology of the Reptiles*. University of Chicago Press, Chicago.

Sanz, J. L., Moratalla, J. J., and Casanovas, M. L. 1985. Traza icnologica de un dinosaurio iguanodontido en el Cretácio inferior de Cornago (La Rioja, España). *Estudios Geológicos* 41: 85–91.

Sarjeant, W. A. S. 1974. A history and bibliography of the study of fossil vertebrate footprints in the British Isles. *Palaeogeography, Palaeoclimatology, Palaeoecology* 16: 265–378.

———. 1990. A name for the trace of an act: approaches to the nomenclature and classification of fossil vertebrate footprints. P. 299–307 in Carpenter, K., and Currie, P. J. (eds.). *Dinosaur Systematics: Approaches and Perspectives*. Cambridge University Press, Cambridge.

Smith, M. J. 1981. *Soil Mechanics*. Longman Scientific and Technical, London.

Tarsitano, S. 1983. Stance and gait in theropod dinosaurs. *Acta Palaeontologica Polonica* 28: 251–264.

Thulborn, T. 1981. Estimated speed of a giant bipedal dinosaur. *Nature* 292: 273–274.

———. 1982. Speed and gaits of dinosaurs. *Palaeogeography, Palaeoclimatology, Palaeoecology* 38: 227–256.

———. 1989. The gaits of dinosaurs. P. 39–50 in Gillette, D. D., and Lockley, M. G. (eds.). *Dinosaur Tracks and Traces*. Cambridge University Press, Cambridge.

———. 1990. *Dinosaur Tracks*. Chapman and Hall, London.

Thulborn, T., and Wade, M. 1979. Dinosaur stampede in the Cretaceous of Queensland. *Lethaia* 12: 275–279.

———. 1984. Dinosaur trackways in the Winton Formation (mid-Cretaceous) of Queensland. *Memoirs of the Queensland Museum* 21: 413–517.

———. 1989. Footprint as a history of movement. P. 51–56 in Gillette, D. D., and Lockley, M. G. (eds.). *Dinosaur Tracks and Traces*. Cambridge University Press, Cambridge.

Tucker, M. E., and Burchette, T. P. 1977. Triassic dinosaur footprints from South Wales: their context and preservation. *Palaeogeography, Palaeoclimatology, Palaeoecology* 22: 195–208.

Vargas, A. O., and Fallon, J. F. 2005. Birds have dinosaur wings: the molecular evidence. *Journal of Experimental Zoology* 304: 1–5.

Viera, L. I., and Torres, J. A. 1995. Anàlis comparativo sobre dos rastros de Dinosaurios Therepodus: forma de marcha y velocidad. *Munibe* 47: 53–56.

Walker, A. D. 1977. Evolution of the pelvis in birds and dinosaurs. P. 319–358 in Andrews, S. M., Miles, R. S., and Walker, A. D. (eds.). *Problems in Vertebrate Evolution*. Linnaean Society Symposium Series 4.

Witmer, L. M. 1995. The extant phylogenetic bracket and the importance of reconstructing soft tissue in fossils. P. 19–33 in Thomason, J. J. (ed.). *Functional Morphology in Vertebrate Paleontology*. Cambridge University Press, Cambridge.

ATLAS OF THE SKULL BONES OF *TYRANNOSAURUS REX*

13

Peter Larson

Introduction

Tyrannosaurus rex was described by Osborn in a series of papers at the beginning of the last century (1905, 1906, 1912, 1916). His work was primarily based on 3 specimens: the holotype (AMNH 973, now CM 9379), the holotype for *Dynamosaurus* (AMNH 5866, now BMNH R7994), which was synonymized with *T. rex* in 1906, and AMNH 5027, which provided much of the description of the skull. Interestingly, the designation of AMNH 5027 has recently come under scrutiny; it may actually represent a second, unnamed species of *Tyrannosaurus* (P. Larson this volume). Molnar's (1991) treatise on the skull added greatly to Osborn's earlier descriptive work, and it also contained a section on arthrology. Molnar studied LACM 23844, AMNH 5027, MOR 008, and SDSM 12047. In 1998 and in this volume, Molnar increased our understanding of skull mechanics and musculature. This was followed by Brochu's (2003) description of FMNH PR2081, including computed tomography of the articulated skull, presented significant new data and interpretations. Several other important contributions to the skull of tyrannosaurids include Carr (1999) on craniofacial ontogeny, Currie (2003) on cranial anatomy of tyrannosaurid dinosaurs (particularly *Gorgosaurus* and *Daspletosaurus*), Currie et al. (2003) on skull structure, Hurum and Sabath (2003) on a comparison of *Tarbosaurus* and *Tyrannosaurus*, Carr and Williamson (2004) on tyrannosaurid diversity, and Carr et al. (2005) on a description of a new tyrannosaurid from Alabama.

Although several new *Tyrannosaurus rex* skulls are now available for study, by far the most significant is the disarticulated, undistorted skull of BHI 3033 (see N. L. Larson this volume). For the first time, the individuals bones of the skull can be described and illustrated for *Tyrannosaurus rex*. This specimen was discussed by Hurum and Sabath (2003) in their comparison of *Tyrannosaurus rex* and *Tarbosaurus bataar*. Smith (2005) also utilized BHI 3033 in his work describing the heterodont dentition of *T. rex*. This subject was also discussed in some detail in Osborn (1916), Molnar (1991), and Brochu (2003). P. Larson (this volume) has utilized this specimen in his discussion of an avianlike kinetic palate in *T. rex*.

Here, I present the first in-depth supplemental description of the skull bones and illustrate them in multiple views online at https://www.iupress.indiana.edu/media/tyrannosaurusrex/, based primarily on BHI 3033, with observations of other specimens.

Dermal Skull Bones

Premaxillae: The premaxillae of BHI 3033 are typical for Tyrannosauridae and for *Tyrannosaurus rex*, bearing 4 alveoli. Details not discussed by Osborn (1905, 1906, 1912), Molnar (1991), Brochu (2003), and Hurum and Sabath (2003), but visible in BHI 3033, include a scalloped region of ridges on the articular surface shared by the right and left premaxillae. This scalloped region is centrally located dorsoventrally and is at the anterior border of the suture. Although nearly 80% of this suture is smooth, ligaments attached to the scalloped area could have restricted lateral movement along this surface, in effect bonding the 2 premaxillae into a single structural unit. Below this scalloped area, the premaxillae separate, leaving a cleft in the articulation. The significance of this cleft is unclear, and it is not seen in the articulated FMNH PR2081 (Brochu 2003), a robust morphotype (P. Larson this volume).

The right premaxilla shows erosion surrounding the anterior portion of the fourth alveolus. This erosion is associated with the deposition of spongy bone. Osteomyelitis is the probable cause of this pathology.

The palatal surfaces of the premaxillae also bear a feature analogous to the interdental plates Osborn 1912) of the maxillae and dentaries. These interdental plates are found at the junctions of the alveoli. They are depressed from the palatal surface of the premaxillae and separated from one another at the base by large nutrient foramina.

The articular surface of the premaxillary-maxillary suture is smooth, interrupted (in the center of the lateral aspect) only by the opening for the subnarialis foramen (Carr 1999; Brochu 2003), whose border is shared by the 2 elements. The premaxillary-nasal articulation is also quite smooth, allowing dorsoventral movement while restricting anteroposterior movement. It is probable that the premaxillae unit was at least passively kinetic, movable relative to the maxillae and nasals, providing shock-absorbing benefits for the premaxillary teeth.

MAXILLAE: The maxillae of BHI 3033 bear 11 alveoli. This number is shared with MOR 980, although to date, all other *Tyrannosaurus rex* specimens have 12 (P. Larson this volume). The 2 maxillae of BHI 3033 articulate near the front of the palate in a series of overlapping ridges not mentioned by Molnar (1991) or Brochu (2003). This is similar to the condition in *Tarbosaurus bataar* (Hurum and Sabath 2003).

In *Tyrannosaurus*, the antorbital fossa is a very deep depression in even the smallest individuals (i.e., BHI 4100 and LACM 2345). The ventral border of the antorbital fossa and the antorbital fenestra are coincident along much of the ventral border of the antorbital fenestra.

In most tyrannosaurs, like *Gorgosaurus* (TCM 2001.89.1; Carr 1999), *Daspletosaurus* (RTMP 91.36.500; Currie 2003), *Appalachiosaurus* (RMM 6670; Carr et al. 2005), and *Nanotyrannus* (CMNH 7541, BMRP 2001.4.1; Bakker et al. 1988; Larson in press), the antorbital fossa is shallower than *T. rex*, and a thin ridge of bone (the ventral antorbital maxillary ridge) rises along the dorsal margin of the posteroventral extension of the maxilla. It extends well past the last alveolus and passes under the jugal where it articulates to the maxilla. In *Gorgosaurus*, *Daspletosaurus*, *Appalachiosaurus*, and *Nanotyrannus*, the ventral border of the antorbital fenestra and the antorbital fossa are not coincident. The constriction at the base of the ventral antorbital maxillary ridge forms the ventral border of the antorbital fossa, and the top of the ridge is the ventral border of the antorbital fenestra (Larson in press).

The maxillary fenestra (the second antorbital fenestra of Osborn 1912) in *Tyrannosaurus* BHI 3033) and *Tarbosaurus* (Hurum and Sabath 2003) contacts the anterior and posterior border of the antorbital fossa. An additional opening between the maxillary and antorbital fenestrae is found on both maxillae of BHI 3033, but is not seen in other specimens of *T. rex*. A similar opening is seen in a *Tarbosaurus* specimen, ZPAL MgD-I/4, described by Hurum and Sabath (2003). This opening may be the result of pre-depositional weathering and breakage of an extremely thin area of bone.

A long, narrow, and deep abrasion is incised into the lateral aspect of the left maxilla. This gouge is oriented dorsoventrally and is directly above the third maxillary tooth near the center of the mass of the maxilla. The abrasion measures 12 cm long by 2 cm wide and is approximately 1 cm deep at the center of the scar. There is evidence of new bone growth, especially near the ventral end of the gouge, demonstrating that this mark was made some days or weeks before death.

Several perforations mar the anterior palatal shelf of the right maxilla. It could be that these are a result of postmortem weathering or a fungal infection. They do not seem to have been caused by osteomyelitis because cancellous bone is not evident.

The nasal-maxillary suture is highly scalloped on BHI 3033. Interlocking fingers of bone prevent any anterioposterior movement along this articulation. There is, however, the possibility of lateral movement, with the nasal-maxillary suture acting as a hinge, allowing the ventral portion of the maxilla to swing outward. If such movement were to occur, the interlocking ridges of the maxilla–maxilla palatal contact would have spread apart. Such kinetic movement would enhance the shock-absorbing abilities of the maxillae and teeth.

Movement is also possible at the jugal-maxillary suture. Here the bones are joined by a tongue (jugal) and groove (maxilla) joint. This joint allows anterioposterior slippage.

LACHRYMALS: One of the characters uniting *Tyrannosaurus* with *Tarbosaurus*, but separating it from other tyrannosaurs, such as *Daspletosaurus*, *Gorgosaurus*, and *Albertosaurus*, is the absence of a corneal process on the lachrymal (Carr 1999; Carr and Williamson 2004). This character also separates *Tyrannosaurus* from *Nanotyrannus* (Larson in press). An additional character is the shape of the lachrymal: it is an inverted L shape in *Tyrannosaurus*, as it is in *Tarbosaurus*, and T shaped in *Albertosaurus*, *Gorgosaurus*, *Daspletosaurus* (Carr et al. 2005), and *Nanotyrannus* (Larson in press).

In addition to the pneumatic "lateral foramen" mentioned by Molnar (1991), the lachrymals of BHI 3033 possess 3 additional pneumatic openings. The first of these pneumatopores lies anterior to the lateral foramen, near the front of the lachrymal. It is on the dorsal margin of the antorbital fenestra and opens ventrally. In medial view, the lachrymals of BHI 3033 expose 2 more pneumatic foramina. The medial lachrymal pneumatopore (Larson in press) opens anterior to and at the dorsal margin of a thin ridge that descends diagonally across the vertical ramus. This ridge terminates at the ventral, anterior border of the vertical ramus. Finally, a fourth pneumatopore (also medial) may be found at the front of the horizontal ramus. This pneumatopore opens anteriorly.

The lachrymals barely touch the ascending ramus of the maxilla, thus allowing movement. However, the nasal-lachrymal articulation locks quite firmly, with a somewhat scalloped surface on the medial aspect of the horizontal ramus of the lachrymal and a cleft on the lateral aspect of the horizontal ramus that receives the lachrymal process of the nasal (Hurum and Sabath 2003). The posterior surface at the junction of the horizontal and vertical rami is somewhat ball shaped. This ball inserts into a socket in the anterior surface of the frontals, suggesting possible prokinesis, or lifting of the muzzle (Larson and Donnan 2002).

NASALS: As is the case with all tyrannosaurids, the nasals of BHI 3033 are fused. This fusion, as well as lateral arching and the consistent thickness of the bone (2 cm or more over nearly the entire length of the nasals), provides an extremely strong structure for the dissemination of stress developed during feeding (see Rayfield 2004). The nasal-frontal suture is a tongue (frontal) and groove (nasal) joint that allows fore-and-aft movement and would not preclude prokinesis.

POSTORBITALS: The medial aspects of the postorbitals show a shallow central depression, or fossa. Other *Tyrannosaurus* specimens (BHI 4812, MOR 555, and MOR 980) also show this feature. Much deeper fossae are seen in *Gorgosaurus* (TCM 2001.89.1), *Daspletosaurus* (Currie 2003), and *Nanotyrannus* (Larson in press).

Two paired, isolated osteoderms were found associated with the skull of BHI 3033 (Larson et al. 1998). These dermal elements are analogous to the "postorbital rugosity" noted as fused to the postorbitals of some robust *Tyrannosaurus* specimens by Molnar (1991), Larson (1994), and Brochu (2003). *Tyrannosaurus* specimens with fused postorbital rugosities include MOR 008, FMNH PR2081, BHI 4182, and UWGM 181 (NS 1565.26). On BHI 3033, these loose postorbital rugosities articulate with the postorbitals on the lateral surface directly above the orbit.

The anterior surface of the postorbitals (above the orbit) and the posterior surface of the horizontal ramus of the lachrymals articulate with pyramidally shaped osteoderms, or horns. These horns were found as isolated elements in BHI 3033. They were noted by Brochu (2003) as fused to the postorbitals in FMNH PR2081 and were thought to be part of the postorbital rugosity of Molnar (1991). It is clear from BHI 3033 (a set of isolated horns was also found with MOR 1125) that they are independent dermal elements. These horns bridge the gap between the postorbital and the lachrymal.

The postorbital-frontal suture is interfingered and allows no movement. The joint between the postorbital and the jugal is smooth and planar, allowing diagonal slippage. The postorbital-squamosal joint is an expanding tongue and groove, which allows fore-and-aft separation, although kinesis at this point is difficult to reconstruct.

SQUAMOSALS: The central portions of the anterior aspect of the squamosals are perforated by a large pneumatopore (most clearly seen in BHI

4100). This pneumatopore is also present in *Tarbosaurus* (Hurum and Sabath 2003), but it is absent in *Gorgosaurus*, *Nanotyrannus* (Larson in press), and *Daspletosaurus* (Currie 2003). Just anterior of center, the lateral ventral borders of the V-shaped squamosals are marked by a deep fold. This fold corresponds to a deep concavity on the squamosal process noted by Molnar (1991) on the quadratojugal. These 2 surfaces could have been the origin and insertion for a ligament connecting the 2 elements (see Quadratojugals, below). The articulation of the squamosal with the exoccipital is fairly flat and may have allowed limited sliding motion. The squamosal-quadrate joint is well developed and is a double ball-in-socket (quadrate-in-squamosal) joint. The 2 balls are side by side and are connected by a saddle that is reversed in the squamosal. This joint gives great flexibility fore and aft, but limits lateral movement.

JUGALS: Both jugals have pathological, healed puncture wounds (Larson 2001). These may have been the result of face-biting behavior described by Tanke and Currie (1998). The injury to the left jugal is a circular hole, 3.5 cm by 2.5 cm, that penetrates at a point near the origin of the cleft that receives the quadratojugal (here the jugal is approximately 4 cm thick). In medial aspect, the perforation emerges, leaving an opening 2.7 cm by 1.5 cm. The injury to the right jugal is located just anterior of center of the ascending ramus, at the ventral termination of the postorbital-jugal joint. The lateral opening measures 2 cm by 2.5 cm. The medial exit is larger (3.5 cm by 6.5 cm) and somewhat triangular in shape. Molnar (1991) showed that there is pneumatic sinus in this area (also seen in BHI 3033) that separates the lateral layer from the medial layer of bone. The larger exit hole may be explained by the thinness of that medial layer.

The jugal-ectopterygoid joint is smooth, allowing palatal kinesis (see Larsson this volume). The articulation of the jugal and the quadratojugal is also smooth, marked by the cleft mentioned above. A horizontal ridge on the lateral aspect of the jugal process of the quadratojugal fits into this cleft. Fore-and-aft movement of the quadratojugal along this joint, with the firmly attached quadrate (see below) rocking on the double ball-and-socket squamosal-quadrate joint described above, would produce streptostyly.

QUADRATOJUGALS: The left quadratojugal is isolated, and the right one is firmly attached to the quadrate. Although Molnar (1991, p. 161) presumed that "no articulation with the squamosal existed," there might have been a ligamental attachment (see Squamosals, above) that joined the 2 elements. This ligament could have acted as a tensile spring, returning the quadrate-quadratojugal to rest position after a streptostylic extension. The quadratojugal-quadrate articulation is somewhat edentate, creating a nonmovable joint. This suture also accounts for the firm attachment of the right quadratojugal to the right quadrate in BHI 3033 and other disarticulated skulls (i.e., MOR 1125).

QUADRATES: The quadrates of BHI 3033 are both preserved with a small notch (1 cm in diameter) on the dorsal edge of the squamosal process. This notch lies just anterior to the articulation with the quadrate. The notch probably corresponds to a much larger notch (several centimeters in diameter) found in *Gorgosaurus*, *Nanotyrannus* (Larson in press), *Albertosaurus*, and *Daspletosaurus* (Currie 2003).

The quadrate-pterygoid joint surfaces are smooth, allowing streptostyly (hinted at by Molnar 1991).The proximal end has a double articular surface (see Squamosals, above). This is similar to the situation in *Gorgosaurus* (TCM 2001.89.1) and *Daspletosaurus* (Currie 2003), whereas *Albertosaurus* (BHI 6234) and *Nanotyrannus* have only a single ball-in-socket articulation. Although all tyrannosaurs exhibit the capacity for streptostyly, it seems that it was accomplished in different ways.

The double condylar surface of the quadrates at the quadrate-articular joint is oriented in such a way that it forms a kind of screw on the surface of the main joint for the opening and closing of the jaws. As the jaws opened, this screw would, in effect, widen the gape of the jaws by forcing the articulars away from each other. Likewise, this screw would bring the rear of the jaws back together again as the jaws closed (Molnar 1991).

Palatal Complex

PTERYGOIDS: Both pterygoids are complete. The articulation with the quadrate is discussed above. The rear of the palatal plate wraps over the basisphenoid, limiting posterior movement. Although the dorsal surface of the palatal plate of the pterygoid is fairly smooth, little movement probably occurred at the ectopterygoid-pterygoid joint, and the folding of the quadrate process of the pterygoid blocked the ectopterygoid from moving posteriorly relative to the pterygoid. The folded posterior pterygoid-palatine joint allowed fore-and-aft movement but restricted lateral movement. The smooth anterior pterygoid-palatine joints allow fore-and-aft movement. The vomerine process of the pterygoid is a smooth vertical blade that touches the palatine on its lateral surface and the vomer on its medial surface The pterygoid-vomer suture is bounded by lateral extensions of the bifurcate stem. These 1-cm-deep lips limit forward movement of the pterygoid relative to the vomer. See Larsson (this volume) for a discussion of palatal kinesis.

EPIPTERYGOIDS: The epipterygoids articulate on the anterior surface of the vertical quadrate process of the pterygoids. The epipterygoids of BHI 3033 are similar to those described for *Daspletosaurus* (Currie 2003). The angle of the inverted V in BHI 3033 is, however, slightly greater than that of *Dasplatosaurus*, and the concavity at the base, mentioned by Currie (2003), is much deeper in *Tyrannosaurus rex* to such an extent that the 2 arms of the V are actually bifurcated. The more medial wing of the epipterygoid butts against the anterior medial edge of the quadrate process of the pterygoid, continuing the curve, established by the medial edge of the quadrate process. This butt joint restricts movement (other than folding) at the epipterygoid-pterygoid joint. The more dorsal epipterygoid-laterosphenoid articulations are smooth and would allow movement during palatal kinesis.

ECTOPTERYGOIDS: The hook-shaped ectopterygoids of BHI 3033 are perforated with a large pneumatopore typical of theropods (Molnar 1991). Carr et al. (2005) noted that the thick lip that bounds the pneumatopore was a character for *Tyrannosaurus*. The anterior limb of the ectopterygoid contacts the ventral medial surface near the center of the jugal. The smooth ectopterygoid-jugal joint is critical for palatal kinesis (see Larsson this volume).

PALATINES: The palatine was incompletely described by Molnar (1991). Although both palatines are complete in BHI 3033, they were somewhat weathered before burial, making a complete assessment difficult. BHI 4100, however, does include an exceptionally well-preserved right palatine that aids in the interpretation of BHI 3033. This palatine has a large pneumatopore on the central lateral surface near the articulation with the maxilla, which opens into a vast chamber. There is also a deep pneumatic fossa directly anterior to this pneumatopore and of near-equal size. The palatines of *Tyrannosaurus* are much more inflated than those of *Daspletosaurus* (Currie 2003; Carr et al. 2005), *Albertosaurus* (Carr et al. 2005), *Gorgosaurus* (TCM 2001.89.1), and *Nanotyrannus* (BMRP 2002.4.1). However, their similarity to *Tarbosaurus* (Hurum and Sabath 2003) could indicate that this inflation is a function of size.

The palatine-maxilla joint is slightly grooved, with corresponding low tongues on the opposing bones. This joint would allow limited movement fore and aft and would not preclude lateral movement of the maxillae. The palatines also contact the anterior medial surface of the jugals and the ventral medial surface of the lachrymals in a smooth joint that would not restrict palatal kinesis or lateral movement.

VOMER: The vomer of BHI 3033 matches Molnar's (1991) reconstructed description. The anterior rhomboid plate in BHI 3033 is convex in dorsal aspect and concave ventrally. This rhomboid plate also sports a small healed puncture wound near the base (Larson and Donnan 2002).

Braincase, Including Skull Roof

BHI 3033 has a relatively complete braincase that is missing only the ventral portion of the basicranium (as a result of predepositional weathering). As in FMNH PR2081, the prefrontals are firmly affixed to the frontals (Brochu 2003). The subadult MOR1125 has loose prefrontals, and a mature individual of *T. rex* A (see N. L. Larson this volume) preserves both frontals as disarticulated elements. A healed puncture wound is located near the center of the supraoccipital crest of the left parietal. This 3-cm-diameter hole perforates what was originally 3 cm of dense bone. Immediately above the perforation, a 12 cm by 5 cm fragment of bone is missing from the edge of the supraoccipital crest, presumably the result of the same incident causing the perforation (Larson 2001; Larson and Donnan 2002).

Mandibles

ARTICULAR: Only the left articular was recovered from BHI 3033; it was tightly articulated to the left surangular. The articular surface of the exposed right articular-surangular suture shows a great deal of interfingering, eliminating the possibility of movement at this suture. Likewise, the nature of the articulation with the prearticulars reveals that the articular was firmly locked into position in life.

PREARTICULARS: The prearticulars thin anteriorly to approximately 2 mm in lateral dimension for the final third of their length, although they measure an average of 8 cm in width. Both of the prearticulars show small (several millimeters), possibly pathologic perforations near their anterior terminations.

ANGULARS: The angulars of BHI 3033 match Molnar's (1991) description. The left angular shows some minor remodeling on the anterior most surface, corresponding to a puncture wound (where it articulates) on the left surangular. The spatulate posterior medial surface articulates with the surangular dorsally and the prearticular ventrally. Anteriorly, the angular thins dorsoventrally and is sandwiched between the dentary (laterally) and the splenial (medially). All these surfaces are smooth and allow fore-and-aft movement.

SURANGULARS: Both clam-shaped surangulars show healed puncture wounds. The right surangular has a large, remodeled perforation just anterior to and the same size as the surangular fenestra; a second perforation is on the anterior edge of this surangular. The left surangular is even more pathologic, with 4 large perforations. All show evidence of healing. They are rounded and thickened at the edges by remodeling. This is not unusual for *Tyrannosaurus*, and perforations of this large, thin element are found in nearly every preserved surangular. FMNH PR2081 (Larson 2001), LACM 23844 (Molnar 1991), AMNH 5027 (Osborn 1916), MOR 555 (Horner and Lessem 1993), MOR 008, MOR 980, BHI 4812, and BHI 6230 all sport extra perforations, thought to be evidence of face biting (Tanke and Currie 1998; Larson 2001; Rothschild and Molnar this volume).

DENTARIES: The dentaries each have 13 alveoli. They also both exhibit pathologies. The right dentary has a healed puncture, and a healed puncture and tear near the posterior margin. There is also a 4.5 cm by 1.5 cm erosion located 5 cm below the fifth alveolus. This erosion on the right dentary shows no healing or spongy bone and may be evidence of a fungal infection. The left dentary shows a 4 cm by 1 cm abrasion located 5 cm below the 13th alveolus. This abrasion does show evidence of new bone growth.

The skull of *Tyrannosaurus* (and other theropods) are often reconstructed with the symphysis of the lower jaws widely separated, sometimes by as much as 30 cm (i.e., LACM 23844). This seems to be an attempt to force the teeth of the lower jaws to occlude with those of the upper. Of course, theropods are not crocodiles, and their teeth did not occlude in life. Rather, the teeth of the lower jaws pass medial to those of the upper jaws as the jaws close. Evidence for this is presented by Molnar (1991, p. 143), who noted, "The medial face of the maxilla bears a number of shallow depressions . . . assumed to have accommodated the tooth crowns [of the dentary] when the mouth was closed, as in alligatoroids." Thus, the symphysis of the lower jaws were joined by a ligamentous attachment that allowed only limited movement for the purpose of absorbing the shock of teeth hitting bone.

SPLENIALS: Both splenials of BHI 3033 are preserved as separate elements showing the lateral surface which articulates against the dentary. A ridge on the dorsal anterior surface of the splenial fits into and becomes part of the Meckelian groove of Osborn (1912). This seems to indicate that there is no movement between the dentary and splenial. The dorsalmost portion of the splenial covers the coronoid, just posterior to the last alveolus of the dentary.

CORONOIDS: Only the right coronoid was recovered, but it was preserved in its entirety. Molnar (1991, p. 155) accepted Osborn's (1912) interpretation of the coronoid as "a small triangular plate laying at the antidorsal angle of the Meckelian fossa." However, because the coronoid passes behind the splenial in AMNH 5027 (and in other articulated *Tyrannosaurus* lower jaws), Osborn's interpretation was incorrect. The coronoid actually continues behind the splenial and overlaps the interdental plates of the dentary in what Osborn (1912) called the supradentary. Currie (2003), noting that the splenial and supradentary were joined, and called the entire structure a "coronoid-superdentary" in his description of *Daspletosaurus*. Hurum and Sabath (2003) referred to it as a "superdentary/coronoid."

Brochu (2003) provides a long discussion about whether this element is a single bone (the coronoid) or 2 separate bones (the coronoid and supradentary) that fused during ontogeny. His argument that no immature tyrannosaur individuals display these as separate elements seems to indicate that this is a single element and there is no supradentary. Certainly in *Tyrannosaurus* (BHI 3033, BHI 4812, MOR 1125, FMNH PR2081, etc.), *Daspletosaurus* (Currie 2003), *Gorgosaurus* (TCM 2001.89.1), and *Tarbosaurus* (ZPAL MgDI/4; Hurum and Sabath 2003) the "coronoid-supradentary" is a single bone. The same is true for other theropods like *Acrocanthosaurus* (NCSM 14345) and the allosauroid *Sinraptor* (Currie and Zhao 1993, p. 2055) where this element is illustrated as a "ceratobranchial." Even in the prosauropod *Plateosaurus* it has been interpreted by Galton (1990) as a single element: the coronoid. Likewise, Romer illustrates the medial aspect of the lower jaws of *Labidosaurus* (1956a), *Kotlassia, Diadectes, Bradysaurus, Python, Peloneustes, Dimetrodon,* and *Edaphosaurus* (1956b) as all possessing a coronoid with a thinning anterior projection that passes over the medial aspect of much of the tooth row—the same situation seen in theropods. It is for these reasons that the "coronoid-supradentary" should simply be referred to as the coronoid. The coronoid for BHI 3033 closely resembles that of *Tarbosaurus*, described by Hurum and Sabath (2003).

Conclusion

The disarticulated skull of *Tyrannosaurus rex* specimen known as BHI 3033 preserves many details not seen in other specimens. A close examination reveals details of the air sac system that invade or leave its impressions on many of the bones of the skull. Pathologies—evidence of disease or healed injuries—found on some of the cranial elements provide insight into behavior, such as intraspecific combat (face biting) and the commonality of certain pathogens such as osteomyelitis. And finally, the exquisite preservation of the cranial joint surfaces on BHI 3033 opens the door to a more complete analysis of the capacity for cranial kinesis in *Tyrannosaurus rex*.

Limited discussions of some aspects cranial kinesis in *T. rex* may be found in Larson and Donnan (2002) and Larsson (this volume).

Acknowledgments

I thank Neal Larson for taking the photographs of the individual elements and Larry Shaffer for photographs of the assembled skull and for turning them all into illustrations. The hours of labor involved in this process are many, and my talents leave much to be desired. Also worthy of my undying gratitude are my colleagues, who provided access to all the *T. rex* specimens involved in this research and who patiently taught me tyrannosaur osteology.

References Cited

Bakker, R. T., Williams, M., and Currie, P. 1988. *Nanotyrannus*, a new genus of pygmy tyrannosaur, from the Latest Cretaceous of Montana. *Hunteria* 1(5): 26.

Brochu, C. A. 2003. Osteology of *Tyrannosaurus rex*: insights from a nearly complete skeleton and high-resolution computed tomographic analysis of the skull. *Journal of Vertebrate Paleontology Memoir* 7: 1–138.

Carr, T. D. 1999. Craniofacial ontogeny in Tyrannosauridae (Dinosauria, Coelurosauria). *Journal of Vertebrate Paleontology* 19(3): 497–520.

Carr, T. D., and Williamson, T. E. 2004. Diversity of Late Maastrichtian Tryrannosauridae (Dinosauria: Theropoda) from western North America. *Zoological Journal of the Linnean Society* 142: 479–523.

Carr, T. D., Williamson, T. E., and Schwimmer, D. R. 2005. A new genus of tyrannosaurid from the Late Cretaceous (Middle Campanian) Demopolis Formation of Alabama. *Journal of Vertebrate Paleontology* 25(1): 119–143.

Currie, P. J. 2003. Cranial anatomy of tyrannosaurid dinosaurs from the Late Cretaceous of Alberta, Canada. *Acta Palaeontologia Polonica* 48(2): 191–226.

Currie, P. J., and Zhao, X. J. 1993. A new carnosaur (Dinosauria, Theropoda) from the Jurassic of Xinjian, People's Republic of China. *Canadian Journal of Earth Sciences* 30: 2007–2081.

Currie, P. J., Hurum, J. H., and Sabath, K. 2003. Skull structure and evolution in tyrannosaurid dinosaurs. *Acta Palaeontologia Polonica* 48(2): 227–234.

Galton, P. M. 1990. Basal Sauropodomorpha. P. 320–344 in Weishampel, D., Dodson, P., and Osm<oacute>lska, H. (eds.). *The Dinosauria*. University of California Press, Berkeley.

Horner, J. R., and Lessem, D. 1993. *The Complete T. rex*. Simon & Schuster, New York.

Hurum, J. H., and Sabath, K. 2003. Giant theropod dinosaurs from Asia and North America: skulls of *Tarbosaurus bataar* and *Tyrannosaurus rex* compared. *Acta Palæontologia Polonica* 48: 2, 161–190.

Larson, N. L., Farrar, R. A., and Shaffer, L. 1998. New information on the osteology of the skull of *Tyrannosaurus rex*. P. 68–69 in Wolberg, D. L., Stump, E., and Rosenberg, G. D. *Dinofest International Proceedings*. Academy of Natural Sciences, Philadelphia.

Larson P. L. 1994. *Tyrannosaurus* sex. P. 139–155 in Rosenberg, G. D., and Wolberg, D. L. (eds.). *Dino Fest*. Paleontological Society Special Publication No. 7.

———. 2001. Paleopathologies in *Tyrannosaurus rex* (in Japanese). *Dino Press* 5: 26–35.

———. In press. *The Case for Nanotyrannus*. Northern Illinois University Press, DeKalb.

Larson P. L., and Donnan, K. 2002. *Rex Appeal: The Amazing Story of Sue, the Dinosaur that Changed Science, the Law and My Life.* Invisible Cities Press, Montpelier, VT.

Molnar, R. E. 1991. The cranial morphology of *Tyrannosaurus rex*. *Palaeontographica Abteilung A* 217: 137–176.

Osborn, H. F. 1905. *Tyrannosaurus*, and other Cretaceous carnivorous dinosaurs. *Bulletin of the American Museum of Natural History* 21: 259–266.

———. 1906. *Tyrannosaurus*, Upper Cretaceous carnivorous dinosaur. *Bulletin of the American Museum of Natural History* 22: 281–296.

———. 1912. Crania of *Tyrannosaurus* and *Allosaurus*. *Memoir of the American Museum of Natural History* (n.s.) 1: 1–30.

———. 1916. Skeletal adaptations of *Ornitholestes, Struthiomimus, Tyrannosaurus*. *Bulletin of the American Museum of Natural History* 35: 733–774.

Rayfield, E. J. 2004. Cranial mechanics and feeding in *Tyrannosaurus rex*. *Proceedings of the Royal Society of London B* 271: 1451–1459.

Romer, A. S. 1956a. *A Shorter Version of the Vertebrate Body.* W. B. Saunders, Philadelphia.

———. 1956b. *Osteology of the Reptiles.* University of Chicago Press, Chicago.

Smith, J. B. 2005. Heterodonty in *Tyrannosaurus rex*: implications for the taxonomic and systematic utility of theropod dentitions. *Journal of Vertebrate Paleontology* 25(4): 865–887.

Tanke, D. H., and Currie, P. J. 1998. Head-biting behavior in theropod dinosaurs: paleopathological evidence. P. 167–184 in Pérez-Moreno, B. P., Holtz, T. J., Sanz, J. L., and Moratalla, J. (eds.). *Aspects of Theropod Paleobiology. Gaia: Revista de Geociencias, Museu Nacional de Historia Natural, Lisbon,* 15.

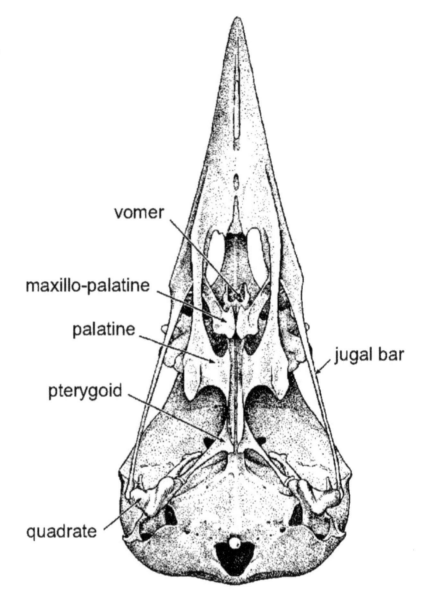

Figure 14.1. *Palatal aspect of **Corvus** with palatal bones labeled. Modified from Bock (1964, fig. 2B).*

PALATAL KINESIS OF *TYRANNOSAURUS REX*

14

Hans C. E. Larsson

Cranial kinesis is widespread throughout Gnathostomata. Although this clade is diagnosed by a cranial-mandibular joint, taxa as diverse as tiger sharks and tiger bitterns have complex cranial articulations that allow for intracranial kinesis. Only a few taxa have completely lost cranial kinesis, such as amphibians, mammals (with the exception of rabbits), turtles, and crocodiles. Cranial kinesis has at least 2 mechanical benefits. Many chondricthyans have hyostylic jaws that have a mobile articulation only with a hyomandibula, which in turn has a mobile articulation with the braincase. These articulations allow for jaw protrusion from the neurocranium. Most osteichthyan fishes have a similar splanchnocranial kinetic jaw suspension. However, their dermatocranial bones augment jaw kinesis with maxillae that are free from the circumorbital bones and mobile with respect to them and, in neoteleosts, a protrusible premaxilla (Westneat 2004). These dermatocranial joints allow the maxillae and premaxillae to be manipulated in complex 3-dimensional excursions and provide the range of jaw mechanics that reflect the high diversity of teleost fishes.

Tetrapods, on the other hand, appear to have started life on land with relatively akinetic skulls. Most Paleozoic and all modern amphibians appear to have little to no cranial kinesis. Synapsida continued this trend to the fused skulls exhibited by modern mammals. The akinetic mammalian skull may have been the result of providing a solid framework for powerful and complex masticatory muscles (reviewed in and references found in Carroll 1988). Presence of complex jaw musculature has been cited as one correlate for the evolution of the complex tooth morphologies present in mammals.

Reptiles are generally more conservative in tooth morphology. Most have simple conelike teeth that range from straight conical teeth to sharply recurved, bladelike teeth. Notable exceptions include the many herbivorous and durophageous taxa. The relatively simple teeth of many squamates contrasts with some of the forms of cranial kinesis that they present. Most squamates have a kinetic quadrate that hinges on the undersurface of the squamosal. Presence of a mobile quadrate is called streptostyly. A classic study on varanids illuminated the complex motions the quadrate makes during jaw opening and closing (Smith 1980). Quadrate rotation is translated into a series of other cranial motions that modify the position of the snout and palatal bones. Extreme versions of streptostyly are present in many snakes, where each quadrate can move independently of the other, providing a walking

Introduction

ratchet motion that helps tractor prey toward their esophagus with palatal teeth (Cundall 1983).

Modern birds also exhibit streptostylic quadrates. Birds use quadrate rotation to provide the forces required to move their upper jaws in a sagittal plane independently from their braincase. The upper jaw is hinged on either mobile joints or over thin bone bridges that can bend. The location and number of these joints vary, but the result is that the upper beak can move in either complex motions, as in nut-eating parrots, or shock absorption, as in woodpeckers (Beecher 1953, 1962; Bock 1964). The diversity of beak kinesis in birds is only beginning to emerge, but research suggests that it is broad (Zusi 1984).

The mechanical driver for bird beak kinesis is a rotary quadrate, similar to the streptostylic condition of most squamates, and a kinetic palate (Bock 1964). The pterygoids contact the quadrates and receive parasagittal forces from the quadrates to slide the pterygoid bones along a mobile joint over an elongate basipterygoid processes. The pterygoids contact the palatines, which in turn contact the vomer to translate this propalinal motion throughout the palate, and in turn to the maxillae and premaxillae (Fig. 14.1). The palatines and vomer have a kinetic contact with the parasphenoid. The distal quadrates also contact thin, elongate jugal bars to translate rotational vectors of the quadrates directly to the lateral margins of the maxillae. The mechanical forces are generated and controlled by a complex set of jaw and palatal muscles (Bock 1964).

Paleognathous and neognathous palates of modern birds have long been used to diagnose ratites from all other birds, respectively (McDowell 1948; Bock 1963; Gussekloo and Zweers 1999). Ratites are characterized by having large, robust palatal bones that appear more similar, in some respects, to a generalized reptile in morphology than the gracile and elongate neognathous palate. Major differences between the 2 palate forms include the presence of elongate posterolateral processes of the palatines to form the only contact with the pterygoid bones laterally in paleognaths. Neognaths have a more typical palatine-pterygoid contact in the midline of the palate. In both forms, the vomer contacts the palatines, but the vomer of paleognaths is relatively larger and more robust. Because of the osteological differences and some in vivo observations, many researchers have long suggested that paleognathous birds have unique cranial kinesis that is associated with their palatal morphology (Hofer 1954; Simonetta 1960; Bock 1963).

Gussekloo and colleagues (2001) examined differences between paleognathous and neognathous bird cranial kinetic patterns by calculating displacements of marked joints within intact bird heads. Displacements were calculated for a set of paleognathous and neognathous birds with closed bills and upper bills that were dorsally flexed approximately 10° from the horizontal plane. This range was used to estimate the linkage arrangements within the normal range of upper bill kinesis. The results suggested the kinetic motions within the entire set of paleo- and neognathous birds were qualitatively similar. The marked differences in palatal morphology between the 2 avian clades was not associated with any obvious difference in the linkages of bones within their skulls during palatal and upper jaw kinesis (Gussekloo et al. 2001). The similar kinetic patterns within paleognathous and neognathes suggest their kinetic skulls evolved from a common ancestor.

The common ancestry of modern bird cranial kinesis has been hypothesized before, and some have suggested that fossil forms, such as *Hesperornis*, exhibited rhynchokinetic skulls that may be ancestral to the diversity of modern bird cranial kinesis (Zweers et al. 1997; Zweers and Vanden Berge 1997).

The evolution of the avian kinetic skull from nonavian ancestors has never been explored. Part of the reason may be that adequately preserved material is rare, and there are significant challenges to reconstructing biomechanical hypotheses for fossil taxa. To date, the possibilities of cranial kinesis in dinosaurs have only been explored in *Iguanodon* (Norman 1984) and hadrosaurs (Cuthbertson 2005). These studies suggested a remarkable mechanism of a lateral excursion of the maxillae to efficiently process tough plant material. The immediate ancestry of birds lies within theropod dinosaurs (Gauthier 1986; Sereno 1999). No nonavian theropod has been hypothesized to have significant cranial kinesis. However, Molnar (1991) did acknowledge that *Tyrannosaurus rex* may have had limited streptostyly.

The purpose of this chapter is to explore a hypothesis that *T. rex* may have been capable of cranial kinesis, and to examine to what degree that kinesis was similar to the kinesis of birds. Previous workers have suggested that *T. rex* may have been capable of limited streptostyly (Molnar 1991). That streptostyly is evinced by the presence of an articular condyle on the proximal end of the quadrate. However, the quadrate is integrated into the lower temporal and palatal bones and requires further examination to describe its possible role in being streptostylic.

Anatomy of *T. rex* Palatal Kinesis

The most complete and well-preserved skull of *T. rex* is housed in the Black Hills Institute (BHI). BHI 3033 has been given the vernacular title of Stan. The head skeleton is lacking only a right articular, is virtually undistorted, and is completely disarticulated. More research casts of this specimen exist in the world than any other theropod, which has led Philip Currie to refer to this specimen as the Stan-dard of theropod cranial osteology (Currie, personal communication 2005).

Examination of the infratemporal region of BHI 3033 reveals a plausible case to begin modeling streptostyly for the specimen. Tyrannosaurs and many other theropods feature an arcuate anteroventral process on the quadratojugal (Fig. 14.2A). This process articulates laterally into a similarly arcuate prong and trough on the posterior ramus of the jugal. Although this type of slot joint is common in reptile skulls, the arcuate shape of the joint is of particular interest. The joint is in close proximity to the quadrate. The dorsal head of the quadrate features a saddled surface that articulates with the ventral surface of the squamosal with probably a synovial joint. A synovial joint between the quadrate and squamosal is present in most squamates, all birds, and even in early-stage crocodylian embryos (Larsson, in preparation). The similar anatomy found in nonavian dinosaurs suggests

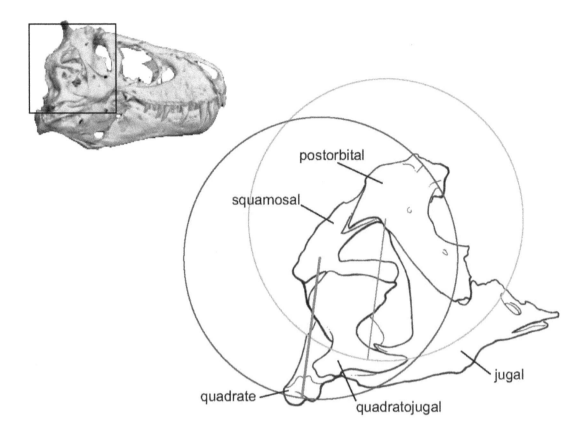

Figure 14.2. Right lateral aspect of *Tyrannosaurus rex* (BHI 3033) as reconstructed from digitized casts (A) and a detail of the right infratemporal region (B). Darker circle and radius represents the length of the quadrate and its distal excursion path. Lighter circle and radius are the same as the darker curves but translated to overlap the approximate axis of the anteroventral ramus of the quadratojugal.

the presence of a synovial joint in these taxa during life as well. If the quadrate is hypothesized to have at least limited streptostyly, then the distal quadrate must swing with a circular excursion with a radius equivalent to the length of the quadrate. If a circular excursion path is constructed with such a radius and shifted to overlap the quadratojugal-jugal contact, then the circle closely follows the arcuate path of the prong-and-trough articulation (Fig. 14.2B). The congruence between the quadrate's circular excursion and shape of the quadratojugal-jugal arcuate articulation suggests that if the quadrate were streptostylic, its motion could be accommodated by a mobile joint with the jugal. Moreover, the articulation between the quadratojugal and squamosal is never fused and is instead a simple broadly overlapping planar joint. The thinness of the quadratojugal in this region suggests that this articulation may also have been mobile. The possible mobility of the quadratojugal with respect to the squamosal and jugal effectively isolates any streptostylic motion of the quadrate to only the quadratojugal on the external surface of the skull.

Internally, the quadrate articulates with the pterygoid (Fig. 14.3). The articular surfaces between the quadrate and pterygoid are broadly overlapping with robust grooves and ridges that would probably have been rather immobile in life. The articulation is further stiffened by a modest groove on the ventrolateral margin of the quadrate flange of the pterygoid to receive the ventral edge of the pterygoid flange of the quadrate. The pterygoid articulates with the braincase on the basipterygoid processes. These processes are

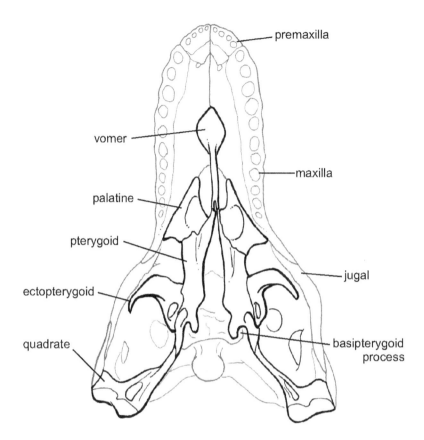

Figure 14.3. *Palatal aspect of **Tyrannosaurus rex** (BHI 3033) as reconstructed from digitized casts. Palatal bones are darkened from remainder of skull.*

elliptical and would have been capped with cartilage. The pterygoids fit over the basipterygoid processes with a mitt-shaped cotyle that is open posteriorly. The open end and loose fit of the articulation suggest some anteroposterior mobility that would have been osteologically unlimited in the anterior direction.

The pterygoids contact the ectopterygoids laterally and palatines anterolaterally. Both these articulations appear to be immobile if the force on the contact was from a posterior vector. The ectopterygoid rests in a saddle on the lateral surface on the pterygoid that would have limited any posterior motion of the ectopterygoid relative to the pterygoid. The posteromedial region of the ectopterygoid exhibits a robust process that butts into a receiving concavity near the anterior surface of the basipterygoid cotyle of the pterygoid. There is no evidence of the articulation retaining the ectopterygoid in place should the pterygoid have been displaced posteriorly. In fact, the articular surface is not well defined, suggesting that the ectopterygoid may have had some degree of mobility, perhaps swinging between somewhat dorsolateral to ventrolateral positions while remaining in contact with the pterygoid medially. Laterally, the ectopterygoids contact the medial surface of the jugals. This contact is does not exhibit any scarring on the jugal and suggests the capacity of some sliding motion between the 2 bones.

The palatines contact the pterygoids in a blind-ended prong in groove joint. The contact provides a buttress from which the pterygoid may have pushed the palatine anteriorly. However, like the ectopterygoid, there are

Palatal Kinesis

appears to be little osteology keeping the palatine in this tight articulation should the pterygoid be displaced posteriorly. Anteriorly, the palatines contact the maxillae in an approximately 30° angle posterolateral to the sagittal plane (Fig. 14.3). This contact is composed of broad striate ridges that extend the length of the contact and are parallel to one another. The delta-shaped profile of the anterior margin of both palatines wedges against the maxillae.

The vomer contacts the anterior ramus of the pterygoids within a tall peg-in-socket joint. The contact is sagittally elongate, with the concave pocket on the vomer receiving the anteriorly convex anterior ramus of the pterygoid. The contact appears braced to have received anteriorly directed forces from the pterygoids to move the vomer anteriorly. However, like the other palatal contacts, the pterygoid-vomer contact does not appear to have a joint with much strength to pull the vomer posteriorly.

Discussion

The palatal bones of *T. rex* do not appear to be rigidly connected to any other bone in the skull, with the exception of the quadrates. The relative isolation suggests some kinesis may have been possible. Palatal kinesis appears to have been limited in a horizontal plane and in a sagittal vector. On the basis of osteological anatomy, anterior motion seems plausible. The pterygoid-basipterygoid process articulation appears to have been capable of propalinal motion. The ectopterygoid-jugal contact shows no indication of sutures and suggests a dynamic articulation. The palatine-maxilla contact is delta shaped, and the surfaces contact along longitudinal striae that do not seem to have hindered a sliding motion. The vomer underlaps the palatal shelves of the maxillae without sutures or any other morphology that would suggest immobility.

Posterior motion would be necessary, of course, to return the palate to its posterior position. This motion does not have strong bone signatures and buttressing. However, if less force was inflicted on the palatal joints, they may have remained in contact with the aid of ligament attachments. If the quadrate was indeed capable of even limited streptostyly, then the anteroposterior excursion of the distal quadrate appears to be accommodated by relatively loose-fitting quadratojugal-squamosal-jugal contacts and a palate that appears to be capable of propalinal kinesis. All modern birds have streptostylic quadrates that drive a palate with propalinal kinesis. Although birds lack an ectopterygoid and have significantly different skeletal anatomies of the pterygoid, palatine, and vomer, their palatal kinesis appears similar to the kinesis hypothesized for *T. rex*. Although this study is in no way a test of palatal kinesis in *T. rex*, it does demonstrate that the osteological anatomy of the quadrate and palate appears to have been capable of accommodating a kinetic palate sliding in a propalinal vector. Further work is under way to examine this model with 3-dimensional kinematic models to determine what effects a kinetic palate may have in the skull (Larsson and Larson, in preparation). The snout probably did not move in a sagittal plane, as in birds, but perhaps other kinetics resulted.

The possibility of palatal kinesis in *T. rex* has significant evolutionary implications. Should at least some of the kinematic functions be homologous (i.e., found to be present in the common ancestral lineage) between *T. rex* and modern birds, then the kinetic bird palate will have had a lengthy history of origin. Such a deep history of bird cranial kinesis has not been proposed before and deserves attention. Previous authors have suggested early birds such as *Hesperornis* and *Parahesperornis* (Biahler et al. 1988) and even *Archaeopteryx* (Versluys 1912; Simpson 1946; Bock 1964) may have had an ancestral form of modern avian cranial kinesis. Others, however, have hypothesized that *Archaeopteryx* possessed an akinetic skull (Simonetta 1960; Beecher 1962). Although some workers had addressed the early stages of avian cranial kinesis within birds (Zweers and Vanden Berge 1997; Zweers et al. 1997), little attention has been paid to examining the evolutionary origin of avian cranial kinesis. Certainly, as is generally true for most paleontological questions, further work on well-preserved fossils will aid reconstructions of the evolutionary origins of avian cranial kinesis.

Acknowledgments

This work has been the product of many conversations on the topic with Pete Larson, Phil Currie, Jorn Hurum, and students in my lab (the Deep Time Specialists). This work was supported by funding from Canada Research Chairs, Natural Sciences and Engineering Research Council of Canada and Fonds québécois de la recherche sur la société et culture.

References Cited

Beecher, W. J. 1953. Feeding adaptations and systematics in the avian order Piciformes. *Journal of the Washington Academy of Science* 43: 293–299.
———. 1962. The bio-mechanics of the bird skull. *Bulletin, Chicago Academy of Science* 11: 10–33.
Biahler, P., Martin, L. D., and Witmer, L. M. 1988. Cranial kinesis in the Late Cretaceous birds *Hesperornis* and *Parahesperornis*. *Auk* 105: 111–122.
Bock, W. J. 1963. The cranial evidence for ratite affinities. P. 39–54 in Sibley, C. G., Hickey, J. J., and Hickey, M. B. (eds.) *Proceedings of the 13th International Ornithological Congress*. American Ornithologists' Union, Baton Rouge, FL.
———. 1964. Kinetics of the avian skull. *Journal of Morphology* 114: 1–42.
Carroll, R. L. 1988. *Vertebrate Paleontology and Evolution*. W. H. Freeman, New York.
Cundall, D. 1983. Activity of head muscles during feeding by snakes: a comparative study. *American Zoologist* 23: 383–396.
Cuthbertson, R. 2005. Reconstructing *Brachylophosaurus canadensis* (Hadrosaurinae): morphological revision and insight into hadrosaurs chewing. P. 18 in Braman, D. R., Therrien, F., Koppelhus, E. B., and Taylor, W. (eds.) *Dinosaur Park Symposium*. Special Publication of the Royal Tyrrell Museum, Drumheller, Alberta, Canada.
Gauthier, J. 1986. Saurischian monophyly and the origin of birds. P.1–55 in Padian, K. (ed.). *The Origin of Birds and the Evolution of Flight*. Memoirs of the California Academy of Science 8.
Gussekloo, S. W. S., Vosselman, M. G., and Bout, R. G. 2001. Three dimensional kinematics of skeletal elements in avian prokinetic and rhynchokinetic skulls determined by roentgen stereophotogrammetry. *Journal of Experimental Biology* 204: 1735–1744.

Gussekloo, S. W. S., and Zweers, G. A. 1999. The paleognathous pterygoid-palatinum complex. A true character? *Netherland Journal of Zoology* 49: 29–43.

Hofer, H. 1954. Neue Untersuchungen zur Kopf Morphologie. P. 104–137 in Portmann, A., and Sutter, E. (eds.). *Acta XI Congressus Internationalis Ornithologici*. Birkhauser Verlag, Basel.

McDowell, S. 1948. The bony palate of birds. I. The Paleognathae. *Auk* 66: 520549.

Molnar, R. E. 1991. The cranial anatomy of *Tyrannosaurus rex*. *Palaeontographica Abteilung A* 217: 137–176.

Norman, D. B. 1984. On the cranial morphology and evolution of ornithopod dinosaurs. *Symposium, Zoological Society of London* 62: 521–546.

Simonetta, A. M. 1960. On the mechanical implications of the avian skull and their bearing on the evolution and classification of birds. *Quarterly Review of Biology* 36: 206–220.

Simpson, G. G. 1946. Fossil penguins. *Bulletin of the American Museum of Natural History* 87: 1–100.

Sereno, P. C. 1999. The evolution of dinosaurs. *Science* 284: 2137–2147.

Smith, K. K. 1980. Mechanical significance of streptostyly in lizards. *Nature* 283: 778–779.

Westneat, M. W. 2004. Evolution of levers and linkages in the feeding mechanism of fishes. *Integrative and Comparative Biology* 44: 378–389.

Versluys, J. 1912. Das Streptostylie—Problem and die Bewegung im Schadel bei Sauropsida. *Zoologia Jahrbuch, Supplement* 15(2): 545–716.

Zusi, R. L. 1984. A functional and evolutionary analysis of rhynchokinesis in birds. *Smithsonian Contributions to Zoology* 395: 1–40.

Zweers, G. A., and Vanden Berge, J. C. 1997. Birds at geological boundaries. *Zoology* 100: 183–202.

Zweers, G. A., Vanden Berge, J. C., and Berkhoudt, H. 1997. Evolutionary patterns of avian trophic diversification. *Zoology* 100: 25–57.

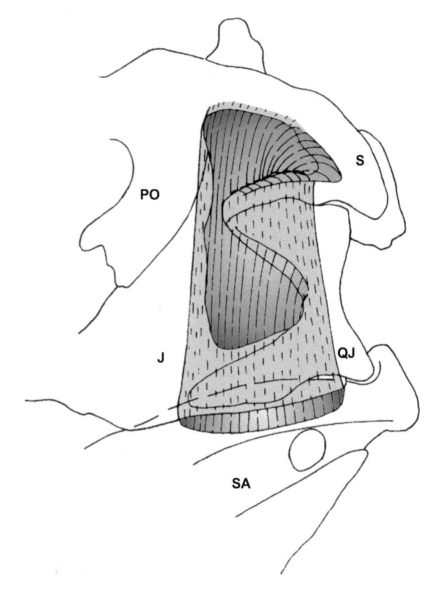

Figure 15.1. *Restoration of the M. adductor mandibulae externus superficialis et medialis of **Tyrannosaurus rex**. Abbreviations: J, jugal; PO, postorbital; QJ, quadratojugal; S, squamosal; SA, surangular.*

RECONSTRUCTION OF THE JAW MUSCULATURE OF *TYRANNOSAURUS REX*

Ralph E. Molnar

Introduction

This study is part of a series on the cranial morphology of *Tyrannosaurus rex* that originally formed my Ph.D. dissertation at the University of California, Los Angeles, in 1973. The descriptive and structural analytical parts have already been published (Molnar 1991, 2000) and present information relevant to this work. Since 1973, a total of 3 reconstructions of the cranial musculature of *T. rex* have appeared: Horner and Lessem (1993), Ushiyama (1995), and Farlow and Molnar (1995), the first 2 likely done independently of the 1973 dissertation. These reconstructions agree with that presented here. But no account of the reconstruction of the cranial musculature of *T. rex* has appeared since 1973, although such a reconstruction had appeared more than 80 years ago (Adams 1919). Paul (1988) published a reconstruction of the superficial musculature of *Allosaurus fragilis*, including the superficial jaw muscles, and briefly discussed the jaw musculature of theropods in general. The only other account of the disposition and function of the jaw musculature of a large theropod is that of Mazzetta et al. (2000) for *Carnotaurus sastrei*.

MATERIAL: Of the specimens of *T. rex* examined for the dissertation (given in Molnar 1991), only AMNH 5027, LACM 23844, and MOR 008 proved useful for this study. AMNH 600 (*Allosaurus fragilis*) and AMNH 5343 (*Gorgosaurus libratus*) also provided significant information. Further specimens were examined for this chapter: BHI 3033, BHI 4100, and BHI 6230, as well as casts of MOR 555, MOR 980, and ZPAL MgD-I/4. The use of the term *fenestra* follows that of Molnar (1991) in referring only to an aperture, not also to the excavation (fossa) that may surround cranial fenestrae.

METHODS: For modern organisms, description of the arrangement of muscles and their bony attachments is a matter of careful dissection: it is conceptually straightforward, if rather less so in practice. For extinct tetrapods, the reconstruction of musculature is more difficult (unless one is so lucky as to have available one of the extremely rare specimens in which some trace of the musculature is actually preserved). Important comments regarding the reconstruction of musculature in extinct organisms are given by Ostrom (1961). Before the 1970s, muscles of extinct organisms were reconstructed by analogy with the arrangement of muscles in selected modern analogs. Basically this involved conceptually locating the muscle attachments on the fossil with reference to selected points or landmarks of

Muscle	Origin	Insertion
Add. mand. ext. sup. med.	Smooth surfaces on squamosal and quadratojugal	Surangular facet
Add. mand. ext. prof.	Supratemporal fossa	Analogy with crocodilians
Add. mand. post.	Analogy with crocodilians and lizards	Analogy with crocodilians and lizards
Pseudotemporalis	Supratemporal fossa: analogy with birds and lizards	None
Pterygoideus ant.	Analogy with crocodilians	Analogy with crocodilians and lizards
Pterygoideus post.	Analogy with crocodilians	None
Intramandibularis	None	None
Dep. mand.	Smooth surfaces on paroccipital processes	Smooth concavities on articulars

Table 15.1. *Kinds of Evidence Used in Establishing Attachment Area*

the morphology of the appropriate bones; muscle scars, where they can be discerned, are the best landmarks (Table 15.1). Where muscle scars could not be discerned, muscles were assumed to attach to corresponding parts of homologous elements. If, for example, a muscle extended from the supratemporal fenestra to the dorsal margin of the surangular in the modern analog, then this was its course when reconstructed in the fossil (Table 15.1).

The critical feature is the selection of appropriate analogs. There seems little difficulty in extrapolating muscle structure among phylogenetically closely related forms—say, *Alligator mississippiensis* to *Crocodylus porosus*, and probably to forms as phyletically distant as *Sebecus icaeorhinus*. However, arbitrary decisions are involved where the muscle scars are indistinct, imperceptible, or substantially different in form. Thus, for morphologically different or more distantly related forms, this method becomes problematic. In attempting to extrapolate from *A. mississippiensis* to the domestic dog, for example, the number of arbitrary decisions becomes unmanageably large, the confidence in the result correspondingly small, so it must be concluded that the one is not an acceptable model for the other. Thus, the choice of a modern analog must rely on having a sufficient number of independent osteological features in common to give confidence that the structure in question—here the cranial skeleton—did in life function as analogs to some desired level of similarity. In the case of tyrannosaurs and crocodilids, these features include sharply pointed teeth, snouts long relative to the postorbital part of the skull, and diapsid postorbital cranial structure (probably more than a single feature).

After choosing a modern analog, the muscle attachments are located, and it is ascertained whether or not muscle scars (i.e., characteristic indications on the surfaces of the bones to which muscles attach) are present, and if present, the positions and forms of these scars. In my experience, muscle scars are often more easily recognized by touch (by changes in the surface texture of bone) than by sight. If scars are not present (or not obvious)—in other words, if the surface of the bone has a texture that is uniform across the area of attachment and continues without change into regions where other muscles or other tissues attach—then one can only approximate the attachment site by reference to nearby landmarks. If the modern analog is judiciously chosen, those scars seen on the fossil should be similar in position and form to those of the analog. Generally, not all of the attachment areas will have easily recognizable scars, and some approximation of the location of the areas by other landmarks will be necessary.

In principle, a second, logically independent course is to compare the forms of the individual muscle scars. To use them in an independent fashion requires comparison of the forms of the scars, rather than the pattern of their placement. For example, a straplike muscle will leave scars that are long and narrow. Therefore, to reconstruct the position of a straplike muscle, one needs to find relatively long, narrow scars for the origin and insertion. If scars of some other form are found, then either the muscle is placed in the wrong position, or it is incorrectly reconstructed as straplike. This method assumes that one already has some notion of the muscular form that can be used to identify the muscles that attached to the scars. Used in isolation, this method would indicate where muscles were present, but it would provide no clear guide as to their arrangement. In other words, attachment sites might be found on different elements, but which site on one element corresponds to which on the other remains unknown. As pointed out by McGowan (1986, and citations therein), it also suffers from the disadvantage that muscle scars are not always present or recognizable, even if the bony surfaces are well preserved. Thus, at best, this method can be used as a partial check on the previous one, but it cannot itself provide a reconstruction of the musculature.

So in reconstructing the musculature of an extinct tetrapod, some guidance from modern analogs is necessary. Lizards and crocodilians were widely considered as models for dinosaurs before the 1970s. The relationship of theropod dinosaurs to birds implies that these, too, may be potential models. However, the cranial structure of birds is much modified from those of large theropods, even those as closely related to birds as tyrannosauroids. The pattern of muscle scars may be compared with the pattern of muscles seen in birds, crocodilians, or lizards to determine which is more similar to that of *Tyrannosaurus rex*. For example, the M. adductor mandibulae externus profundus takes origin in lizards from the region of the posttemporal fenestra, but in crocodilians from the region of the supratemporal fenestra. In theropods, the posttemporal fenestra tends to be closed and lacking clear indication of muscle attachment. However, there are indications of muscle attachment around the supratemporal fenestra. Thus, here, crocodilians were selected as the more appropriate model. On the other hand, the crocodilian M. pseudotemporalis originates from the laterosphenoid ventral to the supratemporal fenestra. In some birds (e.g., *Cepphus grylle, Gallus domesticus*; Lakjer 1926), the M. pseudotemporalis originates anterior to the M. add. mand. prof. from the lateral side of the braincase in the region that is presumably homologous to the medial wall of the channel leading to the supratemporal fenestra in their theropod ancestor. Thus, the M. pseudotemporalis is located anterior to the M. add.

mand. prof. As described in the appropriate section, in *T. rex*, the subdivision of the supratemporal fenestra suggests that the M. pseudotemporalis arose from this region anterior to the M. add. mand. prof., and hence was located anterior to that muscle. Here the pattern of these birds (and some lizards) seems the more appropriate model. By means of such considerations, the jaw muscle pattern of crocodilians was chosen as a generally appropriate guide, although the patterns seen in lizards (and birds) also provide useful comparisons.

In the living crocodilians, as in other tetrapods, the jaw musculature sometimes leaves characteristic muscle scars. Fleshy attachments are generally onto smooth, flat, or nearly flat surfaces of bones. Such scars are sometimes set off from the general surface by a distinct rim or angulation (as in the case of the M. adductor mandibulae externus superficialis et medialis insertion) and sometimes not (as with the M. adductor mandibulae posterior origin). Tendinous attachments tend to arise from ridges along the surface of the bone (as with the crocodilian A tendon; Iordansky 1964) or from sharp ridges of the element (as with the surface tendon of the M. pterygoideus posterior). By using these indications of muscle attachment, together with the patterns of muscle origin and insertion in living crocodilians, the jaw musculature of *T. rex* has been reconstructed. Reconstructions of the individual muscles and the musculature as a whole are presented in a series of figures, whereas the text deals with the rationale for the reconstructions, the comparisons with living forms, and the form and position of the muscle scars.

McGowan (1986, and citations therein) found considerable variation in the pattern of muscles in the wings and hind limbs of birds. Specifically, he found that some muscles were developed to different degrees on different sides of the animal, or in some cases, even absent on one side, as well as differently developed in different individuals of the same species. Examination of lacertilian cranial material in the Northern Arizona University Quaternary Studies Program collection showed that muscle scars of similar form were present in comparable positions across a range of species (*Chamaeleo calyptratus*, *C. melleri*, *C. oustaleti*, *C. parsoni*, *Ctenosaura pectinata*, *Furcifer pardalis*, *Hydrosaurus amboensis*, *Iguana iguana*, *Polychrus gutturosus*, and *Tupinambis teguixin*). However, they may vary in details of form and degree of development, are more prominent on larger specimens, and are not seen on specimens smaller than 50 mm in length. The similarity of the muscle scars of both sides of the cranial skeletons of *T. rex*, examined together with considerations of feeding efficacy, suggests that individual variation was not a problem in this study. However, only 3 specimens of *T. rex* were examined, and so these results should not be extended to other specimens without appropriate examination.

The data on the origin and insertions of lacertilian jaw muscles are taken largely from Lakjer (1926), Bolk et al. (1938), and Oelrich (1956). Some of the attachments were verified in specimens of *Iguana iguana*, *Ctenosaura pectinata*, and *Varanus komodoensis*, but most of the data are from the literature. The data for the crocodilian jaw musculature attachments are taken largely from my own dissections of *Alligator mississippien-*

M. abductor mandibulae externus
M. abductor mandibulae externus supereficialis et medialis (M. add. mand. ext. sup. med.)
M. abductor mandibulae externus profundus (M. add. mand. ext. prof.)
M. abductor mandibulae posterior (M. add. mand. post.)
M. abductor mandibulae internus
M. pseudotemporalis
M. pterygoideus anterior
M. pterygoideus posterior
M. intramandibularis

Table 15.2. *Terminology Used in Description of Jaw Abductors*

sis and *Paleosuchus trigonatus*. This has been supplemented with data on other genera taken from Iordansky (1964).

TERMINOLOGY: The muscle terminology used is that of Tage Lakjer (1926) for crocodilians, but with slight modifications. I was unable to distinguish between the M. adductor mandibulae externus superficialis and the M. adductor mandibulae externus medialis in my dissections of *Alligator mississippiensis* and *Paleosuchus trigonatus*. Even in lizards, these parts fuse and are thus difficult to distinguish (Oelrich 1956; Fisher and Tanner 1970). These muscles do not appear to have separate attachment scars in *T. rex* (although they do in *Allosaurus fragilis*). Hence, I shall refer to this muscle as the M. adductor mandibulae externus superficialis et medialis. The structure of the pterygoid-quadrate region of the skull of *T. rex* differs considerably from that of the crocodilians, and as a result, it is not feasible to treat the M. pterygoideus as divided into 4 portions, as Lakjer did. Thus, only a M. pterygoideus anterior and a M. pterygoideus posterior will be distinguished here.

The terminology that will be used for the jaw adductors (abbreviations used follow the full name) is listed in Table 15.2.

Iordansky (1964) has published a thorough description of the jaw musculature and the associated tendons of the Crocodylia. Previous workers (Lakjer 1926; Anderson 1936) had noted the existence of the zwischensehne (a large tendon sheet associated with the insertion of the M. pterygoideus anterior, M. pseudotemporalis, and M. intramandibularis of crocodilians) but had not conducted the thorough, detailed investigation of the tendinous structure that was undertaken by Iordansky. Although reference will be made to these structures, a description of the tendons and their terminology will not be presented here because there is no evidence for their occurrence in *T. rex*.

Reconstructed Musculature

M. ADDUCTOR MANDIBULAE EXTERNUS SUPERFICIALIS ET MEDIALIS (FIG. 15.1): Among living lizards, these 2 portions are distinguishable (but sometimes with difficulty: Oelrich 1956; Nash and Tanner 1970; Avery and Tanner 1971). They usually arise from the ventral and internal surfaces of the supratemporal arch, and fibers also arise from the quadratojugal and the anterior surface of the quadrate (Lakjer 1926; Oelrich 1956; Fisher and Tanner

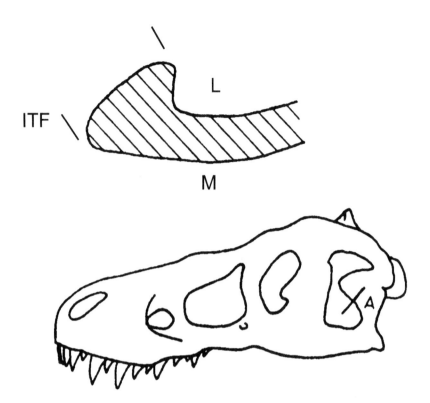

Figure 15.2. Cross section through the anterior margin of the squamosal (dorsal) process of the quadratojugal of **Tyrannosaurus rex** in a plane perpendicular to the anterior margin (as indicated by line A of the lower drawing). Anterior is to the left. The anterolateral-facing surface of the quadratojugal, between the diagonal bars, is that possibly giving origin to a part of the M. add. mand. ext. sup. med. Abbreviations: ITF, region of infratemporal fenestra; L, lateral surface; M, medial surface.

1970; Nash and Tanner 1970; Avery and Tanner 1971). Among living crocodilians, these 2 portions cannot be easily distinguished (cf. Iordansky 1964) and are here considered as a single muscle. In the crocodilians, this muscle originates from the ventral surface of the quadratojugal just above the infratemporal fenestra, with some fibers coming from the posterior wall (formed by the quadrate) of the channel ascending to the supratemporal fenestra.

In *Tyrannosaurus rex* (AMNH 5027 and LACM 23844), the concave anterior margin of the quadratojugal, which bounds the posterior extension of the ventral portion of the infratemporal fenestra, bears along the ventral margin of the upper ramus a smooth surface extending posteriorly from the margin of the fenestra over onto the lateral face of the element (Molnar 1991, pl. 1, fig. 5). This surface is posteriorly bounded by an abrupt rim (Fig. 15.2). In MOR 008, this rim bears rugosities, and in MOR 980, it is raised into a low but distinct lateral shelf. The rim separates the smooth surface adjacent to the fenestra from the roughened external bone surface. Uniquely, the quadratojugal of MOR 980 has a distinct rostroventrally facing facet on the lower margin of the upper ramus, which, like the lateral shelf, is more marked on the right element. Because these features are on the lateral wall of the adductor chamber, similar in position to the area of origin of this muscle in modern crocodilians, it is believed to represent part of the area of origin of the M. add. mand. ext. sup. med.

The central portion of the squamosal also posteriorly bounds the posterior extension of the upper portion of the infratemporal fenestra. Around the apex of this part of the fenestra, there is a smooth surface on the squamosal

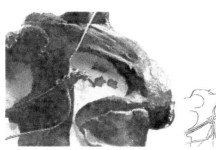

extending from the margin of the fenestra over onto the lateral face of the element (Fig. 15.3), similar to that of the margin of the quadratojugal. This region is bounded posteriorly by marked, low, obtuse ridges on both squamosals in AMNH 5027, as well as on those of MOR 008 and MOR 555. In BHI 3033, BHI 4100, and BHI 6230, the ridge on the dorsal ramus is developed into a sharp crest, forming the surface into a shallow ventrally facing trough. In MOR 555 (and ZPAL MgD-I/4, *Tyrannosaurus bataar*), the crest has become a prominent lateral shelf. Because of the similarity in form to the quadratojugal feature just described, and because of its appropriate location, these features are also considered to be part of the origin of the M. add. mand. ext. sup. med. The internal surface of the squamosal of MOR 4100 has a low, oblique ridge parallel to the ventral edge of the upper ramus, about 1 cm from the edge. This presumably marks the medial edge of the muscle attachment. These features suggest that some fibers, presumably of the M. add. mand. ext. sup. med., may also have originated from the medial surface of the quadratojugal-squamosal flange between the dorsal and ventral posterior extensions of the infratemporal fenestra. Or again, the muscle may have been separated into dorsal and ventral parts. A structure roughly similar to this surface is also found on the quadratojugal of *Allosaurus fragilis* (AMNH 600) and among the other tyrannosaurids, where it is not as marked. (Because distinct scars that can be interpreted as representing both the M. add. mand. ext med. and the M. add. mand. ext. sup. may be seen on the skull of *A. fragilis* in addition to this quadratojugal feature, the latter presumably does not mark the origin of one of these 2 portions.)

The proposed area of origin of the M. add. mand. ext. sup. med. is based on the recognition of features that appear to be muscle scars and are in an approximately homologous position (posterior part of the lateral wall of the adductor chamber) to the area of origin in living crocodilians. Thus, the M. add. mand. ext. sup. med. in *T. rex* is assumed to have arisen from the squamosal at the posterodorsal corner of the infratemporal fenestra, as well as from the internal surface of the supratemporal arch. This situation is essentially similar to that described in hadrosaurs (Ostrom 1961). Additional fibers may have originated from the medial surface of the squamosal-quadratojugal flange and from the anterior margin of the quadratojugal. These fibers would be shorter than those originating from the more dorsal infratemporal region and hence capable of lesser percentage exten-

Figure 15.3. *The origin scar of the M. adductor mandibulae externus superficialis et medialis on the squamosal of **Tyrannosaurus rex**, left and right sides. Insets show the attachment area in diagonal hatching. Some of this region on the right side appears to have been reconstructed in plaster. Photos of the Los Angeles County Museum of Natural History cast of AMNH 5027.*

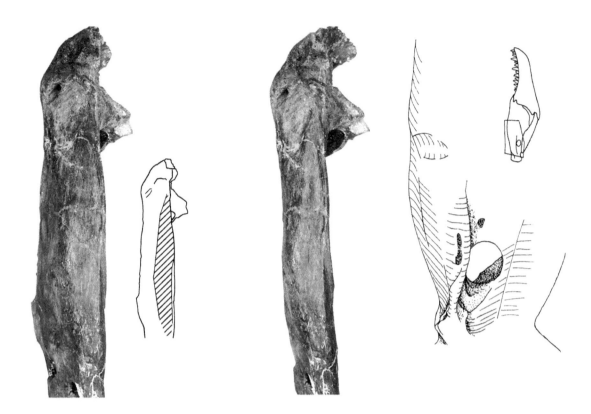

Figure 15.4. (A) Stereophotograph of the insertion scar of the M. adductor mandibulae externus superficialis et medialis on the surangular of *Tyrannosaurus rex*, LACM 23844. Dorsal view of the scar. (B) Lateral view of the same scar of the right surangular of LACM 23844. The small inset shows the location of the scar on the mandible.

sion, in keeping with the decreased distance between the areas of origin and insertion, relative to that of the more dorsally attaching fibers.

In the living crocodilians, the M. add. mand. ext. sup. med. inserts onto a flat facet on the dorsal margin of the surangular. This is located just anterior to the glenoid. Among the living lizards, fibers of these muscles insert both onto the bodenaponeurosis and the dorsal margin of the surangular, as well as onto adjacent bones (Lakjer 1926; Oelrich 1956; Fisher and Tanner 1970; Nash and Tanner 1970; Avery and Tanner 1971). The dorsal margin of the surangular of *T. rex* (AMNH 5027, BHI 3033, BHI 4100, BHI 6230, LACM 23844, and MOR 008) bears an anteroposteriorly elongate facet (Fig. 15.4). This facet is bounded anterolaterally by a low but distinct ridge and is composed of 2 almost-plane subfacets. The medial of these subfacets is almost horizontal in transverse section and slightly convex upward in parasagittal section. The lateral subfacet faces dorsolaterally and is similar to the medial in form. The angulation separating them becomes indistinct posteriorly so that the subfacets fuse together. This feature probably represents the area of insertion for the M. add. mand. ext. sup. med., and the existence of the subfacets suggests that this muscle had a bipartite structure anteriorly. However, the surangulars of BHI 4100 and MOR 008 do not show these subfacets; instead, this region is smoothly convex. In BHI 3033, BHI 4100, and MOR 008, the medial edge of the facet rises into a sharp crest, perhaps indicating the existence of a tendon.

This proposed area of insertion is based on the similarity of both the position and form of the putative muscle scar to those seen in modern crocodilians.

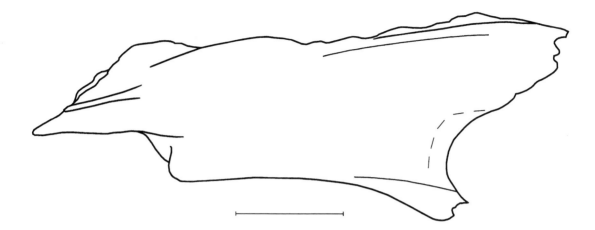

M. ADDUCTOR MANDIBULAE EXTERNUS PROFUNDUS (FIG. 15.5): In the living crocodilians, this muscle originates from around and within the supratemporal fenestra, whereas among the living lizards, it originates from the region of the posttemporal fenestra (Lakjer 1926; Oelrich 1956; Fisher and Tanner 1970; Nash and Tanner 1970; Avery and Tanner 1971). In *Tyrannosaurus rex*, the supratemporal fenestra is surrounded by a roughly bowl-shaped, smooth-surfaced supratemporal fossa, and the posttemporal fenestra is almost completely closed. This is similar to the condition in crocodilians, although there, the fossa is shallower and less bowllike in form. In *T. rex*, the supratemporal fossa is deepest over the parietals, shallowing abruptly in the frontal region. The parietal portion of the supratemporal fossa, as well as the smooth anterior face of the supraoccipital crest, presumably formed the area of origin of the M. add. mand. ext. prof. No clear delimiting mark for this muscle could be found on the supraoccipital crest, although the rugose texture along the dorsal margin suggests that this was not part of the area of fleshy origin. In my dissertation, I suggested that this muscle may also have taken origin from a flat surface on the dorsum of the squamosal. This seems to be correct, but not in the sense intended in the dissertation, where I had believed that the entire dorsal face might have given attachment to the muscle. Most of the dorsum appears to be too rough in texture for a muscle scar. However, Tom Carr (personal communication) observed a crescentic smooth surface medially adjacent to the margin of the supratemporal fenestra on the squamosals of BHI 3033 (and less clearly on BHI 4100) that appears to have been a muscle attachment. Thus, it seems likely that a portion of the M. add. mand. ext. prof. took origin from the dorsum of the squamosal.

Medially, the parietals rise to form a sagittal crest that is much lower than the supraoccipital crest. Hence, it would seem that the antimeres of this muscle probably met across this crest. The lateral surface of the parietal is smooth all the way to its ventral margin. This smooth surface continues down across most of the posterodorsal portion of the lateral surface of the laterosphenoid and the anterodorsal corner of the lateral surface of the prootic (Fig. 15.7). The supratemporal fossa consists of 2 parts: first, the shallower frontal excavation anteriorly, and second, a deeper posterior excavation bounded by the parietal, the supraoccipital crest, and the arch

Figure 15.5. *Sketch of the left squamosal in dorsal view, based on BHI 4100. The proposed area of origin of the M. adductor mandibulae externus profundus, based on BHI 3033 and BHI 4100, is to the right of the dashed line. Anterior is to the right, lateral to the top. Scale bar = 5 cm.*

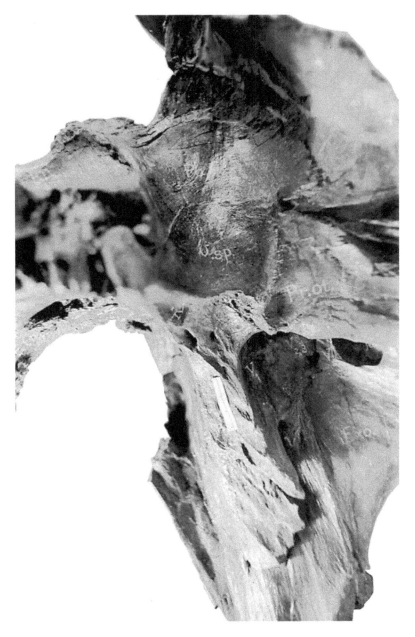

Figure 15.6. Oblique anteroventrolateral view of the braincase of *Tyrannosaurus rex*, AMNH 5117. Most of the lateral surface of the parietal (labeled "Pa." on the specimen), the laterosphenoid (labeled "O. sp."), and the prootic (labeled "pr.ot.") is here believed to have given rise to the M. adductor mandibulae externus profundus. This region is marked by diagonal hatching on the inset. The orthogonally perpendicular hatching on the frontal indicates the proposed origin of the M. pseudotemporalis.

separating the supra- and infratemporal fenestrae. This bipartite structure suggests that 2 muscles took their origin in the fossa. I believe that the M. pseudotemporalis arose from the anterior portion for reasons presented in the section on that muscle. Thus, the M. add. mand. ext. prof. probably originated only from the more posterior portion, from the lateral surfaces of the parietal and laterosphenoid (and possibly from the medial face of the squamosal, which was not available for inspection). There is no discernible feature separating the areas of origin ventrally.

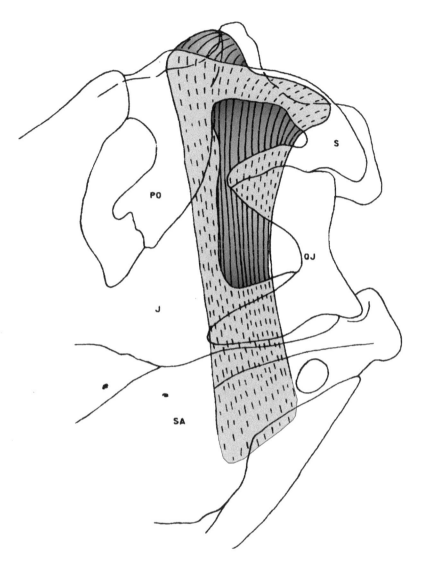

Figure 15.7. *Restoration of the M. adductor mandibulae externus profundus of **Tyrannosaurus rex**. Abbreviations: J, jugal; PO, postorbital; QJ, quadratojugal; S, squamosal; SA, surangular.*

The proposed position of the area of origin of this muscle is based on analogy with its position in modern crocodilians and the inferred position of the M. pseudotemporalis.

The M. add. mand. prof. inserts onto the zwischensehne in the living crocodilians and into the bodenaponeurosis, as well as onto the medial surface of the surangular, and sometimes the coronoid, in living lizards (Lakjer 1926; Oelrich 1956; Fisher and Tanner 1970; Nash and Tanner 1970; Avery and Tanner 1971). The zwischensehne of the living crocodilians inserts in turn into that portion of the angular forming the ventral margin of the Meckelian fossa. The insertion is marked by a prominent, roughly sigmoid ridge along the medial surface of the angular, bordering and extending just below the ventral margin of the Meckelian fossa (Iordansky 1964). The bodenaponeurosis of lizards leaves a similar feature along the anterior margin of the Meckelian fossa and the posterior margin of the coronoid process in those forms examined (*Ctenosaura pectinata,*

Figure 15.8. *Restoration of the M. adductor mandibulae posterior of **Tyrannosaurus rex**. The surficial elements of the skull are rendered as transparent and indicated by dashed lines in the drawing. Abbreviations: AR, articular; BC, braincase; E, epipterygoid; Q, quadrate; QP, quadrate process of the pterygoid.*

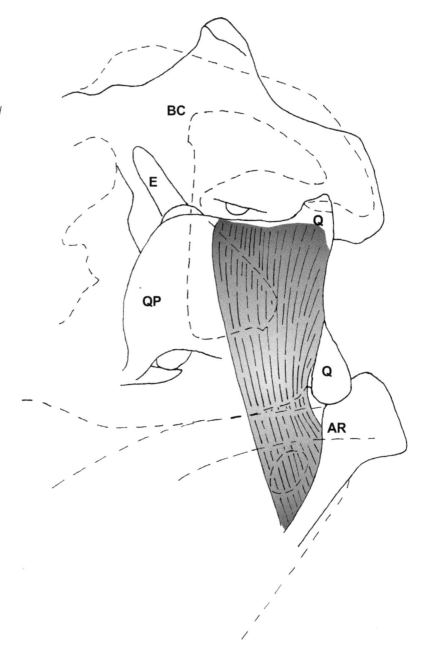

Iguana iguana, *Varanus komodoensis*, and *Varanus* sp.). This margin is continuous from the fossa up onto the coronoid process and shows slight rugosities in both *Varanus* and *Ctenosaura*. No such evidence has been found for a zwischensehne in *T. rex*, and it is assumed to have been absent. The M. add. mand. ext. prof. is thus presumed to have inserted onto the smooth medial surface of the surangular (Molnar 1991, pl. 15, fig. 1), together with the M. pseudotemporalis and the M. pterygoideus anterior. In the living crocodilians *Alligator* and *Paleosuchus*, the Mm. pterygoideus anterior, pterygoideus posterior, and add. mand. post. all fuse together at

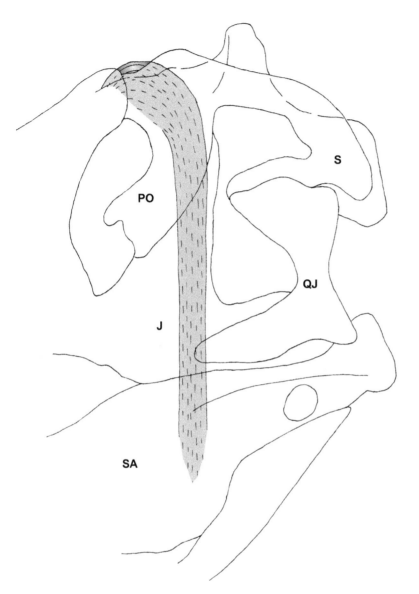

Figure 15.9. *Restoration of the M. pseudotemporalis of **Tyrannosaurus rex**. Abbreviations: J, jugal; PO, postorbital; QJ, quadratojugal; S, squamosal; SA, surangular.*

their insertion onto the mandible. Because there are no separate insertion scars discernible on the medial surface of the surangular, it seems likely that a similar condition existed in *T. rex*.

In view of the absence of evidence for a zwischensehne in *T. rex*, the proposed insertion area of this muscle is based solely on analogy with the insertion on the mandibular bones (as opposed to that on the bodenaponeurosis) in living lizards.

M. ADDUCTOR MANDIBULAE POSTERIOR (FIG. 15.8): In the living crocodilians, the M. add. mand. post. originates from the ventral surface of the posterior portion of the quadrate as well as from 2 tendons (the A and B tendons of Iordansky 1964) originating from the quadrate. Among living lizards, this muscle takes origin from the anterior surface of the quadrate, sometimes via an aponeurosis, and also sometimes additionally from the

Jaw Musculature   267

Figure 15.10. *(Top)* Dorsal view of the braincase of ***T. rex***, AMNH 5117. The abrupt change of level of the inner surface of the fenestra, believed to indicate the boundary between the areas of origin of the M. pseudotemporalis (left) and the M. adductor mandibulae externus profundus (right), is clearly visible. These areas are indicated by diagonal hatching in the inset outline sketch. *(Bottom)* Dorsolateral view of the supratemporal fenestra of ***Tyrannosaurus rex***, AMNH 5027. This also shows the proposed scars of origin of the M. pseudotemporalis (left) and the M. adductor mandibulae externus profundus (right). The outline inset indicates these 2 areas. Photo of Los Angeles County Museum of Natural History cast.

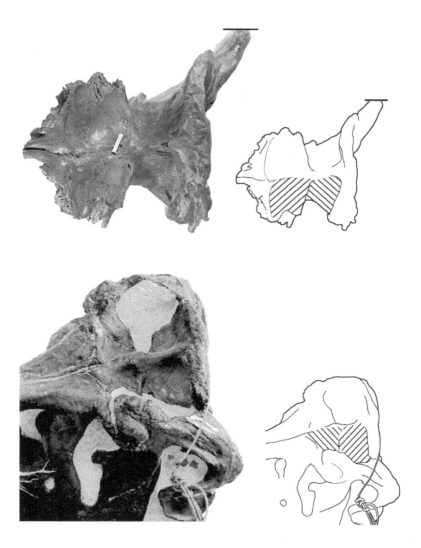

prootic (Lakjer 1926; Fisher and Tanner 1970; Nash and Tanner 1970; Avery and Tanner 1971). The quadrate of *Tyrannosaurus rex* differs from those of the living crocodilians and lizards in that it has a deep, flat, plate-like pterygoid process projecting anteromedially from the corpus quadrati. Unlike the quadrates of modern crocodilians, that of *T. rex* (LACM 23844) does not show any marked ridges of the form from which the A and B tendons take origin. Hence, presumably *T. rex* did not have such tendons. As in modern crocodilians, the quadrate of LACM 23844 does not show any clear marks that might delimit the area of origin of this muscle. Presumably this muscle took origin from the anterior surface of the quadrate and possibly also from the lateral surface of the pterygoid process. The surfaces of both the anterior face of the quadrate and the lateral face of the pterygoid process are smooth, as is the surface of origin for the M. add. mand. post. of living crocodilians.

In both living crocodilians and lizards, this muscle takes origin from the posterior wall of the adductor chamber, and by analogy with this, it is proposed also to take origin there in *T. rex*. In the living crocodilians, this muscle

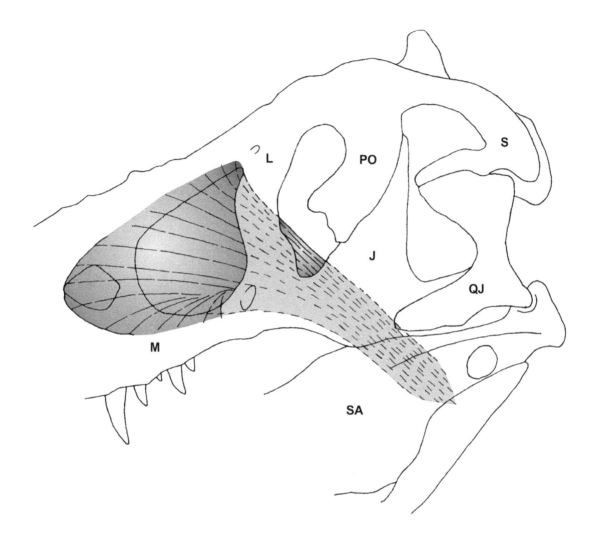

Figure 15.11. *Restoration of the M. pterygoideus anterior of* **Tyrannosaurus rex**. *Abbreviations: J, jugal; L, lachrymal; M, maxilla; PO, postorbital; QJ, quadratojugal; S, squamosal; SA, surangular.*

inserts onto the ventral margin of the Meckelian fossa and the anterior face of the articular. In most living lizards, the area of insertion is basically the same (Lakjer 1926; Fisher and Tanner 1970; Nash and Tanner 1970; Avery and Tanner 1971). The anterior face of the articular of *T. rex* was sheathed by a thin process of bone from the medial surface of surangular. The M. add. mand. post. presumably inserted onto this, probably in addition to inserting onto the adjacent portions of the medial face of the surangular.

As with the origin, the proposed insertion is based solely on analogy with the positions of the insertions in living crocodilians and lizards.

M. PSEUDOTEMPORALIS (FIG. 15.9): In the living crocodilians, the M. pseudotemporalis is a unipartite muscle arising from the external surface of the laterosphenoid. It is a multipartite muscle in living lizards, taking origin from the anterior and medial walls and margins of the supratemporal fenestra and from the upper portions of the epipterygoid (Lakjer 1926; Oelrich 1956; Avery and Tanner 1971), although in some lizards, it is not reported at all (e.g., Fisher and Tanner 1970; Nash and Tanner 1970). The area of origin in birds is similar (Lakjer 1926), with some fibers arising from

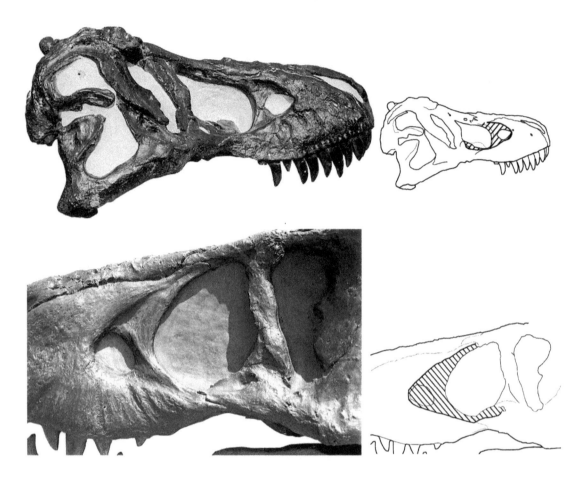

Figure 15.12. Antorbital fossa of **Tyrannosaurus rex**, AMNH 5027 (cast; above) and LACM 23844 (also cast; below). The outline insets indicate the antorbital fossa.

the quadrate (George and Berger 1966). So there are 2 potential guides to reconstructing the M. pseudotemporalis of *Tyrannosaurus rex*, either according to the crocodilian or to the avian-lacertilian pattern. As described in the section on the M. add. mand. ext. prof., the supratemporal fossa is divided dorsally into 2 parts, suggesting that 2 muscles took origin here (Fig. 15.10). This is roughly similar to the condition among hadrosaurs (Ostrom 1961) and some Triassic ornithischians (Thulborn 1971). Medially, the fossa consists of a shallow excavation of the frontals (which forms the "lobes" discussed by Walker 1964), whereas laterally, the area of origin seems to have extended onto a medial shelf of the postorbital sloping medioventrally, and seems to have had a smooth surface texture quite unlike that of the external surface of this element (Molnar 1991, pl. 4, fig. 2). Because it is situated anterior to the M. add. mand. ext. prof. in the living forms, the M. pseudotemporalis presumably originated from the anterior part of the fossa, by analogy with lizards and birds.

The M. pseudotemporalis inserts onto the dorsal surface of the zwischensehne in the living crocodilians, and into the bodenaponeurosis and onto the coronoid in lizards (Lakjer 1926; Oelrich 1956; Avery and Tanner 1971). As mentioned previously, there is no evidence of a zwischensehne in *T. rex* and no indication of muscle attachment on the coronoids, and thus this muscle presumably inserted into the medial surface of the surangular along with the Mm. add. mand. ext. prof. and pterygoideus anterior.

The proposed insertion is hypothetical and is based only on the assumption that in the absence of the zwischensehne, the M. pseudotemporalis would attach to the element to which the zwischensehne attaches in living crocodilians.

M. PTERYGOIDEUS ANTERIOR (FIG. 15.11): The M. pterygoideus anterior of the living crocodilids arises from the internal surface of the maxilla, the dorsal surfaces of the palatines and pterygoids, the posterior surface of the fascial partition separating the orbital region from the nasal capsule, and the interorbital septum. The M. pterygoideus arises from the ventral surface of the ectopterygoids and from the ventral and lateral surfaces of the pterygoids in lizards (Lakjer 1926; Oelrich 1956; Fisher and Tanner 1970; Nash and Tanner 1970; Avery and Tanner 1971). In *Tyrannosaurus rex*, the lateral face of the maxilla is relatively smaller in area than among the crocodilids because the antorbital fenestra occupies a sizable portion of the lateral surface of the snout. This surface of the maxilla is covered with low rugosities that are not continued into the antorbital fossa. At the edge of the depression, the typical surface sculpture abruptly terminates, and low, curved ridges approximately perpendicular to the border of the fossa extend onto the surface of the fossa for about 1 cm, there fading into a smooth surface (Fig. 15.12). This surface may have given origin to some of the fibers of the M. pterygoideus anterior. The dorsal surfaces of the palatal processes were not examined because they are either not exposed or not preserved in any of the specimens available. It may be assumed that the M. pterygoideus anterior took origin from these areas, as it does among crocodilians. The dorsal surface of the palatal plate of the pterygoid seems largely occupied by the ectopterygoid articulation, but the free surface is smooth and may have afforded an area of fleshy origin for this muscle. Such of the dorsal surface of the ectopterygoids as may be examined (in LACM 23844 and MOR 008) is also smooth and may have also contributed attachment area. In addition, those soft areas from which this muscle takes origin in crocodilids may have also served similarly in *T. rex*.

The proposed origin of the M. pterygoideus anterior is based on analogy with its origin from the lateral wall of the rostral cavity and dorsum of the hard palate in living crocodilians.

This muscle inserts chiefly into the zwischensehne in living crocodilians and to a lesser extent onto the anterior face of the articular. As mentioned previously, this muscle fuses with several of the other adductors at its insertion. The insertion area in lizards is on the medial and ventral surfaces of the posterior part of the mandible (Lakjer 1926; Oelrich 1956; Fisher and Tanner 1970; Nash and Tanner 1970; Avery and Tanner 1971). Again, in *T. rex*, there is no clearly defined insertion area, but presumably this muscle attached to the medial surface of the surangular along with the Mm. add. mand. ext. prof. and pseudotemporalis.

The rationale for the proposed insertion is the same as for the M. pseudotemporalis.

M. PTERYGOIDEUS POSTERIOR: This muscle originates from the posterior edge of the pterygoid wing in the living crocodilians and is often not distinguished from the M. pterygoideus anterior for lizards (Oelrich 1956; Fisher and Tanner 1970; Nash and Tanner 1970; Avery and Tanner 1971). In *Tyrannosaurus rex*, the pterygoid wing is made up chiefly by the ectopterygoid, not by the pterygoid, as in the living crocodilians and lizards, and is a relatively smaller structure than in those forms. There is no distinct mark left by the M. pterygoideus posterior on the posterior edge of the wing in crocodilians, nor is any such mark discernible on the edge of that of AMNH 5027. The M. pterygoideus posterior may have originated there, but there is no clear evidence for such an origin other than by analogy with the condition of other reptiles.

The proposed origin for this muscle is based on structural analogy with modern crocodilians, assuming the ectopterygoid part of the pterygoid wing in *T. rex* functioned in a similar fashion to the pterygoid portion of the pterygoid wing in living crocodilians.

Ostrom (1969) has suggested that a portion of the M. pterygoideus posterior in *Deinonychus antirrhopus* took origin from the concave lateral side of the ectopterygoid. The surface of the ectopterygoid of LACM 23844 is quite smooth here, but there is no clear evidence of a muscle origin. The base of the jugal ramus of this ectopterygoid shows a low, obtuse ridge roughly similar to those delimiting the posterior margin of the area of origin of the M. add. mand. ext. sup. med. on the squamosal: this supports Ostrom's suggestion and suggests that a portion of the M. pterygoideus may have originated here in *T. rex* also. However, the quadrate process in *T. rex* is a flat, vertical plate on the pterygoid just posterodorsal to the region of the pterygoid wing (Molnar 1991, fig. 6). This process extends posteriorly to articulate with the quadrate. The anterolateral face of this plate may have served as a muscle origin. The muscle originating here may have been the M. pterygoideus posterior, which may have migrated up posterodorsally from its presumed primitive position. It may also have been part of, or derivative from, the M. add. mand. post. or the M. pterygoideus anterior, or the M. pterygoideus posterior may not have differentiated from the M. pterygoideus anterior in *T. rex*.

In the crocodilians, this muscle wraps around the posteroventral portion of the mandible to insert on both the inner and outer surfaces of the mandible. The lateral insertion forms a smooth surface on the surangular and prearticular, set off from the sculptured lateral mandibular surface by a marked ridge. In *T. rex*, no indication could be found that clearly implied any insertion on the lateral surface of the mandible. This muscle presumably inserted onto the medial surface of the surangular and possibly also onto the ventral margin of the mandible (onto the prearticular). There is, however, no clear indication of this on the prearticular.

Just where this muscle inserted in *T. rex* is unclear: the only rationale for proposing any insertion area is analogy with the condition of living crocodilians. However, in crocodilians, the mandibular bones are in tight contact, sometimes interlocking, permitting no motion between them. In LACM 23844, the only firm contact between postdentary bones is that between the surangular and articular, although in MOR 008, the suran-

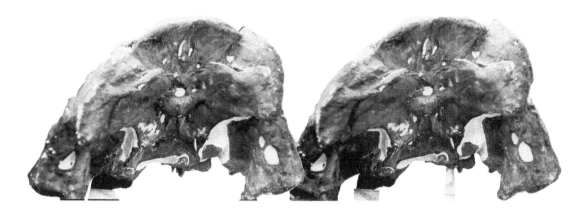

Figure 15.13. Stereophotographs of the area of origin of the M. depressor mandibulae of **Tyrannosaurus rex**, AMNH 5027. The area of origin is visible at the distal end of the paroccipital process, indicated by diagonal hatching on the outline inset. Photo of Los Angeles County Museum of Natural History cast of AMNH 5027.

gular and prearticular are—as far as could be determined without attempting to disarticulate them—fused, suggesting that this contact, too, may have been immobile (at least in some individuals). Other contacts, especially that between the surangular and angular, have left no trace on the surfaces of the bones (Molnar 1991), suggesting that some mobility between these elements may have occurred in life. The implications of this for the insertion of the M. pterygoideus posterior are obscure, but it is clear that the mechanical situation was quite different from that in crocodilians (and lizards). In crocodilians, the M. pterygoideus posterior exerts a force on the angular that is transmitted undiminished via the firm joints to the dentary and on to the tips of the teeth. In *T. rex*, any force exerted on the angular could result in movement of that element on the surangular, thus diminishing the force transmitted on to the other postdentary elements, then on to the dentary (again via a possibly mobile joint; Molnar 1991) and then on to the teeth, a seemingly less effective situation than in crocodilians.

M. INTRAMANDIBULARIS: Crocodilians are the only living tetrapods to have a M. intramandibularis known to Lakjer (1926), although Bolk et al. (1938) mention such in *Struthio camelus*. If this report is correct, then it would seem likely (a type I inference in the terms of Witmer 1997) that this muscle was widespread among extinct archosaurs. However, there is no evidence for such a muscle in *Tyrannosaurus rex*, although this muscle leaves no easily discernible traces in the modern crocodilians, so no evidence of it might be expected in *T. rex* if it occupied a homologous position. The M. intramandibularis, if present, might have acted to operate an intramandibular joint (if such existed in *T. rex*) if, for example, it originated from the surangular and inserted onto the dentary. But such attachments are purely hypothetical: in living crocodilians, it originates from the zwischensehne and inserts onto the walls of the Meckelian canal. Alternatively, it might also have acted to resist anteriorly directed forces on the jaw exerted by the prey. This muscle has been largely ignored by modern functional morphologists and paleobiologists, and so far as I have been to determine, its function in crocodilians and ostriches is unknown.

M. DEPRESSOR MANDIBULAE (FIG. 15.15): In the living crocodilians, the M. depressor mandibulae originates from the ventral portion of the posterior face of the squamosal and the posterior portion of the lateral tip of the exoc-

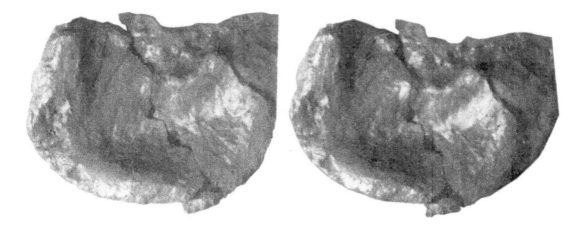

Figure 15.14. Stereophotograph of the posterior surface of the right articular of *Tyrannosaurus rex*, MOR 008. The whole surface presumably represents the area of insertion of the M. depressor mandibulae.

cipital. The area of origin is roughly crescentic. Lizards show a rather different area of origin, along the parietal or from the dorsal cervical fascia (Oelrich 1956; Fisher and Tanner 1970; Nash and Tanner 1970; Avery and Tanner 1971). In *Tyrannosaurus rex*, the distal tips of the exoccipitals bear flat, dorsoventrally elongate facets, wide at the top and narrowing to the bottom (Fig. 15.13). This surface is set off by an obtuse angulation along its medial margin from the rest of the posterior face of the exoccipital. This facet is considered to be the area of origin of the M. depressor mandibulae.

The proposed origin for this muscle is based on analogy of both the position and the form of the muscle scars to what is seen in living crocodilians. In these forms, this muscle inserts onto the dorsal surface of the retroarticular process. In lizards, it also attaches to the retroarticular process (Oelrich 1956; Fisher and Tanner 1970; Nash and Tanner 1970; Avery and Tanner 1971). *T. rex* lacks a retroarticular process. The posterior surface of the articular is a smoothly surfaced concavity and into this concavity the M. depressor mandibulae probably inserted (Fig. 15.14).

The proposed insertion of the M. dep. mandibulae is hypothetical and is based on the consideration that a retroarticular process is absent. However, the presence of a suggestive concavity of the articular in the expected position of that process implies that the muscle would most likely insert as close to the expected position as possible. This would not give significant leverage in depressing the mandible, but given the expected weight of a mandible nearly a meter in length, muscular exertion would not be needed in opening the mouth. Once the mouth was opened, the muscle could act to open it more widely (to about 90°, if that were permitted at the joint), and hence this insertion is functionally sensible.

Discussion

To my knowledge, only one reconstruction of the jaw musculature of a large theropod had been made before the research reported here was carried out. This report was by Adams (1919) for *Tyrannosaurus rex*. His terminology of the jaw adductors has not gained general acceptance. In general terms, Adams's reconstruction is similar to that presented here. Adams, however, did not recognize the existence of the Mm. add. mand. post. or

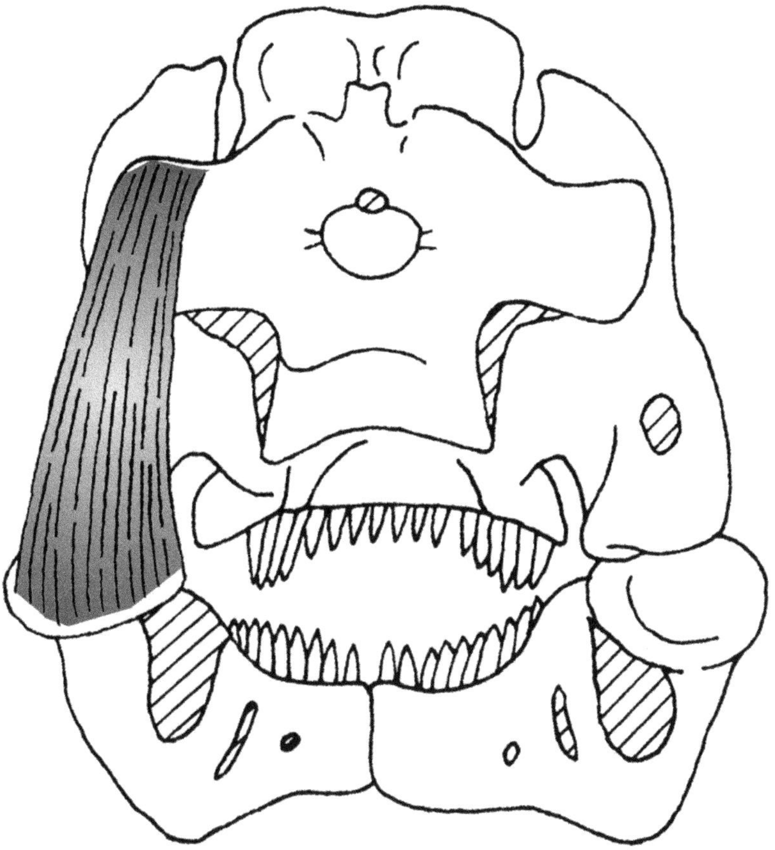

Figure 15.15. *Restoration of the M. depressor mandibulae of the left side of* **Tyrannosaurus rex**.

pseudotemporalis in either *Alligator* or *Tyrannosaurus*. He also assumed that the M. pterygoideus anterior inserted onto the mandible by wrapping around the posteroventral margin. He thus analogized the lateral ridge of the surangular in *T. rex* with the delimiting ridge of the area of the M. pterygoideus insertion on the angular and surangular of the living crocodilians. However, in the crocodilians, it is the M. pterygoideus posterior and not the M. pterygoideus anterior that inserts by wrapping around the jaw (as Adams did indeed recognize). There is no evidence of a muscle insertion onto the lateral face of the surangular in *T. rex*.

Witmer (1997) has argued that the antorbital fossa was occupied by an antorbital sinus cavity rather than a jaw muscle. He reported that reconstruction of a muscle in this region was based on biomechanical hypothesis. Here, it is based on the similarity of the bone surface texture to that seen at muscle scars in crocodilians, and other presumed muscle scars in *T. rex*. Nonetheless, Witmer may be correct. Histological examination of the superficial bone at the proposed muscle attachment in the antorbital fossa might reveal indication of Sharpey's fibers, which would support the hypothesis of muscle attachment, but pending such examination, it seems best to consider both hypotheses as possible. The form of the antorbital fossa varies in different archosaurs, so it may be that different structures occupied the fossa in different taxa.

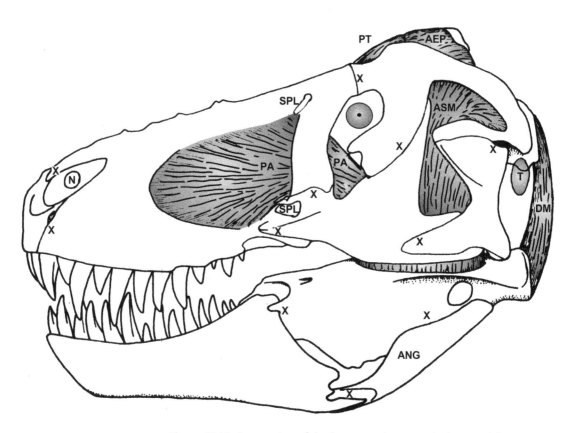

Figure 15.16. Restoration of the jaw musculature and other cranial structures of *Tyrannosaurus rex.* This and the next 5 figures represent the assumed cranial anatomy as it would appear were the head available for dissection. The shape of the tympanum and the shape and exact location of the nares are arbitrary. The tympanum is located so as to give a minimum length to the stapes. Abbreviations: AEP, M. add. mand. ext. prof.; ALP, possible area of origin of M. levator pterygoideus; ANG, angular; AP, M. add. mand. post.; APS, area of origin of M. add. mand. ext. prof.; AS, articular sinus; ASM, M. add. mand. ext. sup. med.; B, brain; BA, basipterygoid articulation; BSS, basisphenoid sinus; C, ventral entrance of canal for internal carotid; DM, M. depressor mandibulae; E, eyeball; EPI, epipterygoid; ET, eustachian tube; FO, fenestra ovalis; IN, internal naris; LD, nasolachrymal duct; LP, M. levator pterygoideus; N, nostril; NC, nasal capsule; OC, occipital condyle; P, pterygoid; PA, M. pterygoideus anterior; PT, M. pseudotemporalis; PTS, area of origin of M. pseudotemporalis; PW, pterygoid wing; Q, pterygoid process of the quadrate; QF, quadrate foramen; QS, quadrate sinus; RS, possible area of insertion of M. levator pterygoideus; S, stapes; SPA, passage to articular and quadrate sinuses; SPB, passage to basisphenoid sinus; SPE, passage to ectopterygoid sinuses; SPJ, passage to jugal sinus; SPL, passage to lachrymal sinus; SPM, passage to maxillary sinuses; T, tympanum; X, mobile joint of skull or mandible.

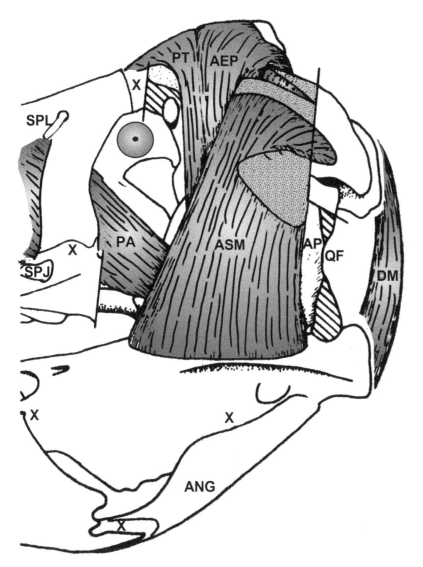

Figure 15.17. *Lateral view of the head of **Tyrannosaurus rex**, with the cheek elements removed to reveal underlying structure. Sections through the musculature are dotted; those through bone are diagonally lined. The posterior squamosal region is not shown because it has not been possible to examine the squamosal. Abbreviations as in Figure 15.16.*

Summary

The reconstruction of the jaw musculature of *Tyrannosaurus rex* presented here is based on examination of muscle scars seen on the various cranial and mandibular elements, interpreted by analogy with the jaw musculature of living crocodilians, lizards, and birds. Reconstructions of the individual muscles are presented in Figures 15.1 (M. add. mand. ext. sup. med.), 15.7 (M. add. mand. ext. prof.), 15.8 (M. add. mand. post.), 15.9 (M. pseudotemporalis), 15.11 (M. pterygoideus anterior), and 15.15 (M. depressor mandibulae). Unequivocal evidence of the existence and positions of the Mm. pterygoideus posterior and intramandibularis was not found.

The jaw muscles, as they are believed to have appeared in the head of T. rex, are presented in a series of 6 figures (Figs. 15.16–15.21). The first of these shows the external form of the skull, with the superficial musculature drawn in place. Rationale for the various details of the figure is presented in the captions. The succeeding figures show the musculature at increasingly deeper levels of the head, in addition to the major kinetic joints of the skull.

Jaw Musculature 277

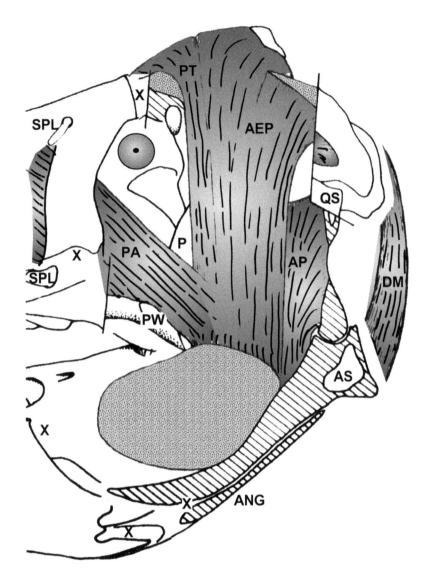

Figure 15.18. *The same view as in Figure 15.17, but with the surangular and M. adductor mandibulae externus superficialis et medialis removed. The section through the quadrate is slightly medial to that in Figure 15.16. Abbreviations as in Figure 15.16.*

Acknowledgments

I am grateful to the following people for their assistance in various aspects of this study: the late E. C. Olson (University of California, Los Angeles), P. P. Vaughn (then at UCLA), L. Drew (then at the Museum of the Rockies), E. S. Gaffney (American Museum of Natural History), W. Langston Jr. (University of Texas, Austin), N. L. Larson and P. Larson (Black Hills Museum of Natural History), J. A. Madsen Jr. (then at University of Utah), B. J. K. Molnar, M. J. Odano (then at the Los Angeles County Museum of Natural History), J. H. Ostrom (then at Yale University), the late S. P. Welles (University of California, Berkeley), T. Carr (Carthage College, Kenosha), J. Meade, and S. L. Swift (Laboratory of Quaternary Paleontology, Northern Arizona University, Flagstaff).

References Cited

Adams, L. A. 1919. A memoir on the phylogeny of the jaw muscles in recent and fossil vertebrates. *Annals of the New York Academy of Sciences* 28: 1–166.

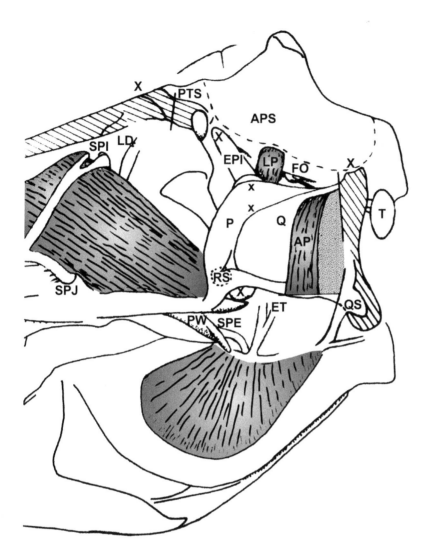

Figure 15.19. *The same view as in Figure 15.18, but with the Mm. pseudotemporalis and adductor mandibulae externus profundus, and the posterior portion of the mandible removed. The M. adductor mandibulae posterior is cut at the level of the ventral border of the pterygoid process of the quadrate. The possible M. levator pterygoideus scar (RS) marked with a dotted line is on the medial surface of the pterygoid. A M. pterygoideus posterior has not been included. The dashed line represents the boundary of the area of origin of the M. adductor mandibulae externus profundus. The posterior squamosal structure and the central portion of the quadrate sinus have not been included because these areas could not be examined. Abbreviations as in Figure 15.16.*

Anderson, H. 1936. The jaw musculature of the phytosaur, *Machaeroprosopus*. *Journal of Morphology* 59: 549–589.

Avery, D. F., and Tanner, W. W. 1971. Evolution of the iguanine lizards (Sauria, Iguanidae) as determined by osteological and myological characters. *Brigham Young Science Bulletin* 12: 1–79.

Bolk, L., Göppert, E., Kallius, E., and Lubosch, W. 1938. *Handbuch Der Vergleichenden Anatomie Der Wirbeltiere*. Vol. 5. Urban & Schwarzenberg, Berlin.

Colbert, E. H. 1946. The eustachian tubes in the Crocodilia. *Copeia* 1946: 11–14.

Colbert, E. H., and Ostrom, J. H. 1958. Dinosaur stapes. *American Museum Novitates* 2076: 1–20.

Farlow, J. O., and Molnar, R. E. 1995. *The Great Hunters*. Franklin Watts, New York.

Fisher, D. L., and Tanner, W. W. 1970. Osteological and mylogical [sic] comparisons of the head and thorax regions of *Cnemidophorus tigris septentrionalis* Burger and *Ameiva undulata parva* Barbour and Noble (family Teiidae). *Brigham Young University Science Bulletin* 11: 1–41.

Figure 15.20. *The same view as in Figure 15.19, but with the epipterygoid, pterygoid, quadrate, and associated structures removed. The eustachian tube, middle ear cavity, and tympanum have been restored by analogy with those of crocodilians. The stapes is taken after that of* **Dromaeosaurus albertensis** *as figured by Colbert and Ostrom (1958). The ocular muscles have been omitted. The eyeball is positioned dorsally in the orbit because it would presumably have been near the nasolachrymal canal, which opens high on the lachrymal. Abbreviations as in Figure 15.16.*

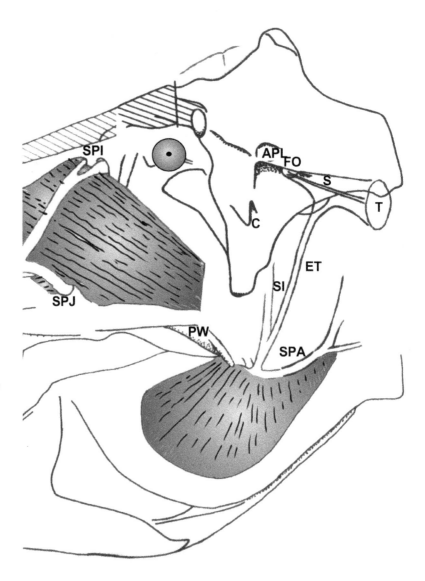

George, J. C., and Berger, A. J. 1966. *Avian Myology*. Academic Press, New York.

Horner, J. R., and Lessem, D. 1993. *The Complete T. rex*. Simon & Schuster, New York.

Iordansky, I. I. 1964. The jaw muscles of the crocodiles and some relating structures of the crocodilian skull. *Anatomischer Anzeiger* 115: 256–280.

Lakjer, T. 1926. *Studien über die Trigeminus-versorgte Kaumuskulatur der Sauropsiden*. C. A. Reitzel, Copenhagen.

Mazzetta, G. V., Fariña, R. A., and Vizcaíno, S. F. 2000. On the palaeobiology of the South American horned theropod *Carnotaurus sastrei* Bonaparte. P. 185–192 in Pérez-Moreno, B. P., Holtz, T. J., Sanz, J. L., and Moratalla, J. (eds.). *Aspects of Theropod Paleobiology. Gaia: Revista de Geociencias, Museu Nacional de Historia Natural, Lisbon*, 15.

McGowan, C. 1986. The wing musculature of the Weka (*Gallirallus australis*), a flightless rail endemic to New Zealand. *Journal of Zoology* A 210: 305–346.

Molnar, R. E. 1973. The Cranial Morphology and Mechanics of *Tyrannosaurus rex* (Reptilia: Saurischia). Ph.D. diss., University of California, Los Angeles.

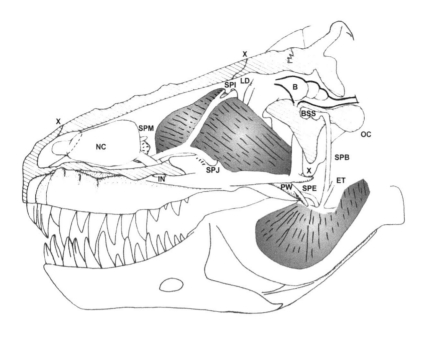

Figure 15.21. *Sagittal section through the head of **Tyrannosaurus rex**. The cervico-occipital musculature has not been included. The brain is essentially that of an alligator modified to conform to the endocranial cavity of **T. rex**. The posterodorsal corner of the splenial and the pterygoid-vomerine contact are not known and have hence been omitted. The nasal capsule represents the general form described in the discussion section of Parsons (1959) fitted into a tyrannosaur skull. The passages from the respiratory tract to the various cranial sinuses are sketched as following the shortest routes to the sinuses, with some guidance from the system of **Alligator mississippiensis** (Colbert 1946). Abbreviations as in Figure 15.16.*

———. 1991. The cranial morphology of *Tyrannosaurus rex*. *Palaeontographica* A 217: 137–176.

———. 2000. Mechanical factors in the design of the skull of *Tyrannosaurus rex* (Osborn, 1905). P. 193–218 in Pérez-Moreno, B. P., Holtz, T. J., Sanz, J. L., and Moratalla, J. (eds.). *Aspects of Theropod Paleobiology*. Gaia: Revista de Geociencias, Museu Nacional de Historia Natural, Lisbon, 15.

Nash, D. F., and Tanner, W. W. 1970. A comparative study of the head and thoracic osteology and myology of the skinks *Eumeces gilberti* van Denburgh and *Eumeces skiltonianus* (Baird and Girard). *Brigham Young University Science Bulletin* 12: 1–32.

Oelrich, T. M. 1956. *The Anatomy of the Head of Ctenosaura pectinata (Iguanidae)*. Museum of Zoology, University of Michigan Miscellaneous Publication 94.

Ostrom, J. H. 1961. Cranial morphology of the hadrosaurian dinosaurs of North America. *Bulletin of the American Museum of Natural History* 122: 33–186.

———. 1969. Osteology of *Deinonychus antirrhopus*, and unusual theropod from the Lower Cretaceous of Montana. *Bulletin, Peabody Museum of Natural History* 30: 1–165.

Parsons, T. S. 1959. Studies on the comparative embryology of the reptilian nose. *Bulletin of the Museum of Comparative Zoology* 120: 101–277.

Paul, G. S. 1988. *Predatory Dinosaurs of the World*. Simon & Schuster, New York.

Thulborn, R. A. 1971. Tooth wear and jaw action in the Triassic ornithischian dinosaur *Fabrosaurus*. *Journal of Zoology* 164: 165–179.

Ushiyama, T. 1995. *Tyrannosaurus*. P. 14–43 in NHK Special Project Team (ed.). *This Is the Dinosaur! New Aspects of Dinosaurs Based on the Scientific Approach* (in Japanese). Shinchosha, Tokyo.

Walker, A. D. 1964. Triassic reptiles from the Elgin area: *Ornithosuchus* and the origin of carnosaurs. *Philosophical Transactions of the Royal Society of London* B 248: 53–134.

Witmer, L. M. 1997. The evolution of the antorbital cavity of archosaurs: a study in soft-tissue reconstruction in the fossil record with an analysis of the function of pneumaticity. *Society of Vertebrate Paleontology, Memoir* 3: 1–73.

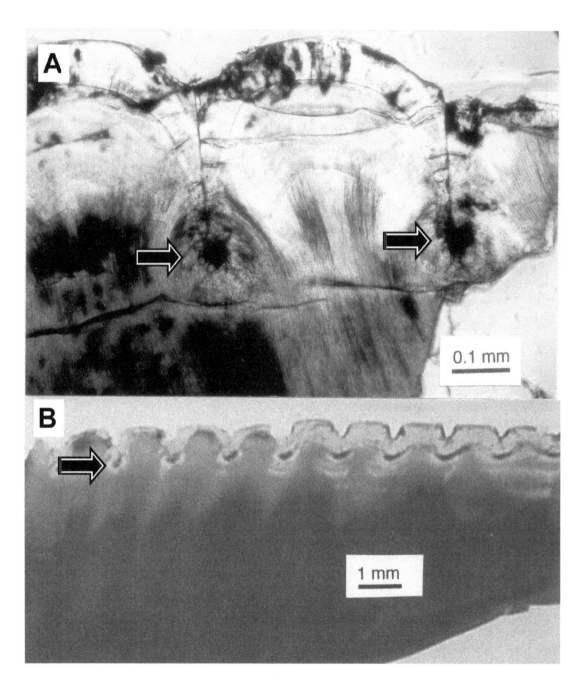

Figure 16.1. Ampullae in the serrations of Tyrannosauridae. (A) **Albertosaurus**. Two ampullae are visible. Note that the ampullae (arrows) are round, clearly defined, and nearly as large as their serrations. Each is connected to the surface by a straight, well-defined gap that is accurately centered on its ampulla. (B) **Tyrannosaurus rex** (FMNH PR2081). Ten dark ampullae are visible beneath the serrations of **T. rex**. The ampulla at left is nearly round (arrow) and is connected to the surface by a gap, confirming some relationship to **Albertosaurus**. But the gaps are reduced to cracks which are also visible in other ampullae, are not centered, and do not appear to have retained any protective function. The ampullae themselves show a clear trend from round and well defined (left), to comma shaped and diffuse (right). See text for interpretation.

VESTIGIALISM IN A DINOSAUR

William L. Abler

16

Introduction

Vestigialism is a well-known mechanism in biology. It occurs when some structure in a living organism is no longer used and has become reduced in size, differentiation, or function, over evolutionary time. The most famous example of vestigialism is the sightless eyes of cave-dwelling fishes and salamanders. In the total absence of light, the once-functional eyes have become small and have lost the power of sight. Another example is the immobile dew claw high on the leg of a dog. In a remote ancestor of the dog, a structure having the same embryonic source as the dew claw was a functioning toe on a much shorter foot. Vestigialism allows us to directly see the consequences of evolutionary processes. Darwin (1859) interpreted vestigial structures as evidence of descent with modification. Theropod tooth serrations, with their linear rows of discrete, repetitive objects, are as close to controlled conditions as is possible in nature. In them we can see the developing process of vestigialism frozen in time in the ampullae beneath the serrations in a tooth of *Tyrannosaurus rex*.

The findings reported here are based on a thin section of a tooth, a destructive procedure. Because of the precious nature of all *T. rex* material, the resulting observations and conclusions are based on a single sample. The observed vestigialism might thus result from any of 3 causes. First, it might represent pathology. Second, it might vary depending on the location along the tooth row. Last, it might be typical of all *T. rex* teeth. The observations that follow assume the vestigial ampullae were typical and the comments apply to vestigial structures. But the precise importance of the observations reported here must remain tentative until more material can be examined.

Serrations

Serrations in the teeth of *Albertosaurus* exhibit flask-shaped voids called ampullae (Abler 1992). I have suggested (Abler 2001) that the ampullae are a stress-relief mechanism similar to that used in technology to prevent the propagation of a crack through a hard material such as an airplane surface or a telescope lens. In *Albertosaurus*, neighboring serrations do not touch one another. Instead, they are separated by a narrow gap that extends down into the interior of the tooth and terminates in the flask-shaped ampulla. This structure distributes stress around its perimeter rather than concentrating it at a single point. In *Albertosaurus*, the serrations, the gaps between them, and the ampullae at the ends of the gaps are large and remarkably uniform in shape and in size.

By contrast, serrations in the teeth of *Tyrannosaurus rex* are comparatively crude. The external surfaces of the serrations are similar to those of

Albertosaurus, but *T. rex* teeth lack some of the fine detail, such as lobes on the serration bodies, or the elevated tail that in some cases continues onto the surface of the tooth. But it is in the interior structures seen in the tooth sample that the real differences appear. Instead of systematic gaps between adjacent serrations, the enamel serration caps form a continuous surface. Instead of gaps between serrations, there are cracks in the enamel caps of the serrations. The cracks may extend perpendicular to the tooth, as in *Albertosaurus*, or may be angled. In some cases, there is no crack at all. The ampullae, in the interior of the tooth below and between adjacent serrations, form a clear progression from round and centralized, as in *Albertosaurus*, to lenticular, diffuse, and eccentric (Fig. 16.1).

Vestigialism Compared with Differentiation

The progressive loss of a precise shape and function in the slots and ampullae of the *T. rex* serrations contrasts sharply with the progressive differentiation of other functional structures seen in the tooth set of a carnivorous mammal such as the dog (*Canis familiaris*). The serrations on theropod teeth are discrete, linearly arranged structures, where each successive structure tends to resemble the preceding. Thus, teeth and serrations offer a unique opportunity to see in a single compound object the kind of structural transformations that might take place in a sequence of species over evolutionary time, or that would otherwise require us to compare a number of fossils and species whose relationship may be in doubt.

For example, the teeth on a single jaw of a dog form an anatomical progression (Thompson 1966) from front to back. Posteriorly along the tooth row, each more posterior tooth can be formed from the one ahead of it by enlarging the structure at the anterior end of that tooth (Abler 2005). The canine tooth is a spike, simple except for a small ridge that circles the base of the tooth, and ascends to the point. The premolar is formed from the canine by enlarging the anterior part of its basal ridge. The carnassial is formed from the premolar by enlarging a small ridge on the anterior part of the premolar. And the molar is formed from the carnassial by enlarging a small anterior table on the carnassial. The details are more complicated, but the teeth have a relationship that is sequential and substantially taxonomic in character. The individual serrations in the *Tyrannosaurus rex* tooth have a similar relationship that is sequential and quasi-taxonomic. But instead of showing sequential differentiation, the serrations, and specifically the ampullae, show sequential deterioration. If it is typical, vestigialism in the ampullae of the present *Tyrannosaurus rex* tooth confirms their functional status in *Albertosaurus*.

The presence in *Tyrannosaurus rex* of a structure otherwise known only from the earlier *Albertosaurus* tends to confirm a taxonomic relationship between the 2 species. But in *Albertosaurus*, the ampullae have such functional value that their simple presence in the later *T. rex* might have been a case of evolutionary convergence. It is the vestigial nature of the ampullae in this *T. rex* specimen that may rule out a functional connection and confirm the taxonomic one.

Conclusions

The overall function of tooth serrations seems to be a shifting balance between cutting, the collection and delivery of infectious bacteria, and protection against breaking (Abler 1992, 2001). For its part, *Tyrannosaurus rex* appears to have abandoned metabolically expensive structural specialization and refinement in favor of obtaining maximum size (Erickson et al. 2004). More generally, because vestigial structures do not arise as a result of their selective advantage, they are free from suspicion of convergence with earlier, similar structures and are a more reliable indicator of common inheritance.

Acknowledgments

I thank P. Currie and P. Makovicky for access to specimens; and W. Simpson, L. Bergwall-Herzog, I. Glasspool, and N. Hancoca for technical assistance.

References Cited

Abler, W. L. 1992. The serrated teeth of tyrannosaurid dinosaurs, and biting structures in other animals. *Paleobiology* 18: 161–183.

———. 2001. A kerf- and drill model of tyrannosaur tooth serrations. P. 84–89 in Tanke, D. H., and Carpenter, K. (eds.). *Mesozoic Vertebrate Life*. Indiana University Press, Bloomington.

———. 2005. *Structure of Matter, Structure of Mind*. Pensoft Scientific Publishers, Sofia.

Darwin, C. 1859. *On the Origin of Species*. John Murray, London.

Erickson, G. M., Makovicky, P. J., Currie, P. J., Norell, M. A., Yerby, S. A., and Brochu, C. A. 2004. Gigantism and comparative life-history parameters of tyrannosaurid dinosaurs. *Nature* 430: 772–775.

Thompson, D.W. 1966. *On Growth and Form*. Abridged ed. Bonner, J. T. (ed.). Cambridge University Press, Cambridge.

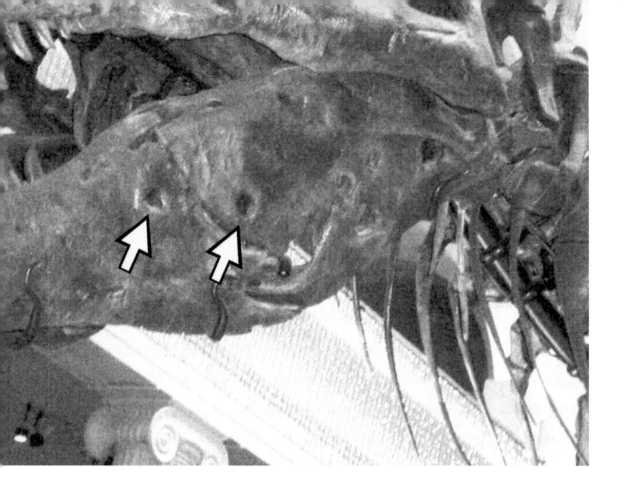

Figure 17.1. *Lateral view of jaw of **Tyrannosaurus** (FMNH PR2081). Multiple holes with partially healed margins.*

TYRANNOSAURID PATHOLOGIES AS CLUES TO NATURE AND NURTURE IN THE CRETACEOUS

17

Bruce M. Rothschild and Ralph E. Molnar

Introduction

Pathology may be more common in tyrannosaurids than any other dinosaurs. Larson (2001a, 2001b) stated that healed fractures or disease was invariably present in *Tyrannosaurus rex* if more than 10% of the skeleton was present. The current analysis will be divided into 4 components: cranial (skull and mandible) pathology, postcranial pathology, nontyrannosaurid evidence for predation, and metabolism and physiology. We will not discuss theropods in general but instead limit ourselves to those tyrannosaurids where sufficient skeletal material was available for assessment. Brochu (2003) lists tyrannosaurids as *Alectrosaurus olseni, Eotyrannus lengi, Siamotyrannus isanensis, Nanotyrannus lancensis, Daspletosaurus torosus, Albertosaurus sarcophagus, Gorgosaurus libratus, Tarbosaurus bataar,* and *Alioramus remotus.* Paul (1988) adds *Chingkankousaurus, Deinodon, Dinotyrannus,* and *Maleevosaurus,* and places *Alioramus* in the Aublysodontidae. Holtz (2001) lists *Aublysodon molnari, Alectrosaurus olseni, Alioramus remotus, Shanshanosaurus, Gorgosaurus libratus, Albertosaurus sarcophagus, Daspletosaurus torosus, Tyrannosaurus bataar, Tyrannosaurus rex,* and the Kirkland Aubylosodontine and Two Medicine Tyrannosaurine, listing *Siamotyrannus isanensis* outside the Tyrannosauridae. *Appalachiosaurus, Eotyrannus, Dilong, Aviatyrannis,* and *Stokesosaurus* are more or less recently described plesiomorphic tyrannosaurids. Of the Tyrannosauridae, only *Albertosaurus, Daspletosaurus, Gorgosaurus, Tarbosaurus,* and *Tyrannosaurus* have a large enough sample size to be consider here. *Dryptosaurus,* although plausibly regarded by some to be a plesiomorphic tyrannosaurid, is too incomplete a specimen to be considered here.

Paleopathologies similar to modern pathologies may result from injury or disease. Here we consider only the former category. In autecological terms, injuries may conceivably result from several categories of causes: aggression, prey capture, courtship and mating, and accident. The kinds of injuries resulting from these would not all be the same. Aggression here results from 2 causes of injury: conflict with conspecific individuals over food, territory, and mates; and conflict with other predators, usually over food. These kinds of injuries may involve trauma from being struck with a blunt instrument, such as a kick or blow from a tail; being bitten; or being injured by claws.

Injuries acquired during prey capture would presumably include only the first of these (being struck) because the prey would not have teeth or

Taxon	Fractures	Bite marks	Infection	Stress Fractures	Abnormal Teeth
Tyrannosaurid	X	X			
Albertosaurus	X	X		X	X
Alectosaurus					X
Daspletosaurus	X	X	X		X
Gorgosaurus	X	X			
Tarbosaurus		X		X	
Tyrannosaurus	X	X	X	X	X

Table 17.1. *Distribution of Paleopathologies among the Tyrannosauridae*

Note.—Data from Anonymous (1997); Currie (1997); Erickson (1995); Harris (1997); Lambe (1917); Larson (2001); Molnar (2001); Rothschild (1997); Rothschild et al. (2001); Tanke (1996a, 1996b); Tanke and Currie (1995, 1998); Webster (1999); Williamson and Carr (1999).

claws capable of inflicting serious injury (although ornithomimosaurs may be an exception; this awaits discovery of pedal material). Injury from prey ceratopsian horns could be considered but are highly characteristic in appearance (Rothschild and Tanke 1997). Injury may also be experienced during courtship. Bears, walrus, and raccoons frequently have baculum fractures. However, in oras (*Varanus komodoensis*), females will not mate with males that cannot physically subdue them (Auffenberg 1981). Injury may be inflicted during this process. Presumably this would not be a serious injury because it would not benefit the female's reproductive prospects to be seriously injured, and we have seen no reports of injuries related to prey capture in oras.

How much this might apply to tyrannosaurids, if at all, is unclear, but this type of mate selection does select for large body size. It could conceivably be involved in the increase of body size in the tyrannosaurid lineage. Injuries due to accident would be predominantly the result of falls and should be distinguishable from those that result from the other categories of injury.

The distribution of pathologies among the tyrannosaurids is shown in Table 17.1.

Skull Paleopathologies

CRANIAL: Dentigerous wounds were described in *Albertosaurus* (Molnar and Currie 1995). A punctured, infected *Daspletosaurus* ectopterygoid was reported by Williamson and Carr (1999). Molnar (2001) described a dentary fracture in *Gorgosaurus libratus* (RTMP 91.36.500). Larson (2001a) illustrated mandibular pathology in *Tyrannosaurus rex* (FMNH PR2081, aka Sue). These holes (Fig. 17.1) were mistakenly attributed by Rega and Brochu (2003) to a fungal infection. The holes actually are smooth walled with ingrowth of new bone. That appearance is indistinguishable from healing trephination and actually represents partial healing of bone-penetrating injuries. Molnar (1991) reported a possible pathological angular in MOR 008, as well as a tooth puncture of its surangular. A similar pathology also occurs in the surangular of LACM 23844. Keiran (1999) reported a right dentary of *Gorgosaurus* (RTMP 91.36.500) with a midlength, dorsoventral yellowish discoloration. It contrasts with the normal medium brown color. This halo of color surrounds a face-bite lesion, but the cause of this discoloration is unknown. Unilateral loss of the occipital crest in *T. rex* Stan (BHI 3033; Larson and Donnan 2002) appears unique and may be related to mating or intraspecific combat: dominance behavior or love bite?

DENTAL ABNORMALITIES: Tooth abnormalities are common, with 47% of broken premaxillary teeth having wear facets, in contrast to 8.4% of broken lateral teeth (Mongelli et al. 1999). Abler (1992) figures a tooth mark occasionally found on teeth of Late Cretaceous tyrannosaurid dinosaurs from Alberta, Canada. He suggested it possibly occurred during aggressive intraspecific interactions.

Abler (1997) reported an unidentified tyrannosaurid tooth (RTMP 83.98.90) from the Campanian Dinosaur Park Formation of Alberta, Canada, with normal shape but multiple rows of supernumerary serrations on the sides of the crown, parallel to the long axis. Erickson (1993a) reported a tyrannosaur (presumably albertosaurid) tooth in Upper Cretaceous sediments (Prince Creek Formation) from the North Slope of Alaska. The anterior carina row was split, and an extra carina segment was present on the posterior face of the tooth. Darren Tanke (Tanke and Rothschild 2002) examined hundreds of isolated Late Cretaceous tyrannosaur teeth and several intact dentitions in hopes of documenting further instances of such anomalies. Approximately 10% had some degree of carina splitting; 0.4% had extra serration rows. Tyrannosaurids with split carina included *Tyrannosaurus*, *Daspletosaurus* sp., *Albertosaurus* sp., *Alectrosaurus olseni*, and possibly 3 other tyrannosaurs. Similar examination of nontyrannosaurid carnosaur taxa, with exception of *Allosaurus fragilis* (Erickson 1995), revealed no such pathologies. Trauma, aberrant tooth replacement, and genetic factors are possible causes.

Erickson (1993b) reported 12% of 993 shed tyrannosaur teeth from Montana, Alberta, and Alaska, with pathologies affecting *Tyrannosaurus*, *Albertosaurus* sp., *Daspletosaurus* sp., and *Alectrosaurus olsoni*. Split carinas were found in both premaxillary and lateral teeth, and were randomly distributed in intact dentitions. Lack of wear on the extra carina segments suggests this anomaly probably had no functional significance. Erickson (1995) reported split carinae as more common in *Albertosaurus* than in *Tyrannosaurus*. He also reported rates of 10% in Campanian albertosaurs (in both highland and lowland environments) in contrast with 27% in the Maastrichtian specimens. The pathologies may be due to environmental stresses or a population bottleneck causing this congenital pathology because such observations have been made on recent animals (Rothschild and Martin 1993).

Postcranial Pathology

SPONDYLOSIS DEFORMANS: Moodie (1919) and Brochu (2003) accurately recognized spondylosis deformans in *Tyrannosaurus*, a phenomenon that is often mislabeled as osteoarthritis. Osteoarthritis has been erroneously diagnosed in dinosaurs because of semantic confusion (Rothschild 1990; Rothschild and Martin 2006). Osteoarthritis is diagnosed on the basis of

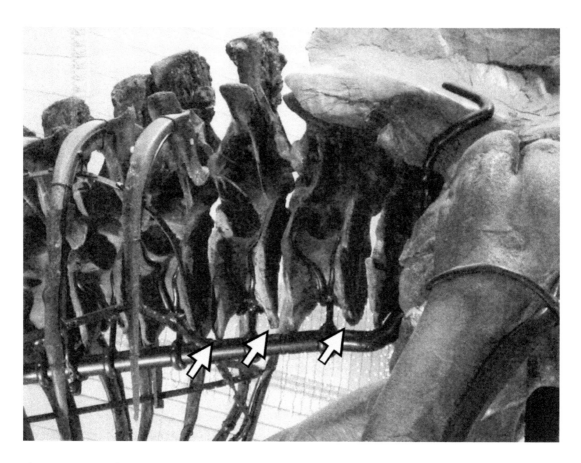

Figure 17.2. Lateral view of thoracolumbar spine of *Tyrannosaurus* (FMNH PR2081). Overgrowth of vertebral margins form osteophytes, spondylosis deformans.

osteophytes at the articular surfaces. Confusion arises because the term *osteophyte* is also used to describe overgrowth of vertebral centra, a condition known as spondylosis deformans (Fig. 17.2) that is unrelated to osteoarthritis (Resnick 2002; Rothschild and Martin 1993). Although spondylosis deformans is occasionally noted in large theropod dinosaurs, osteoarthritis has not (Rothschild 1990). Diffuse idiopathic skeletal hyperostosis (DISH, a disorder characterized by ossification of spinal longitudinal ligaments; Rothschild and Berman 1991) has yet to be reported.

INJURIES: Fractures are commonly reported (Anonymous 1997; Currie 1997; Dingus 1996; Lambe 1917; Larson 2001a, 2001b; Tanke and Currie 1998) and have been quantified by Tanke (1996b), who reported healing tyrannosaur fibula fractures, similar to that found in *Tyrannosaurus*, FMNH PR2081 (Larson and Donnan 2002), in 10%–15% of specimens in Royal Tyrrell Museum collections.

Currie (1997) reported a very large (>8 m long) *Gorgosaurus* (TMP94.12.602) with fractured and healed right fibula and well-healed dorsal ribs, as did Russell (1970) in *G. libratus* and as seen in *T. rex* (FMNH PR2081, Fig. 17.3). Roughened and thickened bone near the distal end of the left fibula of a *Gorgosaurus* (Lambe 1917) suggests a healed bone fracture. Healed midshaft fracture of the right fibula with good alignment was also noted in TMP91.36.500 (*G. libratus*). A possible well-healed unilateral dentary fracture and a fractured and well-healed right fibula were

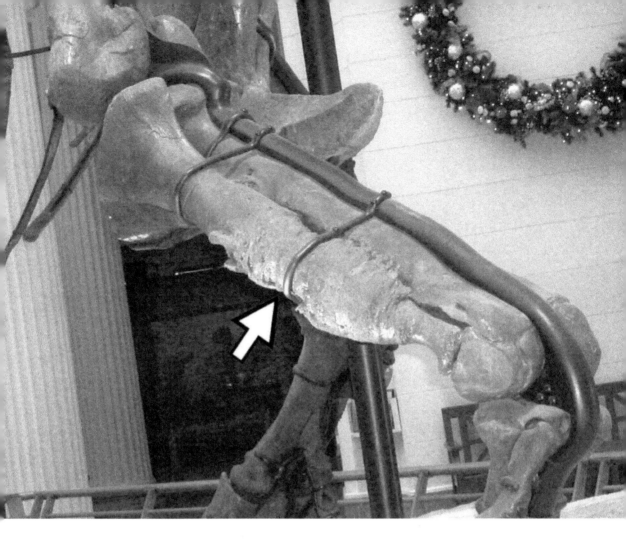

Figure 17.3. Oblique anterior view of leg of ***Tyrannosaurus*** (FMNH PR2081). Partially healed fracture with possible infection.

also found in that specimen (Keiran 1999; Tanke 1996a). The frequency of fibular fractures suggests that these are not from falls, unless the animals were less coordinated than at least one of use (R.M.). More likely, the fractures resulted from conspecific interactions, but are less likely from tail impacts. Tyrannosaurid tails are too high off the ground, although possible injury from a prey animal (e.g., sauropod tail whip) could be considered.

INTERACTIONS: Dingus (1996), Brochu (2002), and Rega and Brochu (2003) reported rib fractures in *Tyrannosaurus* (Fig. 17.4) and by Molnar (2001) in an undescribed tyrannosaurid (which also had a humeral fracture). Fractures have been reported in gastralia by Lambe (1917) in *Gorgosaurus* and by Grierson (1998) in an unspeciated tyrannosaur (TMP97.12.229) from Dinosaur Provencial Park, Alberta. Although fractures are by definition trauma related, the specific source of the trauma in FMNH PR2081 (T. rex) was revealed by the presence in an injured rib of a tooth fragment from a conspecific (Larson 2001a, 2001b).

Fracture healing may result in limb element shortening with malpositioning of elements, as exemplified by shortening in *Albertosaurus* (MOR 3079) and the specimen reported by Molnar (2001) mentioned above. Pseudoarthro-

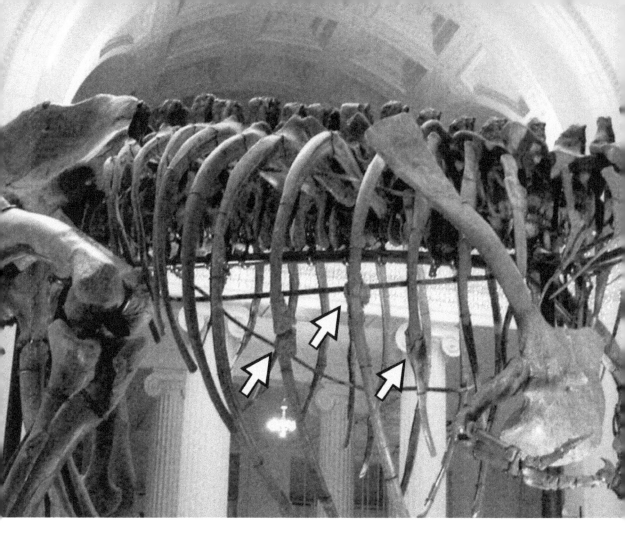

Figure 17.4. *Lateral view of **Tyrannosaurus** (FMNH PR2081) thorax. Rib fractures.*

sis, wherein the fracture components do not fuse but rather form a false joint, was noted in a *Gorgosaurus* gastralium (RTMP 91.36.500; Keiran 1999).

Other evidence of injury includes exostoses (Fig. 17.5), wherein a portion of the bone is spalled free at one end. Growth from the retained base produces an external bony overgrowth, the exostosis. Rothschild and Tanke (2005) reported humeral exostosis on FMNH PR2081. The damage has been variously attributed to avulsion of the teres major or triceps humeralis (Larson 2001a, 2001b; Carpenter and Smith 2001). Exostoses are found in the scapulocoracoid FMNH PR2081 (Larson 2001a, 2001b). Although they were originally attributed to osteoarthritis, these have been reclassified as exostoses because they are not at the joint surface. Rather, they are at sites of muscle attachment. Molnar (2001) described exostosis in metatarsal IV of *A. sarcophagus*. Russell (1970) reported a distal humeral pathology in *Daspletosaurus torus*.

The roughened surface of the left metatarsal IV in *Gorgosaurus* (RTMP 91.36.500) suggests trauma, perhaps prey interaction from an event similar to that which resulted in a small, mushroomlike hyperostosis on a right pes, amorphously shaped, digit III ungual (Lambe 1917). This pathology, referred to as an osteochondroma, was also found in a *Gorgosaurus* pedal phalanx (RTMP 91.36.500, Rothschild and Tanke 2005). Rothschild

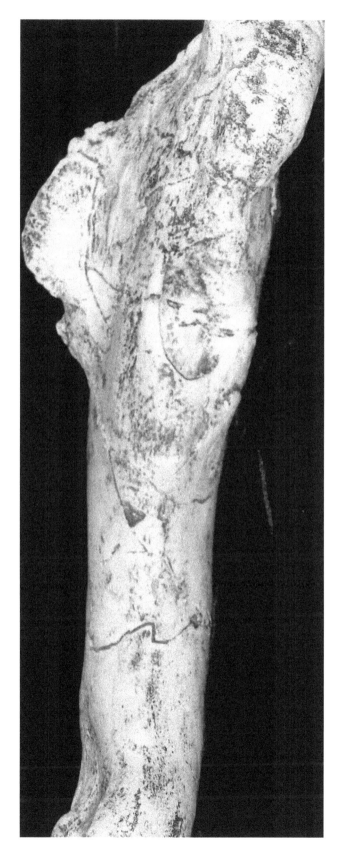

Figure 17.5. *En face* view of ***Tyrannosaurus*** *(FMNH PR2081) humerus. Surface disruption with bone spur.*

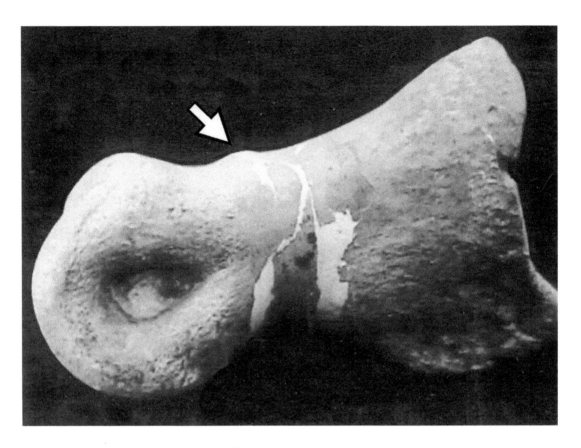

Figure 17.6. Anterior oblique view of **Tyrannosaurus rex** proximal pedal phalanx. Bump reveals site of stress fracture.

and Tanke (2005) described pedal phalangeal osteochondroma in *Gorgosaurus* (RTMP 91.36.500), as was found in the Lancian tyrannosaurid Jane (P2002.4.1) by one of us (B.M.R.). A hole, 2.5 cm by 3.5 cm, in the *A. sarcophagus* (ROM 807) iliac blade may represent trauma or infection (Molnar 2001).

In FMNH PR2081, scarring on the humerus (Fig. 17.5)—suggesting torn tendons, a badly broken and healed fibula, crushed and healed caudal vertebrae, a broken and improperly healed cervical rib, and the injuries present in the skull—suggest that Sue "led a vigorous, very active, and somewhat nasty life" (Anonymous 1992, p. 80). Currie (1997, p. 3) described articulated distal caudal vertebrae of FMNH PR2081 showing "exostotic evidence of a healed break," although only the 2 anteriormost are actually involved. Brochu (2003) described fractures in the 15–22 presacral, diffusely among gastralia, in the right coracoid, and in the right fibula of FMNH PR2081.

STRESS FRACTURES: Stress or fatigue fractures are caused by strenuous repetitive activities (Resnick 2002; Rothschild and Martin 1993), in contrast to the fractures listed above that were the result of acute trauma. Stress fractures have a highly characteristic appearance (Resnick 2002), seen as an osseous bump (Fig. 17.6). The normal plasticity of bone in responding to repetitive stresses is responsible for the bone remodeling. When the resorptive response to stresses is greater or faster than the remodeling component, disruption occurs in the form of stress fractures (Griffith et al.

2005; Resnick 2002). Such fractures are present across the spectrum of theropod size (Rothschild et al. 2001). They were easily distinguished from osteomyelitis (bone infection) because of lack of bone destruction (Resnick 2002; Rothschild and Martin 1993). Stress fractures affecting the manus are common (Rothschild and Tanke 2005), providing further evidence that tyrannosaurs, at least at times, pursued predatory behavior. Active resistance of prey is required to overstress the manus. Perhaps separation of scavenging and predation is more a matter of degree and semantics (Holtz this volume; Paul this volume).

Stress fractures in the feet and even the furcula (Lipkin and Carpenter this volume) suggest running in pursuit of prey rather than of competitors or rivals. Catching food is necessary and may involve prolonged chases, whereas scaring off rivals requires short chases only sufficiently long enough to indicate that the rival is not welcome. Another consideration is theropod copulation. However, considering that FMNH PR2081 is a female, injuries compatible with mating are hard to identify. This contrasts with the occipital defect of BHI 3033 (Stan), which may suggest a bite during mounting. Copulation can probably be eliminated on theoretical grounds, in that creatures that consistently injured themselves during copulation would either become extinct or would modify their sexual activities so that the injury is eliminated. The exception is of course bears, for which fractured baculum is allegedly the most common pathology (Bartosiewic 2000). Repetitive stress activities in this case are to blame. The behavioral question is then to identify those activities.

PLASTIC DEFORMATION: The term *plastic deformation* is shared by the medical and paleontological worlds, but with wholly different implications. For the paleontologist, the term is often applied to postmortem taphonomic alteration of bones. The medical, and hence pathological, implication is quite different: Griffith et al. (2005) describes the response of immature human bone to the removal of excessive force. The return to normal appearance is referred to as elastic deformation. If the excessive force does not actually fracture the bone and it does not return to its pre-forced condition, the third alterative pertains: persistence of excessive force-induced deformation, or plastic deformation. Plastic deformation occurs when the elastic limits of bone are not exceeded sufficiently to actually fully fracture, but it appears to be the result of tubular bone longitudinal compression. Slip lines or microfractures appear at an angle of 30° to the degree of bowing (Chamay 1970). This gives rise to what has been called extrinsic-toughening along and bridging the crack. It allegedly increases the compliance of the region surrounding the crack (Nalla et al. 2005). Such mechanisms are the main source of toughening in brittle materials The most commonly affected bones in humans are the radius and ulna, followed by the femur (Griffith et al. 2005; Resnick 2002).

Femoral angulation in adult *T. rex* has perhaps been overlooked as an example of plastic deformation (Fig. 17.7). Surface alterations at the site of angulation perhaps have been considered sites of muscle-tendon attachments (Fig. 17.8). Examination of the postcranial skeleton of Samson (currently being prepared at the Carnegie Museum), however, revealed focal

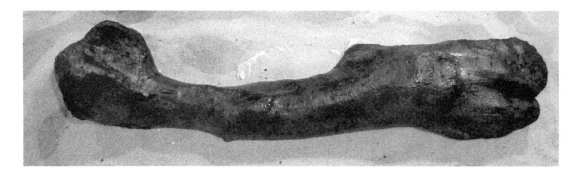

Figure 17.7. Anterior oblique view of Samson's (*Tyrannosaurus*) femur. Deformation of distal portion.

bilateral alterations of femoral surface bone more suggestive of stress fractures. Examination of ontogenetic series will be required to assess the significance of this alteration, its species specificity, and biomechanical implications.

VERTEBRAL CENTRA FUSION: The fused appearance of vertebrae in tyrannosaurids (Dingus 1996; Rothschild 1997) appears to represent 5 phenomena: normal; homeobox congenital abnormality; trauma; infection; and ligamentous fusion. Dingus (1996) described fused *Tyrannosaurus* neck vertebrae, subsequently documented as a normal phenomenon (B.M.R., unpublished observations) resulting in more superior positioning of the head. It is responsible for the acute angulation at the cervical-thoracic junction. The vertebral end plates lack the otherwise normal orientation, producing the convex (with respect to the corpus) orientation of the vertebral column at that point as seen in cervical 10–dorsal 1 of AMNH 5027 (*Tyrannosaurus*; see Osborn 1916, pl. 27). Vertebral centra fusion is occasionally noted at other locations in tyrannosaurids. Radiologic examination of dorsals 7–8 of AMNH 5027 reveals no sign that the 2 vertebrae ever existed independently, and it confirms that this pathology represented a failure of vertebral segmentation in ontogeny, a homeobox phenomenon. It is distinguishable from DISH because of absence of ligamentous fusion, and from spondyloarthrop-

Figure 17.8. Close-up of area at deformation of Figure 17.7. Plastic deformation versus stress fracture.

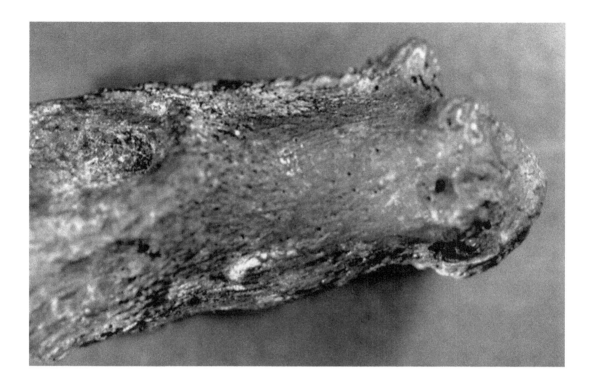

Figure 17.9. Oblique view of tyrannosaurid manus phalanx. Lytic area with overhanging edge.

athy because of absence of segmentation (Rothschild and Martin 1993). DISH ligamentous fusion, reported by Larson (2001a, 2001b) in FMNH PR2081, was also noted in RTMP 81.10.1 (*Tyrannosaurus*).

Archer and Babiarz (1992) reported damage in tyrannosaurid caudals, perhaps related to trauma. Brochu (2000) and Rega and Brochu (2003) described caudal fusion in the FMNH PR2081, with surface reaction documenting infection, perhaps related to a bite.

INFECTION: Infections are predominantly reported as isolated phenomena at bite sites, including skull, vertebrae, scapula, ilium, ischium, fibula, humerus, and pedal and manus phalanges (e.g., infected phalanx; RTMP 71.17.1) (Larson and Donnan 2002; Rothschild 1997; Tanke and Rothschild 2002; Welles 1984; Williamson and Carr 1999). Infection in the FMNH PR2081 was more generalized (B.M.R., personal observation), with Brochu (2000) and Webster (2002) specifically commenting on left fibular osteomyelitis and caudal 26–27 fusion (which B.M.R. attributes to osteomyelitis). Rega and Brochu (2003) described mandibular holes originally attributed them to a fungal infection. These mandibular holes have subsequently have been classified simply as healing bone-penetrating wounds. Their appearance is indistinguishable from healing trephination. The 2.5 cm by 3.5 cm hole in the iliac blade of *A. sarcophagus* (ROM 807) may represent trauma or infection (Molnar 2001).

GOUT: Gout is a metabolic disorder in which uric acid crystals accumulate in and around the joint as masses, producing bone erosion that is characterized by overgrowth at its margins (Resnick 2002; Rothschild and Martin 1993). Such lesions were found in metacarpal I and II of FMNH PR2081 and in an unspeciated tyrannosaurid pedal phalanx from RTMP

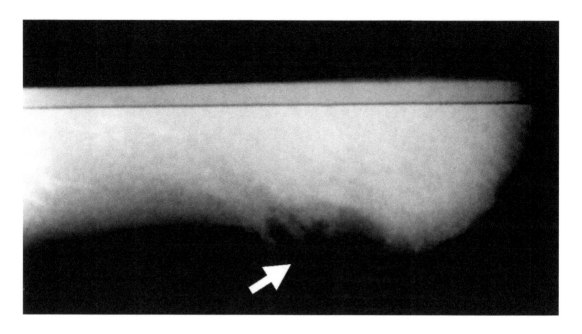

Figure 17.10. *Tangential X-ray view of phalanx in Figure 17.9. Spherical lytic area with overhanging edge.*

92.36.328 (Figs. 17.9, 17.10, Rothschild et al. 1997). The latter represents 1 of 82 *Daspletosaurus* or *Albertosaurus* phalanges examined (Rothschild and Martin 2006).

Nontyranosaurid Pathologies in Support of Predatory Activity

A series of healed fractures in *Edmontosaurus* caudal neural spines suggests attack by a large predator believed to have been *Tyrannosaurus* (Carpenter 1998). *Triceratops* healed horn and squamosal injuries have also been attributed to *Tyrannosaurus* (Happ this volume). Healing documents predatory rather than scavenging activity, and tooth puncture marks identify the predator (Carpenter 1988, 1998).

Metabolism and Physiology

ENDOTHERMY OR HOMEOTHERMY? Surface alterations and those recognized radiologically offer a limited perspective of the occasionally phenomenal quality of tissue preservation. Mary Schweitzer has documented extraordinarily preserved pristine bone with fully intact, nonfossilized microscopic structure (Schweitzer 1993; Schweitzer and Cano 1994; Schweitzer et al. 1996, 2005a, 2005b; Schweitzer and Horner 1999; Schweitzer et al. this volume). One apparently intravascular structure may even be a red blood cell on the basis of iron content. Despite this extraordinary histologic preservation, chemical structure preservation, at least of DNA, has yet to be demonstrated. The report of extraction of ancient DNA from a dinosaur (Woodward et al. 1994) proved actually be the result of contamination by humans handling the specimen (Hedges and Schweitzer 1995).

Although DNA might not be preserved, less complex molecules such as calcium phosphate apparently sufficiently preserved to allow intriguing insights. The isotopic form of oxygen preserved in bone is determined by that of the ingested water (in equilibrium with the individual's body water) and by the body temperature during bone formation. Examination of isotope values of bone from central (e.g., thoracic vertebra) and peripheral (e.g., tibia) body locations seems to distinguish endotherms (e.g., mammalian metabolism) from ectotherms (e.g., reptilian metabolism). Barrick and Showers (1994) found less than a 4°C difference in isotope content in well-preserved *Tyrannosaurus rex* bones, suggesting that this species was at least homeothermic, implying a relatively high metabolic rate similar to that noted in endotherms.

LOCOMOTION PHYSIOLOGY—THEORETICAL: Alexander (1996), Farlow et al. (1995), and Erickson et al. (1995, 2004) suggested that *Tyrannosaurus* probably did not run fast, speculating that a large and heavy animal such as *Tyrannosaurus* would seriously injure or accidentally kill itself if it fell while running at high speed (however, see Paul this volume). Predisposition to falling would be exacerbated by the cylindrical hip structure, which provides less ability to adjust for uneven ground (Hotton 1980).

The literature on speed estimation is confusing, at least to one of us (B.M.R.). Hutchinson and Garcia (2002) suggested a speed of 11–20 m/s (40–72 km/h) because muscle force is proportional to body mass and 43% of body mass is required in extensor muscles to run quickly. Farlow et al. (1995) suggested 7.5–11 m/s (27–40 km/h), as in the white rhinoceros. Erickson et al. (1995), analyzing a gracile *T. rex* (MOR 555), estimated body mass as approximately 6000 kg and muscle strength as 7.5–9.0 m^2/giganewton. Proposed dinosaur speeds, however, are difficult to interpret in view of work by Bramble and Lieberman (2004). They reported that human elite sprinters can achieve burst (less than 15 seconds) speeds of 10.2 m/s (36.7 km/h), contrasted with several-minute bursts of 15–20 m/s (54–72 km/h) for horses, greyhounds, and pronghorn antelopes. Gait in humans switches to running at the intersection of walking-running metabolic cost of transport curves (Bramble and Lieberman 2004). This occurs at 2.3–3.5 m/s (8.3–12.6 km/h) because of the mass-spring mechanism of running in which legs flex more. Human speeds are 2.3–6.5 m/s (8.3–23.4 km/h), depending on elite running status. This is high relative to mass, compared with quadrupedal trotting (3.1 m/s) and the trot-gallop transition (4.4 m/s) speeds of 110–500-kg horses. Galloping speed is 7.7 m/s for a 65-kg quadruped, ranging to 8.9 m/s for up to 10 km by elite horses. Canidae and hyaenidae frequently travel 10–19 km per day.

Bramble and Lieberman (2004) noted that human and canid running requires approximately twice as much energy per muscle mass as other mammals. This appears to be in spite of confounding factors: the springlike leg tendons of humans (compared with apes) are more efficient in generating force, reducing metabolic cost by 50%, and the palmar arch contributes an additional 17%. Another factor that affects speed is stride length. Humans increase stride length, whereas most quadrupeds increase rate (Bramble and Lieberman 2004). Long legs also increase ground contact time, which inversely correlates with energy costs. As distal muscle mass is reduced, the energy required in endurance running is reduced by its square root of the distance to the hip. Limb orientation also affects muscle mass requirement.

More posterior center of balance (of the leg relative to the trunk), shorter extensor muscle fibers, and increased moment arms reduce that required for a given increase in speed. Coombs (1978) noted that faster-running animals have longer lower (tibia + metatarsal) than upper (femur) limb bones. That is opposite of tyrannosaurids (Farlow et al. 1995; but see Holtz this volume).

LOCOMOTION PHYSIOLOGY—PATHOLOGIC EVIDENCE: Many assumptions used in predicting tyrannosaurid speed from normal bones have not yet been adequately analyzed to allow definitive conclusions. However, pathologies provide another perspective. Darren Tanke found 2 healing fractures with prominent callus development positioned about 23 and 29 cm distal to the midline articulation of gastralium in unspeciated tyrannosaur TMP97.12.229 (Grierson 1998). The pathology suggests that falls by tyrannosaurids did indeed occur, in contrast with the suggestion of Farlow et al. (1995). Healed fractures are present on the right third dorsal rib, and the right 13th and 14th gastral ribs of *Gorgosaurus* (Lambe 1917). Thus, whatever the speed, it would appear that it was sufficient for tyrannosaurids to make belly landings, given their short forelimbs.

Conclusion

The study of tyrannosaurid bones provides an opportunity not only flesh out the body morphotype, but also to understand their physiology and environmental interactions. The future of our understanding appears limited only by our ability to develop testable hypotheses. The frequency of gout and the rarity of tyrannosaurids precludes statistical assessment of changes over time. In contrast, isolated teeth are common enough for statistical studies of tooth carina pathologies. Bony manifestations of gout occur only rarely among humans with gout (Rothschild and Heathcote 1995), so many more individuals of tyrannosaurids are required for the epidemiology of gout in that taxon. It will be interesting to see teeth carina pathology related to placement in the jaw and to other jaw pathology. The patterns of pathology must be mapped out in all specimens to understand the behavior and environment of tyrannosaurids.

If bridge is according to Hoyle, then paleopathology is according to Doyle's Sherlock Holmes: once the impossible has been eliminated, whatever remains, no matter how improbable, must be the answer.

References Cited

Abler, W. L. 1992. The serrated teeth of tyrannosaurid dinosaurs, and biting structures in other animals. *Paleobiology* 18: 161–183.

———. 1997. Tooth serrations in carnivorous dinosaurs. P. 740–743 in Currie, P. J., and Padian, K. (eds.). *Encyclopedia of Dinosaurs*. Academic Press, San Diego.

Alexander, R. M. 1996. *Tyrannosaurus* on the run. *Nature* 379: 121.

Anonymous. 1992. Update on the world's largest known *T. rex*. *Geology Today* 7(3): 79–80.

———. 1997. *Tyrannosaurus rex*—a highly important and virtually complete fossil skeleton. Sotheby's Auction House, Sale 7045 auction catalog. Auction, Saturday, October 4, 1997, New York.

Archer, B, and Babiarz, J. P. 1992. Another tyrannosaurid dinosaur from the Cretaceous of northwestern New Mexico. *Journal of Paleontology* 66: 690–691.

Auffenberg, W. 1981. *The Behavioral Ecology of the Komodo Monitor.* University of Florida Press, Gainesville.

Bartosiewic, L. 2000. Baculum fracture in carnivores: osteological, behavioural and cultural implications. *International Journal of Osteoarchaeology* 10: 447–450.

Barrick, R. E., and Showers, W. J. 1994. Thermophysiology of *Tyrannosaurus rex*: evidence from oxygen isotopes. *Science* 265: 222–224.

Bramble, D. M., and Lieberman, D. E. 2004. Endurance running and the evolution of *Homo. Nature* 432: 345–352.

Brochu, C. A. 2000. Postcranial axial morphology of a large *Tyrannosaurus rex* skeleton. *Journal of Vertebrate Paleontology* 20: 32A.

———. 2003. Osteology of *Tyrannosaurus rex*: insights from a nearly complete skeleton and high-resolution CT analysis of the skull. *Journal of Vertebrate Paleontology* 22(Suppl. 4): 1–138.

Carpenter, K. 1988. Evidence of predatory behaviour by *Tyrannosaurus* (abstract). International Symposium on Vertebrate Behaviour as Derived from the Fossil Record, Museum of the Rockies, Bozeman, Montana, p. 6.

———. 1998. Evidence of predatory behavior by carnivorous dinosaurs. P. 135–144 in Pérez-Moreno, B. P., Holtz, T. J., Sanz, J. L., and Moratalla, J. (eds.). *Aspects of Theropod Paleobiology. Gaia: Revista de Geociencias, Museu Nacional de Historia Natural, Lisbon,* 15.

Carpenter, K., and Smith, M. 1995. Osteology and functional morphology of the forelimbs in tyrannosaurids as compared with other theropods (Dinosauria). *Journal of Vertebrate Paleontology* 15: 21A.

Carpenter, K., and Smith, M. 2001. Forelimb osteology and biomechanics of *Tyrannosaurus rex*. P. 90–116, in Tanke, D. H., and Carpenter, K. (eds.). *Mesozoic Vertebrate Life.* Indiana University Press, Bloomington.

Chamay, A. 1970. Mechanical and morphological aspects of experimental overload and fatigue in bone. *Journal of Biomechanics* 3: 263–270.

Coombs, W. P., Jr. 1978. Theoretical aspects of cursorial adaptations in dinosaurs. *Quarterly Review of Biology* 53: 393–418.

Currie, P. J. 1997. *Gorgosaurus? Hip and Tail.* Royal Tyrrell Museum of Palaeontology Field Experience 1996 update, p. 3.

Dingus, L. 1996. *Great Fossils at the American Museum of Natural History—Next of Kin.* Rizzoli, New York.

Erickson, G. M. 1993a. The mystery of the "split-toothed" tyrannosaurs. *PaleoBios* 14(Suppl. 5): 5.

———. 1993b. The evolution of split carinas in tyrannosaur teeth from the Late Cretaceous Western Interior. *Journal of Vertebrate Paleontology* 13(3): 34A.

———. 1995. Split carinae on tyrannosaurid teeth and implications of their development. *Journal of Vertebrate Paleontology* 15(2): 268–274.

Erickson, G. M., Smith, M. B., and Robinson, J. M. 1995. Body mass, "strength indicator," and cursorial potential of *Tyrannosaurus rex*. *Journal of Vertebrate Paleontology* 15(4): 713–725.

Erickson, G. M., Makovicky, P. J. Currie, P. J., Norell, M. A., Yerby, S. A., and Brochu, C. A. 2004. Gigantism and comparative life-history parameters of tyrannosaurid dinosaurs. *Nature* 430: 772–775.

Farlow, J. O., Smith, M. B., and Robinson, J. M. 1995. Bone mass, bone strength indicator, and cursorial potential of *Tyrannosaurus rex*. *Journal of Vertebrate Paleontology* 15: 713–725.

Grierson, B. 1998. Writing down the bones. *Western Living* 23(8): 107–117.

Griffith, J. F., Tong M. P., Hung H. Y., and Kumta S. M. 2005. Plastic deformation of the femur: cross-sectional imaging. *American Journal of Roentgenology* 184: 1495–1498.

Harris, J. D. 1997. A Reanalysis of *Acrocanthosaurus atokensis*, Its Phylogenetic Status, and Paleobiogeographic Implication, Based on a New Specimen from Texas. Master's thesis, Southern Methodist University, Dallas.

Hedges, S. B., and Schweitzer, M. H. 1995. Detecting dinosaur DNA. *Science* 268(5214):1191–1192.

Holtz, T., Jr. 2001. The phylogeny and taxonomy of the tyrannosauridae. P. 64–83 in Tanke, D. H., and Carpenter, K. (eds.). *Mesozoic Vertebral Life*. Indiana University Press, Bloomington.

Hotton, N., III. 1980. An alternative to dinosaur endothermy: the happy wanders. P. 311–350 in Thomas, R. D., and Olsen, E. C. (eds.). *A Cold Look at the Warm-Blood Dinsoaurs*. American Association for the Advancement of Science Selected Symposium 28. Westview Press, Boulder, CO.

Hutchinson, J. R., and Garcia, M. 2002. *Tyrannosaurus* was not a fast runner. *Nature* 415: 1018–1021.

Keiran, M. 1999. *Albertosauus: Death of a Predator* (Discoveries in Paleontology). Raincoast Books, Vancouver, British Columbia, Canada.

Lambe, L. 1917. *The Cretaceous Theropodous Dinosaur Gorgosaurus*. Geological Survey of Canada Memoire 100.

Larson, P. L. 2001a. Paleopathologies in *Tyrannosaurus rex*. In *Dinopress: Dinosaurs, Pterosaurs, Marine Reptiles and Extinct Animals*, 5: 6–9.

———. 2001b. Pathologies in *Tyrannosaurus rex*: snapshots of a killer's life. *Journal of Vertebrate Paleontology* 21: 71A–72A.

Larson P. L., and Donnan, K. 2002. *Rex Appeal: The Amazing Story of Sue, the Dinosaur that Changed Science, the Law, and My Life*. Invisible Cities Press, Montpelier, VT.

Molnar, R. E. 1991. The cranial morphology of *Tyrannosaurus rex*. *Paleontographica A* 217: 137–176.

———. 2001. Theropod paleopathology: a literature survey. P. 337–363 in Tanke, D. H., and Carpenter, K. (eds.). *Mesozoic Vertebral Life*. Indiana University Press, Bloomington.

Molnar, R. E., and Currie, P. J. 1995. Intraspecific fighting behavior inferred from toothmark trauma on skulls and teeth of large carnosaurs (Dinosaurs). *Journal of Vertebrate Paleontology* 15: 55A.

Mongelli, A., Jr., Varricchio, D. J., and Borkowski, J. J. 1999. Wear surfaces and breakage patterns of tyrannosaurid (Theropoda: Coelurosauria) teeth. *Journal of Vertebrate Paleontology* 19: 64A

Moodie, R. 1919. New observations in paleopathology. *Annals of Medical History* 2: 241–247.

Nalla, R. K., Stölken, J. S., Kinney, J. H., and Ritchie, R. O. 2005. Fracture in human cortical bone: local fracture criteria and toughening mechanisms. *Journal of Biomechanics* 38: 1517–1525.

Osborn, H. F. 1916. Skeletal adaptations of *Ornitholestes*, *Struthiomimus*, *Tyrannosaurus*. *Bulletin of the American Museum of Natural History* 35(43): 733–771.

Paul, G. S. 1988. *Predatory Dinosaurs of the World*. Simon & Schuster, New York.

Rega, E. A., and Brochu, C. A. 2003. Paleopathology of a mature *Tyrannosaurus rex* skeleton. *Journal of Vertebrate Paleontology* 21: 92A.

Resnick, D. 2002. *Diagnosis of Bone and Joint Disorders*. W. B. Saunders, Philadelphia.

Rothschild, B. M. 1990. Radiologic assessment of osteoarthritis in dinosaurs. *Annals of the Carnegie Museum* 59: 295–301.

——— 1997. Dinosaurian paleopathology. P. 426–448 in Farlow, J. O., and Brett-Surman, M. K. (eds.). *The Complete Dinosaur*. Indiana University Press, Bloomington.

Rothschild, B. M., and Berman, D. 1991. Fusion of caudal vertebrae in Late Jurassic sauropods. *Journal of Vertebrate Paleontology* 11: 29–36.

Rothschild, B. M., and Heathcote, G. M. 1995. Characterization of gout in a skeletal population sample: presumptive diagnosis in Micronesian population. *American Journal of Physical Anthropology* 98: 519–525.

Rothschild, B. M., and Martin, L. D. 1993. *Paleopathology: Disease in the Fossil Record*. CRC Press, London.

Rothschild, B. M., and Martin, L. D. 2006. *Skeletal Impact of Disease*. New Mexico Museum of Natural History Bulletin 18.

Rothschild, B. M., and Tanke, D. H. 1997. Thunder in the Cretaceous: interspecies conflict as evidence for ceratopsian migration? P. 77–81 in Rosenberg, G. D., and Wolberg, D. L. (eds.). *Dino Fest*. Paleontological Society Special Publication 7.

Rothschild, B. M., and Tanke, D. H. 2005. Theropod paleopathology: state of the art review. P. 351–365 in Carpenter, K. (ed.). *The Carnivorous Dinosaurs*. Indiana University Press, Bloomington.

Rothschild, B. M., Tanke, D. H., and Carpenter, K. 1997. Spheroid erosions in tyrannosaurs: Mesozoic gout. *Nature* 387: 357.

Rothschild, B. M., Tanke, D. H., and Ford, T. 2001. Theropod stress fractures and tendon avulsions as a clue to activity. P. 331–336 in Tanke, D., and Carpenter, K. (eds.). *Mesozoic Vertebrate Life*. Indiana University Press, Bloomington.

Russell, D. A. 1970. Tyrannosaurs from the late Cretaceous of Western Canada. *National Museum of Natural Science Paleontology* 1: 1–34.

Schweitzer, M. H. 1993. Biomolecule preservation in *Tyrannosaurus rex*. *Journal of Vertebrate Paleontology* 13(3): 56A.

Schweitzer, M. H., and Cano, R. J. 1994. Will the dinosaurs rise again? P. 309–326 in Rosenberg, G. D., and Wolberg, D. L. (eds.). *Dino Fest*. Paleontological Society Special Publication No. 7.

Schweitzer, M. H., and Horner, J. R. 1999. Intrasvascular microstructures in trabecular bone tissues of *Tyrannosaurus rex*. *Annales de Paleontologie* 85: 179–192.

Schweitzer, M. H., Marshall, M., Carron, K., Bohle, S., Arnold, E., Buss, S., and Starkey, J. 1996. Identification of possible blood-derived heme compounds in *Tyrannosaurus rex* trabecular tissues. P. 99 in Wolberg, D. L., and Stump, E. (eds.). *Dinofest*. Program and abstracts, April 18–21. Arizona State University, Tempe.

Schweitzer, M. H., Wittmeyer J. L., and Horner, J. R. 2005b. Gender-specific reproductive tissue in ratites and *Tyrannosaurus rex*. *Science* 308: 1456–1460.

Schweitzer, M. H., Wittmeyer, J. L., Horner, J. R., and Toporski, J. K. 2005a. Soft-tissue vessels and cellular preservation in *Tyrannosaurus rex*. *Science* 307: 1952–1955.

Tanke, D., and Currie, P. J. 1995. Intraspecific fighting behaviour inferred from toothmark trauma on skulls and teeth of large carnosaurs (Dinosauria). *Journal of Vertebrate Paleontology* 15(3): 55A.

Tanke, D., and P. J. Currie. 1998. Head-biting behavior in theropod dinosaurs: paleopathological evidence. P. 167–184 in Pérez-Moreno, B. P., Holtz, T. J.,

Sanz, J. L., and Moratalla, J. (eds.). *Aspects of Theropod Paleobiology. Gaia: Revista de Geociencias, Museu Nacional de Historia Natural, Lisbon,* 15.

Tanke, D., and B. M. Rothschild. 2002. *An Annotated Bibliography of Dinosaur Paleopathology and Related Topics, 1838–1999.* New Mexico Museum of Natural History and Science Bulletin 20.

Webster, D. 1999. A dinosaur named Sue. *National Geographic* 195(6): 46–59.

Webster, D. 2002. Debut Sue. *National Geographic* 197(6): 24–37.

Welles, S. P. 1984. *Dilophosaurus wetherilli* (Dinosaura, Theropoda), osteology and comparisons. *Palaeontographica* A 185: 85–180.

Williamson, T. E., and Carr, T. D. 1999. A new tyrannosaurid (Dinosauria: Theropoda) partial skeleton from the Upper Cretaceous Kirtland Formation, San Juan Basin, NM. *New Mexico Geology* 21(2): 42–43.

Woodward, S. R., Weyland, N. J., and Bunnell, M. 1994. Sequence from Cretaceous period bone fragments. *Science* 266: 1229–1232.

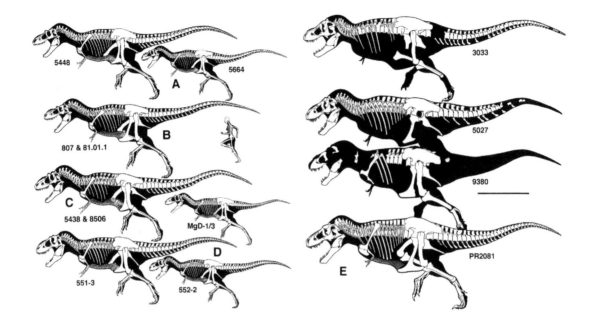

Figure 18.1. Tyrannosaurid skeletons to same scale. Scale bar = 2 m. (A) *A. libratus* AMNH 5448 (2.3 tonnes), juvenile AMNH 5664 (700 kg). (B) ***Albertosaurus sarcophagus*** ROM 807, RTMP 81.01.1, etc. (2.5 tonnes). (C) ***Daspletosaurus torosus*** AMNH 5438 and NMC 8506 (2.4 tonnes). (D) Subadult ***T. bataar*** PIN 551–3 (2.1 tonnes), juveniles ZPAL MgD-1/3 (760 kg), and PIN 552-2 (510 kg). (E) ***Tyrannosaurus rex*** largely as preserved, from top to bottom; BHI 3033 (gracile, 5.6 tonnes), AMNH 5027 (gracile, 5.7 tonnes), holotype CM 9380 (robust morph, ~5.7 tonnes), FMNH PR2081 skull distorted (robust, 6.1 tonnes).

THE EXTREME LIFESTYLES AND HABITS OF THE GIGANTIC TYRANNOSAURID SUPERPREDATORS OF THE LATE CRETACEOUS OF NORTH AMERICA AND ASIA

Gregory S. Paul

Introduction

The 2- to more than 6-tonne tyrannosaurids, which appeared near the end of the dinosaur era and were restricted to North America and Asia, were in many respects the culmination of approximately 100 million years of evolution of gigantic predatory dinosaurs. No other group matched their advanced combination of size, killing power, and speed. Although the classic tyrannosaurids were anatomically sophisticated, they were conservative in sharing a consistent body form (Fig. 18.1). Heads were large and massively constructed by normal avepod (bird-footed theropods, sensu Paul 2002) standards, temporal boxes were enlarged, fields of vision overlapped to varying degrees, and teeth were large and tended to be more conical than those of their relatives. The S-curved necks were stout, bodies were compact, and distally, tails were reduced relative to other large theropods. The arms were reduced. The pelvis was large; the legs were elongated, especially distally, and laterally compressed. This unique suite of features gave the tyrannosaurids a progressive, intricate appearance that other giant avepods lack; the latter appear crude in comparison.

Tyrannosaurids lived in a world of similarly gigantic potential prey. Unarmed and unprotected hadrosaurs and armored ankylosaurs, some with tail clubs, were everywhere. Horned ceratopsids were limited to western North America, whereas therizinosaurs and deinocheirians were limited to Asia. Gigantic sauropods existed over much of the superpredator's range, but they were largely absent in the eastern coastal plain of the western North America. A major question is whether tyrannosaurids, and by general similarity other gigantic predatory avepods, actively hunted the large prey they lived among. A few have argued that big tyrannosaurids were slow-moving scavengers that largely or entirely avoided active hunting (Lambe 1917; Colinvaux 1978; Halstead and Halstead 1981; Barsbold 1983; Horner and Lessem 1993; Horner 1994; Horner and Dobb 1997). But most researchers have concluded or presumed that they were active hunters (Osborn 1916; Russell 1970, 1977, 1989; Farlow 1976, 1994; Bakker 1986; Paul 1987a, 1988a, 2000; Farlow et al. 1991, 1995, 2000; Molnar 1991, 2000; Abler 1992, 1999; Holtz 1994, 2002, 2004, this volume; Coombs 1995; Larson 1997; Carpenter 2000; Currie 2000, 2003; Hurum and Currie

Tyrannosaurus is the most superb carnivorous mechanism among the terrestrial Vertebrata, in which raptorial destructive power and speed are combined.

Osborn (1916, p. 762)

Table 18.1. *Glossary of Terms*

Mega-avepod	Avepods whose adult forms exceen approximately 1 tonne
Bradyenergetic	Basal and resting metabolic rates, aerobic capacity, and energy budgets do not exceen pertilian maximums
Tackyenergetic	Basal and resting metabolic rates, aerobic capacity, and energy budgets exceed reptilian maximum.

2000; Carpenter and Smith 2001; Christiansen 2000; Farlow and Holtz 2002; Hutchinson and Garcia 2002; Larson and Donnan 2002; Meers 2002; Van Valkenburgh and Molnar 2002; Snively and Russell 2003; Hurum and Sabath 2003; Henderson and Snively 2004; Hutchinson 2004b; Rayfield 2004; Currie et al. 2005; Sampson and Loewen 2005; Therrien et al. 2005). Within this majority, there is considerable diversity of opinion on the details of tyrannosaurid predation. In one view, giant adults were high-risk predators that attacked prey at speeds matching or exceeding those of rhinos and possibly approaching those of field horses, engaging equally enormous running herbivores in battles to the death on a size scale not seen on land today. In another view, the adults were safety-conscious killers that moved much slower than their own progeny, and perhaps not much faster than elephants. Perhaps giant tyrannosaurids only attacked prey that was small, weak, or disabled by injury, disease, or age. In all models, tyrannosaurids scavenged when the opportunity arose.

Like other giant avepods, tyrannosaurids were radically different in form and size from any living predators (Paul 1988a; Van Valkenburgh and Molnar 2002). The only Cenozoic predator analogs are the phorusrhacid birds, but they were much smaller and had beaks rather than teeth. They are also extinct. It would therefore seem to be difficult to restore the life mode of tyrannosaurids. However, in recent years, an astonishing amount of information has become available regarding tyrannosaurid biology. Within the limitations inherent to paleobiology, these data allow us to confidently restore the dinosaur's lifestyles and habits to a degree difficult to imagine at the end of the 20th century. It is now understood that tyrannosaurids grew rapidly and apparently died remarkably young. Such a life strategy fits with and is consistent with tyrannosaurids leading such dangerous lives that they had to reach sexual maturity and breed rapidly before they died. The most plausible agent for causing early death on such a regular basis is frequent, intense combat with large, well-armed prey at dangerous speeds.

Direct evidence for active predation by adults on strong adult prey is found in the form of healed bite marks on mature hadrosaurs and ceratopsids, which were healthy enough to escape or even fend off the attack. Tyrannosaurids were capable of velocities far in excess of those of elephants, and similar to those of rhinos and perhaps horses. Such speeds were necessary because hadrosaurs and ceratopsids could run as fast as rhinos. Vision and olfaction were important sensors. The sole killing weapon consisted of the jaws, which were specialized for delivering deep bite wounds. Forearms did not play a significant role. The energetics of tyrannosaurids and their prey is most similar to those of birds in featuring high aerobic exercise capacity. But behavior was reptilian in being more stereotypical and less complex and variable than in bigger-brained birds. Major questions, some probably unanswerable, surround many of these issues.

Table 18.1 provides a glossary of terms used in this chapter.

Biogeography and Habitats of Tyranosaurids

Gigantic tyrannosaurids, which evolved in the last stages of the Mesozoic, were restricted to North America and east-central Asia; elsewhere, abelisaurs were dominant. Albertosaurines were probably limited to the western North American peninsula. This enormous area ranged from lower temperate to polar latitudes, and from coastal floodplains to highland basins. Weather varied from seasonally hot and dry to temperate, winterlike conditions. In all cases, rainfall was sufficient to support enough vegetation to feed a suitable large herbivorous prey population. In some areas, vegetation may have been dense. There is no evidence that individual tyrannosaurids migrated long distances (summary based on Russell 1977, 1989; Paul 1988a, 1988b; Horner and Lessem 1993; Farlow 1994; Holtz 2004; Lehman 2001; Larson and Donnan 2002; Sampson and Loewen 2005).

Adult Masses, and the Juvenile Versus Small Adult Problem

Several researchers have concluded that the remains of a number of small individuals represented those of lesser-sized, adult tyrannosaurid taxa (Russell 1970; Bakker et al. 1988; Paul 1988a; Currie 2003; P. Larson this volume). Others have determined that all small specimens (Fig. 18.1A, D) from North America and from the tarbosaur bearing Nemegt and related beds are the juveniles of large taxa (Rozhdestvensky 1965; Carr 1999; Henderson and Harrison this volume). The failure to find specimens that are clearly assignable both to *Nanotyrannus* and to *T. rex* in the same size range in the same beds suggests that *Nanotyrannus* is the juvenile of the latter. It is questionable whether *Alectrosaurus* and *Alioramus*, which appear to have been medium sized as adults, belong to the Tyrannosauridae (Holtz 2004). If they did not, and if all small specimens within the family are juveniles, then all adult tyrannosaurids were gigantic.

The use of proximal limb bone diameters (as used by Anderson et al. 1985; Erickson et al. 2004; Bybee et al. 2006; P. Larson this volume) to estimate the mass of extinct animals is inherently unreliable when applied to extinct forms whose anatomy does not closely match living forms (Paul 1988a, 1997; Seebacher 2001). Body mass varies by a factor of 2 among animals of differing forms with the same leg element diameters because of greatly varying locomotory adaptations, anatomical configurations, tissue composition, and safety factors. The use of bone diameters as the primary means of estimating mass is highly inappropriate because wide differences in mass–bone dimensions and in limb loading are suppressed. Nor has it been shown by large samples of wild ratites and kangaroos that mass consistently correlates closely with body mass in modern bipeds either in general, or even within populations of single species. It is plausible that bone strength and robustness-mass relationships vary within a species in different locations and/or times, or between the sexes, as a result of differing prey preferences or other factors. Therefore, bone dimensions cannot be relied on to make straightforward body mass estimates that assume that gracile taxa or morphs are correspondingly lighter in total mass than robust examples. It is more probable that the gracile morphs have more slender bones relative to total mass than the robust examples, with the 2 forms possessing similar total masses when overall dimensions are similar.

When restoring exotic extinct animals, actual or high-resolution virtual volumetric models based on rigorously restored skeletons in multiple views have, despite their limitations and the effort involved in producing them, a much smaller range of error than inherently unreliable bone dimension–based estimates (Paul 1988a, 1997, 2002; Henderson 1999; Seebacher 2001). When scientifically restored models are consistently sculpted following the same standards, they are especially valuable in producing a set of mass estimates that can be compared relative to one another, providing an independent test of mass–bone dimension relationships. The possible presence of air sacs does not greatly affect such models because these respiratory spaces are always a modest minority of total volume even in modern birds, and there is a narrow zone of plausible variation in tyrannosaurids. Their skeletons were significantly less pneumatic than those of birds, so tyrannosaurid air sacs were probably less capacious (Paul 1988a, 1997, 2002). Because the air sacs were integral to respiratory capacity and performance, it is unlikely that their volume varied substantially between individuals. The dedication of large portions of total volume to locomotor muscles also limited the potential for variation in internal air volume. It is improbable that the specific gravity of tyrannosaurids was much above or below 0.85–0.80 (the first value is used here), and it is essentially impossible for it to have been below 0.7, the minimum observed in flying birds (Paul 2002).

Volumetric models indicate that adult *Albertosaurus* (=*Gorgosaurus*), *Daspletosaurus*, *T. bataar*, and *T. rex* specimens exceeded 2 tonnes in total mass, and none exceeded the 6-tonne range (Fig. 18.1; Paul 1988a, 1997). Significantly lower estimates (as per Bakker 1986; Christiansen 2000; Erickson et al. 2004) are not possible because such low tissue volumes relative to the correctly sized skeletal framework require that the subjects either be emaciated to starvation levels, or to assume impossibly low specific gravities due to unrealistically large air sac capacity. *Albertosaurus* was no more massive than living estuarine crocodiles, which reach about 1 tonne (Matthews and McWhirter 1992). Because the femora of ~1.5-tonne *Allosaurus* were more robust than those of *Albertosaurus*, femur strength–based mass estimates result in *Allosaurus* as being heavier than the *Albertosaurus* (Erickson et al. 2004; Bybee et al. 2006), even though albertosaurine skeletons are abut 20% larger than those of allosaurids (Paul 1988a, 1997). In reality, *Albertosaurus* and *Daspletosaurus* were in the 2- to 3-tonne range. There is no particular evidence that the total volumes and masses of the robustly constructed *Daspletosaurus* were significantly greater than those of the more lightly built *Albertosaurus* (Fig. 18.1A–C). Rather, the bones of *Daspletosaurus* appear to have been more heavily constructed relative to their mass than those of gracile *Albertosaurus*.

T. bataar skulls approach in size, but do not quite match, those of *T. rex*, and fully mature skeletons cannot be compared because postcrania of the largest Asian specimens have not yet been described. *T. bataar* and *T. rex* were 2 to 3 times heavier than *Albertosaurus* and *Daspletosaurus*. Some high estimates for the mass of the largest *T. rex* (Paul 1988a) were based on incorrect data on the size of the largest specimens. Adult *T. bataar* and especially *T. rex* were robustly constructed, although again, the exact degree of stoutness of the Asian form cannot be assessed because no descriptions of the largest specimens have been made. The strong divergence in body mass scalings estimated by P. Larson (this volume) for the 2 *T. rex* morphs are not supported by the skeletal restorations of large specimens and their volumes (Fig. 18.1E; Paul 1988a, 1997). That the gracile femora are sometimes longer than much more robust femora suggests that the latter individuals were not heavier in total mass than the former, but were stronger boned. The volumetric mass estimates of big adults cluster in a narrow zone of a little below and above 6 tonnes, regardless of the morph. This pattern suggests that tyrannosaurids, whose growth appears to have been determinate (Erickson et al. 2004), were genetically programmed to reach a consistent adult mass.

It is not clear whether *T. bataar* and *T. rex* were scaled-up versions of lithe *Albertosaurus* or stouter *Daspletosaurus*. The increase in robustness is associated with the sheer increase in mass, but the exact manner is which dimensions scale with increasing size to maintain constancy of locomotor and predatory performance is still not entirely understood (Paul 2000). Nor are the 2 tyrannosaurid taxa of the 4- to 6-tonne range, but that have differing skeletal robustness similar to that seen in *Albertosaurus* and *Daspletosaurus*, available for comparison to the smaller morphs.

Growth, Life Spans, and Their Implications

The growth and adulthood of tyrannosaurids was unusual in a number of key respects.

Bradyenergetic continental reptiles always grow slowly, including the largest extinct freshwater crocodilians (Case 1978a; Paul 1994, 2002; Erickson and Brochu 1999). Tachyenergetic land animals are capable of much faster growth and can achieve great bulk in correspondingly shorter time spans (Paul 1994; Paul and Leahy 1994; Erickson et al. 2001; Padian et al. 2001). Growth rates as measured by counting lines of arrested growth show that tyrannosaurids and allosaurids grew as fast as mammals of similar size (Erickson et al. 2004; Horner and Padian 2004; Bybee et al. 2006). If anything, the modest growth rates Erickson et al. (2004) estimated for *Daspletosaurus* and especially *Albertosaurus* are too low as a result of the mass underestimates they used, as they acknowledged was possible. When corrected adult mass values are used, all tyrannosaurids as well as allosaurids exhibit maximum sustained growth rates similar to those of giant mammals and well above those seen in equally large nonmarine reptiles (Fig. 18.2). For the purposes of this study, I assume that the tachyenergetic tyrannosaurids and other predatory mega-avepods had high aerobic exercise capacity broadly similar to those of ground birds and large mammals,

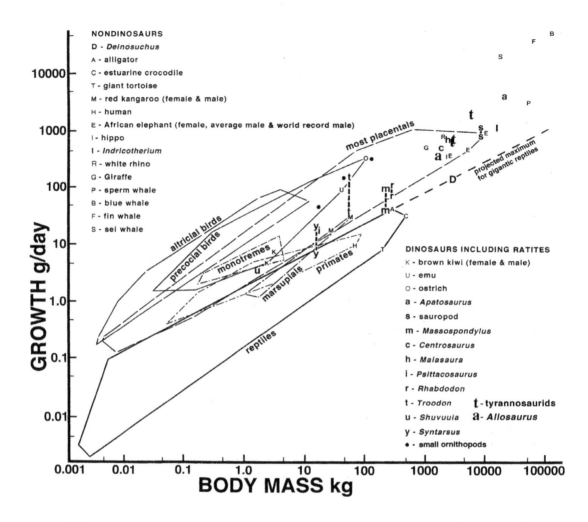

Figure 18.2. Daily growth rate as a function of adult body mass in non-marine amniotes. Data sources include Case (1978a), Erickson and Brochu (1999), Erickson et al. (2001, 2004), Padian et al. (2001), Paul (2002), Horner and Padian (2004), and Bybee et al. (2006).

as well as total yearly energy budgets well above reptilian levels, although their energy budgets and resting metabolic rates may have been in the lower avian-mammalian range (Paul and Leahy 1994; Paul 1998, 2002).

Modern giant animals have long life spans in the wild (Case 1978a; Nowak 1999). The longest living large land animals are bradyenergetic tortoises, which may survive for a century or more if they live on predator- and disease-free islands. Among tachyenergetic continental land giants—hippos, rhinos, and elephants—normal life spans are about 4 to 6 decades (Owen-Smith 1988). The mature period of life after the completion of growth is itself a number of decades.

Past speculations on the life spans of giant dinosaurs ranged from many decades to centuries (Case 1978b). For megapredators (mega is more than 1000 kg, as per Owen-Smith 1988) living on danger-filled continents, it is inherently unlikely that they would live much longer than elephants. Few would have predicted that age estimates would indicate that no tyrannosaurid, including the biggest known, lived past about 3 decades according to 2 independent studies (Erickson et al. 2004; Padian et al. 2001), or that death occurred only a decade or less after the completion of growth. In other words, tyrannosaurids grew fast and reached full adult mass at about the same time as equally large mammals, but then consistently died without experiencing extended adulthoods. The same pattern is observed in allosaurids (Bybee et al. 2006), but the growth pattern in other mega-avepods is not yet documented. The combination of short lives and the absence of extended maturity is unknown in large living animals (Owen-Smith 1988; Nowak 1999) and is aberrational, at least by modern standards. Such a radical termination of adulthood after a short juvenile period indicates an extreme lifestyle that evolved under exceptional selective pressures. The dangers faced by tyrannosaurs are recorded in their skeletons as trauma-induced injuries (Hanna 2000; Rothschild et al. 2001; Tanke and Currie 2000; Larson and Donnan 2002; Brochu 2003; Rothschild and Tanke 2005; Rothschild and Molnar this volume).

Many animals have short adult reproductive periods followed by death, but these are mostly invertebrates, such as squid and butterflies, and certain fish, such as salmon. It is more common for large tetrapods to have extended sexual maturity in order to maximize reproductive success, whether in numbers of offspring created or care that can be given to the young. That tyrannosaurids were more like salmon than elephants indicates that they were prevented from following a more typical life pattern for land giants. The most plausible causal agent is that adult mortality was consistently so high that few individuals would have lived to old age. In this situation, it would be reproductively advantageous to dedicate resources toward maximizing replication in the brief period before death and to forgo placing resources into maintaining health into old age. The statistical likelihood that tyrannosaurids would die young should therefore have imposed selective pressures on tyrannosaurids to be genetically preprogrammed to do so. That the small-brained tyrannosaurids lived fast and died young is compatible with their chronically living closer to the edge of danger and death, and having correspondingly lower anatomical safety factors, such as leg bone strength relative to running performance, than do big-brained, slow-breeding giant herbivorous mammals.

As tyrannosaurids grew up, they experienced significant, fairly straightforward allometric changes, with juveniles being gracile and adults significantly stouter (Fig. 18.1; Russell 1970; Paul 1988a, 2000; Christiansen 2000). This differed from some other dinosaurs, such as hadrosaurs and sauropods, which were much more isometric in trunk and leg dimensions regardless of life stage (Christiansen 1997). But increasing robustness with growth allometry is common in large land tetrapods. The allometry in leg proportions is broadly similar in tyrannosaurids and ungulates, with the distal elements moderately less elongated relative to the femur with increased size. A degree of skull strengthening by increasing the robustness of the elements occurred as size increased. The jugal process of the postorbital tended to invade the orbit with an ossified projection.

Growing tyrannosaurids exhibited an unusual change in skull proportions in which the rostrum deepened with maturity. Juveniles had low, long-snouted, gracile skulls compared with the deeper rostrums of adults (Paul 1988a; Carr 1999; Currie 2003). This is the opposite of the usual ontogenetic pattern, in which the snout begins short and deep and elongates

Figure 18.3. Adult mega-avepod skulls in lateral and dorsal view, reproduced to a constant length to facilitate comparison of element proportions. (A) *Allosaurus*. (B) *Albertosaurus libratus* AMNH 5458/RTMP 85.62.1. (C) *Daspletosaurus torosus* holotype NMC 8506. (D) *Tyrannosaurus bataar* holotype PIN 551-1 with some details from ZPAL MgD-1/4. (E) *T. rex* AMNH 5027. (A) and (E) include ventral view of upper tooth arcades with cross section of a premaxillary tooth.

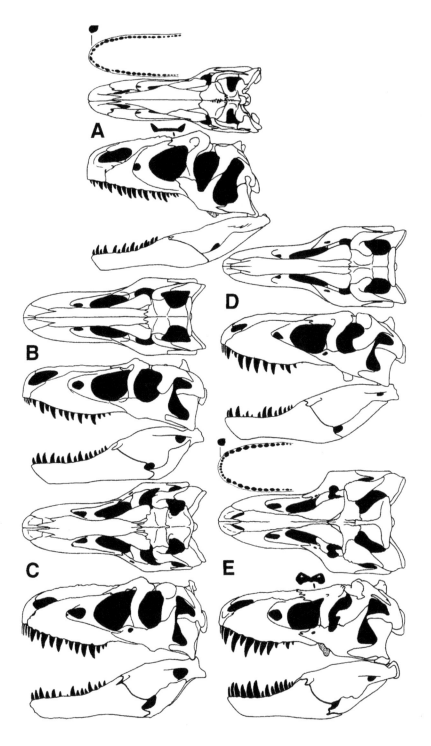

with growth. The teeth of juveniles were more bladelike than those of adults and may have been more numerous, in at least some cases (Carr 1999) but not others (Currie 2003). The changes appear to have been most extreme in *T. rex*. The ontogeny of skull proportions is therefore another unusual feature of the tyrannosaurids. The most dramatic growth changes appear to have occurred in *T. rex* in part because the species experienced the largest size change. However, *T. bataar* did not undergo comparably extensive alterations in form, especially in the teeth (Hurum and Sabath 2003), even though it almost matched its American relation in adult size.

Anatomical Assessment and Comparisons

Tyrannosaurids retained the basic head and body form characteristic of other predaceous mega-avepods (e.g., allosauroids, megalosaurs, abelisaurs, spinosaurs, and ceratosaurs, and distinct from the herbivorous therizinosaurs; Figs. 18.3, 18.4). The skull was large relative to the body (Van Valkenburgh and Molnar 2002) and long and low, with exceptions to the standard being the shorter, deeper, relatively smaller skulls of abelisaurs and the hyperelongated, shallow skulls of spinosaurs. Eyes were large in absolute terms (the illusion they were small being an allometric scaling effect; Paul 1988a; Holtz this volume). Expanded olfactory bulbs of the brain and large nasal posterior nasal passage capacity indicate that olfaction was well developed (Brochu 2000; Franzosa and Rowe 2005). Brains were simple and reptilian in form and organization (Brochu 2000; Larsson et al. 2000; Paul 2002; Franzosa and Rowe 2005). Necks were moderate in length, flexibility, and S curvature. The body was not elongated and was fairly rigid, as indicated by partial ossification of interspinal ligaments. The tail was long and supple. Arms were much too short for use in locomotion. The pelvis was large, and it anchored long, birdlike, tridactyl legs, including a small, unreversed hallux. Allometric scaling of body proportions, in particular abbreviation of the body and tail, assisted large theropods in maintaining reasonable turning ability despite their increasing inertial mass and horizontal body posture (Henderson and Snively 2004; Paul 2005). Extensive pneumaticity of the skeleton and a number of features of the rib cage indicate the presence of a well-developed, preavian air sac lung ventilation system (Paul 1988a, 2001, 2002; Leahy 2000; Perry 2001; O'Connor and Claessens 2005).

In most avepods, the rims of the bones immediately lateral to the tooth rows were sharp edged, and the teeth were closely spaced. This arrangement is more similar to lip-bearing amphibians and lepidosaurs than to crocodilians, which are unusual in lacking tooth coverings. In crocodilians, the rims of the jaw margin tend to be rounded, and most teeth are set in widely spaced sockets. In most theropods, the maxillary teeth were moderate in size, and the lateral row of foramina on the dentary is set immediately below the tooth row. In ceratosaurs and tyrannosaurids, the teeth were longer and the lateral row of foramina set lower on the dentary. The relationship between maxillary tooth length and dentary foramina position is not explicable if lips were absent, but is logical if the mandibular lip pocket was deeper in ceratosaurs and tyrannosaurids in order to accommodate their longer teeth (Fig. 18.5). The crocodilian-like dentition of spinosaurs suggests they were lipless, at least along the anterior tooth row.

Bakker (1986), Paul (1988a), Molnar (1991), Larson (1997), Larson and Donnan (2002) and Larsson (this volume) restore tyrannosaurids and other mega-avepods with varying degrees of cranial kinesis. Even in mature

Figure 18.4. *Skeletons of large avepods reproduced to a constant body mass to facilitate comparison of skeletal proportions. (A) Ostrich. (B) Ornithomimid* **Struthiomimus**. *(C) Juvenile* **Albertosaurus**. *(D)* **Daspletosaurus**. *(E) Abelisaur* **Carnotaurus**. *(F) Charcharodontosaurid allosauroid* **Giganotosaurus**. *(G)* **Allosaurus**. *(H) Sinraptorid allosauroid* **Yangchuanosaurus**. *(I)* **Ceratosaurus**.

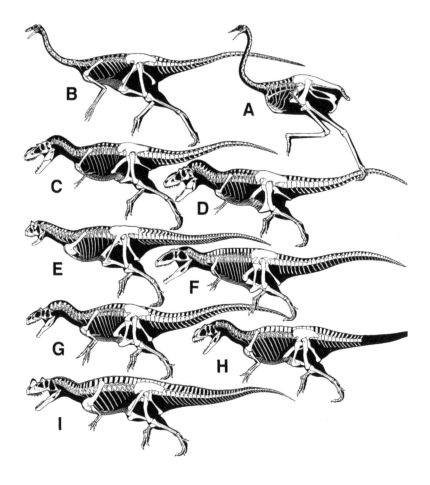

Figure 18.5. *As in all tyrannosaurids, killing power was concentrated entirely in the enormous head of* **Tyrannosaurus rex**, *which took the group's cranial adaptations to an extreme. The posterior temporal box was greatly expanded laterally, enlarging the volume of jaw and neck muscles, and rotating to orbits anteriorly so the eyes had large visual overlap. The teeth were large and stout, and formed D-shaped anterior arcade.*

skulls, many of the sutures are sufficiently open that some skulls were preserved disassembled or have been easily disassembled after preparation, leaving open the possibility of considerable movement in life. However, if skull mobility was present in tyrannosaurids, it may have been less developed than in mega-avepods with more lightly built skulls. The tyrannosaurids' more robust skull roof, extensive squamosal-quadratojugal contact, and broad vomer-premaxilla-maxilla contact may have inhibited or prevented cranial mobility. Tyrannosaurids, like other theropod dinosaurs, lacked the push-pull quadrate-pterygoid articulation integral to avian kinesis (Paul 2002). But the same was true of *Archaeopteryx* (Paul 2002), whose reduced or absent postorbital-jugal contact suggests the presence of some degree of kinesis. How the cranial musculature might have operated any such preavian kinetic system also remains poorly understood. Possibly intracranial mobility was largely or entirely passive.

Small basal tyrannosauroids possessed simple feathers (Xu et al. 2004), but patches of mosaic scales are preserved on some large tyrannosaurid specimens (Currie, personal communication). There is no evidence that any predatory mega-avepod sported an extensive feather pelage (Paul 2002).

In many respects, tyrannosaurids were among the most distinctive mega-avepods. Other giant predaceous avepods shared the following with tyrannosaurids: skulls were more lightly constructed because cranial fenes-

trae were large, being set between rather slender intervening bars (Figs. 18.3A, 18.4E-1), and the roof of the skull above the posterior antorbital fossa was remarkably weakly constructed (Fig. 18.3A, E). Skulls were narrow, with the temporal box not much broader than the rostrum. As a consequence, the eyes had at most limited forward vision. Teeth were modest in size and strongly bladed, except for spinosaurs. The trunk was fairly long, and the tail was quite long and heavy (Fig. 18.4E–I). Arms were usually well developed in most cases, and had 3 large, clawed fingers. The exception was the abelisaurs' atrophied arms with degenerate hands. The pelvis was modest in size, the legs were moderately long, distal elements were moderately elongated, and the feet were only moderately laterally compressed. Abelisaurs are again the exception: their pelves were proportionally larger than the norm, and the legs were longer.

Adult tyrannosaurids shared the following compared with other mega-avepods: the skull was heavily constructed (Fig. 18.3B–E), and the interfenestral bars were strengthened at the expense of the fenestrae, whose size was reduced. In particular, the squamosal-quadratojugal bar was enlarged until it almost, but not quite, split the lateral temporal fenestra into 2 small openings. The skull roof above the antorbital fenestra was stronger in tyrannosaurids than in equal-sized relatives (Fig. 18.3A, E). Tyrannosaurid palates were unusually strongly built. The parietal crests between and posterior to the superior temporal fenestrae were enlarged. The temporal box was broader, both relative to the size of the skull and relative to the rostrum (Fig. 18.5). The mandibles were deeper, both at the dentary and especially posteriorly. The tyrannosaurid dentary was constructed more strongly (Therrien et al. 2005). The overall result was to make the skull markedly stronger and more resistant to impacts as well as to torsional and other bending forces (Paul 1987a, 1988a; Molnar 1991, 2000, this volume; Farlow et al. 1991; Abler 1992, 1999; Erickson and Olson 1996; Erickson et al. 1996; Hurum and Currie 2000; Farlow and Holtz 2002; Rayfield et al. 2002; Holtz 2002; Meers 2002; Rayfield 2004; Therrien et al. 2005). The same researchers concur that the lateral expansion of the temporal box, the enlargement of parietal crests, deepening of the posterior mandible, and the massive construction of the skull indicate that the jaw musculature was significantly larger and more powerful than that of other mega-avepods. The enlargement of the posterior face of the broadened braincase is indicative of enlarged muscles in the neck. Tyrannosaurid teeth tended to be fewer in number, larger, and more conical (Paul 1987a, 1988a; Molnar 1991, 2000, this volume; Farlow et al. 1991; Abler 1992, 1999; Erickson and Olson 1996; Erickson et al. 1996; Holtz 2002, 2004; Meers 2002; Rayfield 2004). A particularly distinctive feature is the D-shaped cross section of the premaxillary teeth, which collectively formed a more rounded arcade that may have formed a scooplike cookie-cutter wound-inflicting device (Figs. 18.3A, E; 18.5; Paul 1987a, 1988a); it may also have been used to help clean flesh from bone (Holtz 2002). The skull, tooth, neck, and muscle adaptations combined to give tyrannosaurids a powerful punch-pull biting action. These features also seem to have rendered tyrannosaurids better able to hold onto and manipulate prey with their mouths (Therrien et al. 2005).

Tyrannosaurid orbits face more anteriorly than those of other mega-avepods (Figs. 18.3, 18.5), a structure that has often been presumed to indicate a greater degree of binocular vision (Stevens in press). However, the lateral expansion of the temporal box forced the anterior rotation of the orbits, and it is possible that the tyrannosaurids' overlapping fields of vision were an incidental side effect of enlargement of the jaw musculature rather than a primary selective adaptation. Alternately, the expansion of the posterior part of the skull may have allowed tyrannosaurids to enjoy the benefits of overlapping visual fields denied other, narrower skulled mega-avepods in which the orbits had to face laterally. On the other hand, birds with overlapping fields of vision often have markedly less stereoscopic depth perception than the external position of the eyes suggests (Molnar 1991; Martin and Katzir 1995). The tyrannosaurids' limited neural capacity casts

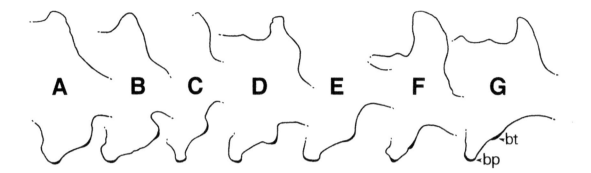

Figure 18.6. Profiles of braincases in left lateral view drawn to a constant height to facilitate comparison of the position of the basitubera (bt) relative to the basipterygoid (bp). (A) *Sinoraptor* IVPP 10600. (B) *Allosaurus* UUVP 5583. (C) *Albertosaurus libratus* TCM 2001.89.1. (D) *A. sarcophagus* RTMP 81.10.1. (E) *Daspletosaurus* sp. RTMP 94.143.1. (F) *Tyrannosaurus bataar* ZPAL MgD-1/4. (G) *T. rex* AMNH 5117 with some details from 5207. Sources include Osborn (1912), Madsen (1976), Bakker et al. (1988), Currie and Zhao (1993), Hurum and Sabath (2003), and Currie (2003).

some doubt on whether their optic lobes could process true stereovision. The tyrannosaurid brain was about 50% larger than those of other mega-avepods, suggesting it had slightly better mental capabilities (Brochu 2000; Larsson et al. 2000; Paul 2002; Franzosa and Rowe 2005). Whether tyrannosaurid brains were in the upper reptilian or lower avian size ranges is not clear because of uncertainties in extrapolating the avian and reptilian data to the size of the megadinosaurs (Larsson et al. 2000; Paul 2002). The olfactory lobes appear to have been better developed than in other mega-avepods (Brochu 2000; Franzosa and Rowe 2005), although comparative data on the olfactory apparatus of various tyrannosaurids are not currently available. It is possible that the enlargement of the brain was partly or largely due to the needs of processing the information received from the olfactory lobes. Conversely, the expansion of brain size may have allowed the improvement in olfaction.

In mega-avepods, the orientation of the occiput from the dorsal rim to the basitubera ranges from moderately dorsal to moderately ventral (Fig. 18.6; Coria and Currie 2002). Tyrannosaurids fit into the latter category (Fig. 18.6C-G). The basitubera tend to be more anteriorly placed relative to the basipterygoids in the tyrannosaurids, but the difference from other giant theropods is modest (Fig. 18.6). These features indicate that the tyrannosaurid head was often held somewhat ventroflexed at the end of the S-curved neck.

In the adult tyrannosaurid postcranial skeleton (Figs. 18.1, 18.4C, D), the neck was stouter, the trunk was more compact, and the distal part of the tail shorter and more gracile than in equal-sized avepods. Tyrannosaurids are famed for having arms that were severely reduced, with just 2 flexible fingers bearing modest claws. The reduction of the forelimbs progressing distally indicates developmental atrophy (Lockley et al. this volume), but the arms were not withered to the point of total nonfunction (Carpenter and Smith 2001; Lipkin and Carpenter this volume). Tyrannosaurid pelves were proportionally large, including the area of the proximal ischial plate. The legs were more elongated than those of most other mega-avepods, especially the distal elements (Coombs 1978; Paul 1987a, 1988a; Holtz 1994, 2004; Farlow et al. 2000). Even *T. rex* had a tibia/femur ratio similar to that of horses. The metatarsus was unusually compressed laterally as a result of the arctometatarsalian condition (Holtz 1994; Snively and Russell 2003). The avepod dinosaurs with trunks, tails, and legs most similar to those of tyrannosaurids were the ratitelike ornithomimids. Tyrannosaurids therefore differed from all other large predatory dinosaurs by focusing mass, especially that of muscles, into the head and the legs at the expense of the arms and distal portions of the tail. As a result, killing power was concentrated in the head, to the point that the arms and hands no longer played an offensive role to the degree seen in bigger-armed mega-avepods. The other group that abandoned arms as important weapons while emphasizing leg power was the abelisaurs. Their arms appear to have been even less functional, but the skull was not proportionally as large, robust, or large toothed, the distal part of the tail was not reduced, and the legs were not as well adapted for speed because the distal elements were not as laterally compressed. Nor were the tyrannosaurids' feet major killing weapons because the claws were blunted and more suitable for running. The concentration of muscle mass into the hind limbs and the elongation and lateral compression of the distal segments mean that tyrannosaurids were better adapted for running that any other gigantic avepods. Only some abelisaurs approached them in these regards.

Tyrannosaurid Leg Power, Posture, and Speed

It is not possible to volumetrically model the precise percentage of body mass dedicated to a given body part (Paul 1997, 2002), so estimates are limited to probable ranges compatible with the anatomical proportions of extinct taxa and extrapolated from modern analogs. The leg muscles of ratites are one-quarter to four-tenths of total body mass (Alexander et al. 1979; Patak and Baldwin 1993; Abourachid and Renous 2000; Hutchinson 2004a), about 70% of which are leg extensors. Most of the musculature is concentrated proximally in the thigh and the calf bundle immediately below the knee. The ilial plates of tyrannosaurids are not proportionally as large as those of ground birds (Fig. 18.4A, C, D). However, ratites lack the dinosaurs' large tail-based caudofemoralis, which retained a stout caudal base for anchoring this large femoral retractor muscle. Although difficult to quantitatively measure in dimensional terms, the area of the combined ilium and proximal part of the tail is approximately comparable in ratites and tyrannosaurids. The length of the femur relative to the mass of the body is similar in ratites and tyrannosaurids according to volumetric models. Therefore, the combined proximal limb musculature of the tyrannosaurids should have been comparable in proportional volume to those of ratites. The large cnemial crest of tyrannosaurids is indicative of a well-developed "drumstick."

In both tyrannosaurids and ratites, overall mass is or was reduced by the presence of extensive air sacs, and little mass occurred in the forelimbs. Ratites constantly carry a load of gastrolithes, consumed fodder, and feces that can exceed 10% of total mass (Herd and Dawson 1984; Wings 2004, personal communication). The belly of a hungry, hunting tyrannosaurid would have been empty. The reduced distal part of the tail made up only a few percent of the mass of a tyrannosaurid. Very probably at least a fifth of the total mass of tyrannosaurids consisted of leg muscles. Otherwise, the hind limb muscles would be atrophied relative to the large legs and the available area of locomotor muscle attachments. It is more probable that leg muscles made up one-quarter to one-third of the total mass, and four-tenths is plausible.

Paul (1987b, 1988a, 1998) and Christiansen (2000) restored giant tyrannosaurids with flexed knees on anatomical grounds. Hutchinson and Garcia (2002), Hutchinson (2004b), Hutchinson et al. (2005), and Hutchinson and Gatesy (2006) argued that the knee had to be essentially straight as a result of the demands of bearing great mass. As knees become increasingly straight in mammals, knee orientation changed dramatically in order to accommodate the change in posture; sauropods and stegosaurs have adaptations that allow their knees to fully straighten (Paul 1987b, 2000). The knees of avepod dinosaurs, regardless of size, are remarkably uniform in morphology, and are essentially avian. As in modern birds, straightening the knee disarticulates the wedge-shaped lateral femoral condyle from the tibia and fibula. This leaves the condyle without any function at this stage of the step cycle, and increases the knee's vulnerability to dislocation by longitudinal rotation along the long axis of the limb.

Christiansen (2000) observed a modest decrease in knee flexion orientation progressing from small to gigantic avepods. However, his sample included many small avepods that were not as speed adapted as tyrannosaurids, and whose posture may have been more crouching as a result of factors not related to size. A comparison limited to more anatomically uniform juvenile and adult ornithomimids and tyrannosaurids found little, if any change, regardless of size (Paul 2000), with the knees unable to straighten much beyond 120°. A detailed restoration showing how a giant avepod knee could remain fully articulated when straight has yet to be published (Farlow et al. 2000, fig. 10, was generated at too low a resolution to qualify, and the figures in Hutchinson et al. 2005 and Hutchinson and Gatesy 2006 are grossly defective, as discussed below). A cartilage extension of the lateral femoral condyle that remained articulated with the tibia-fibula would not provide adequate bracing against dislocation because cartilage is weak when subjected to shear loads. The need to brace the knee with proper hard bone articulations was, if anything, greatest in giants, so if adult tyrannosaurids had straight knees, then the morphology of the joint should have been modified to allow the posture that occurs in all other vertical-limbed tetrapods.

Gatesy et al. (2006) corrected the above errors in knee articulation. By a process of exclusion based on sensitivity analysis, they calculated the only viable leg posture for *Tyrannosaurus*. The result included a strongly flexed knee similar to the posture arrived at by Paul (1987b, 1988a, 2000). This is not surprising when it is realized that horizontal bipeds with a center of mass significantly anterior to the acetabulum, like theropods, must place the foot similarly forward of the hip joint, and that requires that the femur slope anteroventrally and that the knee be flexed. Vertical-bodied bipeds like humans can have a vertical femur and straight knee because the center of mass is directly over the acetabulum.

If low knee flexion reduces the need for large leg muscles, then straight-kneed humans should have smaller leg extensors than flexed-kneed ostriches. However, the data in Hutchinson (2004a) show that the leg extensor/total body mass ratio in humans and ostriches is the same. As explained below, even leaving aside critical anatomical errors, the uncertainties surrounding calculating force requirements for locomotion are still too extensive to allow the method to be the critical factor in restoring limb posture. There is no question that tyrannosaurids had the highly flexible ankles required to achieve a running gait (Paul 1987b, 1988a, 2000).

Thulborn (1982) and Coombs (1978) concluded that young, gracile ornithomimids, and by inference juvenile tyrannosaurids, were not as swift as ratites. Hutchinson and Garcia (2002) estimated that a fast-running juvenile tyrannosaurid needed far larger leg muscles than are present in some similar-sized ratites to run at the same speed. This serious overestimate casts doubt on the study's methodology. Otherwise, there is wide agreement that the small, gracile tyrannosaurids were high-velocity runners able to approach or match the speeds of the similarly designed ornithomimids, and by inference ratites (Russell 1977, 1989; Bakker 1986; Paul 1987a, 1987b, 1988a, 2000; Holtz 1994, 2004; Christiansen 2000; Currie 2000; Farlow et al. 1995, 2000; Hutchinson 2004b). Many researchers presume that bigger tyrannosaurs ran slower (Russell 1977; Coombs 1978; Christiansen 2000; Currie 2000; Farlow et al. 1995, 2000; Hutchinson and Garcia 2002; Hutchinson 2004b; Hutchinson et al. 2005). But McMahon and Bonner (1983, p. 151) observed that speed potential rises with increasing size, so it is as if the largest "animals had the opportunity to be the fastest as well as the biggest but didn't care to try." Bakker (1986) and Paul (1987a, 1988a, 2000) suggested that tyrannosaurids maintained consistently high speeds as they grew gigantic.

The influence that proportional leg muscle mass has on running performance is ambiguous. Even though humans and ostriches of similar mass appear to share similar leg muscle/total body mass ratios, the bird can run much faster. The bird with the largest leg muscles observed to date is the tridactyl-footed emu, not the ostrich (Hutchinson 2004a), even though the semiunidactyl ratite is probably a faster runner (Alexander et al. 1979). It is therefore possible that other anatomical adaptations allow some animals to better exploit the power potential of locomotor muscles in terms of speed, and that such adaptations are at least as important for restoring top attainable speed as is computer analysis of locomotor muscle mass.

Animals often evolve extreme adaptations in order to achieve extreme performance, such as the flight of the biggest pterosaurs and teratornes, the speed of cheetahs, and the deep diving ability of some marine birds and mammals. If giant tyrannosaurids ran at high velocities, and if extreme adaptations were necessary for them to do so, then evolutionary selective pressures would have had pushed biology to the feasible limits needed to achieve the performance, and it cannot be assumed that they possessed average levels of biological performance. The problems of reliably extrapolating and restoring the quantitative factors necessary for restoring the locomotion of gigantic tyrannosaurids are being explored by Sellers and Paul (2004) by using evolutionary robotics. On the basis of genetic algorithms that learn to walk and run (Sellers et al. 2003), initial results suggest that if the leg muscles consisted predominantly of fast-twitch fibers optimized for producing maximal burst anaerobic power, then the largest tyrannosaurids were capable of a full, suspended phase run at speeds similar to those of rhinos, and well in excess of those elephants. These estimates are conservative in that energy-storing and -releasing tendons and bones and other speed-enhancing adaptations have not yet been factored into the simulation. It is possible that giant tyrannosaurid leg tendons and ligaments were stretched so taut that when the legs were folded to sit down, the animals' mass was barely sufficient to hold the body down. Such a springlike loading of the legs could have reduced the work needed to hold the body up when running, allowing more of the locomotory muscle power to be converted into speed and endurance. In order to fully investigate the athletic potential of flexed-limbed dinosaurs, future simulations should include a range of plausible values for biological factors that range from average to the highest that are viable. That will increase the probablility of capturing the maximum speed potential, and the latter is likely to be closest to representing the performance of tyrannosaurids because their skeletons appear to have been better adapted for running than living mammals of the same size.

Modern rhinos evolved from terrestrial running ancestors. They retain flexed legs with flexible ankles that, despite being quite short and not very laterally compressed, allow them to gallop. Columnar, unflexed legs with inflexible ankles and short feet prevent elephants from reaching a full run, thus limiting them to a top speed of about 25 km/h (Paul 2000; Hutchinson et al. 2003). Elephants are slow not because they are gigantic—even juvenile elephants are no faster than the adults—but because they are anatomically adapted to be slow, a result of having descended from semiaquatic herbivores and living in a world free of gigantic predators. A major portion of the mass of mammalian megaherbivores (sensu Owen-Smith 1988) is concentrated in the plant-fermenting gut. About 10% to 20% of a large, healthy herbivore's mass consists of forage and feces (Robertson-Bullock 1962; Short 1963). Because rhino legs are so short, the small locomotor muscle mass is apparently a large portion of total mass. The limited available data suggest that elephants do not have large leg muscles (Robertson-Bullock 1962), and their limb anatomy is not able to effectively convert their power into speed.

Adult tyrannosaurids descended from and grew up from fast runners, and they were probably empty-bellied, flexed-limbed predators with mobile ankles that emphasized nonlocomotor weight reduction in favor of expanded leg muscles. The 1- to 3-tonne *Albertosaurus* and *Daspletosaurus* had much longer, more distally gracile, birdlike legs than similar-sized rhinos, and they were the most speed adapted tetrapods in their size class. It is improbable that they were slower, and may have been actually faster, than rhinos. *T. bataar* and *T. rex* were only 2 or 3 times heavier than *Albertosaurus* and *Daspletosaurus*, and they possessed the same ratitelike running adaptations. Hence, the speed potential of giant tyrannosaurids was not lower than their lesser relatives on a morphological basis, and they were uniquely speed adapted for their size class. The anatomy of *T. rex* was adapted to exploit the maximum practical power production of its enormous leg muscles, weighing between 1.2 and 2.4 tonnes, to produce speed. There is simply no comparison between the locomotor apparatus of the similar-sized giants.

The conclusion by Halstead and Halstead (1981), Thulborn (1982), Alexander (1989, 1996), and Hutchinson and Garcia (2002) that the most speed adapted of giant animals, *T. rex*, was little, if any, faster than one of the least speed adapted of animals, elephants, is correspondingly illogical. If *T. rex* was a "slow runner, at best" (Hutchinson and Garcia 2002, p. 1021), then tyrannosaurids of increasing size should have become increasingly and dramatically anatomically adapted for slow speeds (Paul 1988a). Instead, tyrannosaurids of all sizes were remarkably uniform in having been the most anatomically speed adapted land animals in their size classes. What changes there are are modest, and are limited largely to changing proportions in terms of increasing robustness and distal shortening of distal segments with increasing bulk. Lower limb gracility and segment ratios do not necessarily indicate slower speed outside the same size class. The tyrannosaurids' size-related changes in limb proportions are of the order observed in ungulates that maintain a constant absolute top speed as they mature. Therefore, the stouter construction of *T. rex* does not, a priori, indicate a loss of speed, any more than the heavier build and shorter distal limb elements of an adult zebra indicate that it is slower than the much more gracile foal, which is no faster than it parent. It is therefore not a given that giant tyrannosaurids lost speed as they matured or evolved greater size.

Hutchinson and Garcia (2002) and Hutchinson (2004b) considered ~40 km/h to be a plausible peak velocity for *T. rex* (in broad agreement with Coombs 1978 and Christiansen 2000), a pace similar to that of galloping rhinos and far faster than that of elephants. Farlow et al. (1995) allowed for speeds up to ~55 km/h, which probably matches tridactyl-footed ratites. The most comprehensive field survey of ungulate speeds found that none exceeded ~50 km/h (Alexander et al. 1977). To boost giant tyrannosaurid speed 20% to the 50 km/h considered plausible by Paul (1988a, 2000) would require boosting muscle power by the same amount, a factor within the error zone of quantitative biomechanical estimates (Sellers and Paul 2004). It is probable that other gigantic avepods were able to run at speeds well above those of elephants (Paul 1988a; Blanco and Mazzetta 2001), albeit perhaps somewhat lower than those reached by the more speed-adapted tyrannosaurids.

Hutchinson and Garcia (2002) and Hutchinson (2004b) cited the slow speeds of potential adult tyrannosaurid prey as a reason for the latter being similarly slow. However, Alexander (1989), Paul and Christiansen (2000), and Christiansen and Paul (2001) demonstrated that giant ceratopsids of all sizes were anatomically adapted for rhinolike speeds in terms of limb anatomy and function, skeletal strength, and available muscle power. Although ceratopsids were probably not as fast as the more gracile tyrannosaurids, their area for limb muscle attachment is remarkably large. They had longer limbs than rhinos, suggesting that they possessed an even greater charging power than the latter. A detailed locomotor study is not available yet for hadrosaurs, but they appear have been at least as fast as the more robust ceratopsids. The area for leg muscle attachment is less in hadrosaurs than in tyrannosaurids, suggesting that hadrosaurs were not as swift. Slow prey cannot be used to limit the maximum speed of their potential killers (see Holtz this volume).

The central concentration of mass due to the abbreviation of the neck, body, and tail and reduction of the arms should have improved the maneuverability of tyrannosaurids compared with longer-bodied, longer-armed mega-avepods (Henderson and Snively 2004; Paul 2005). However, the more massive head of tyrannosaurids may have countered this mass concentration somewhat, although how much is difficult to assess because of uncertain differences in pneumaticity in mega-avepod heads. Snively and Russell (2003) concluded that the arctometatarsalian foot also improved the agility of tyrannosaurids.

Similarities and Differences between Tyrannosaurid Taxa

Tyrannosaurids were strikingly uniform (Figs. 18.1, 18.3), to the point that the variation within the family was similar to that seen within modern carnivore genera such as *Canis*, *Panthera*, and *Felis*, and is less than that present in *Varanus* and *Ursus* (Paul 1988a). Even so, there are important differences between the known taxa. The more robust skeleton of *Daspletosaurus* compared with *Albertosaurus* of similar mass is well known. The skull is increasingly more strongly constructed from *Albertosaurus*, *Daspletosaurus*, *T. bataar*, to *T. rex*. The difference in skull robustness in *Albertosaurus* versus similar-sized *Daspletosaurus*, and in *T. bataar* versus *T. rex* indicate that this robustness was not a size-related feature. It is in the skull that the greatest differences are found within the group.

Among tyrannosaurids (Fig. 18.1, 18.3), the temporal box was narrowest in *Albertosaurus*, so that the eyes faced less strongly forward, and the breadth of the posterior face of the braincase was less than in other members of the group. In *T. bataar*, and even more so in *Daspletosaurus*, the posterior of the skull was broader and the eyes faced more anteriorly. This condition was taken to an extreme in *T. rex*. A dorsal view restoration (based on my photographs) of the skull of the holotype of *T. bataar* (Fig. 18.3D), which is almost as large as those of adult *T. rex*, agrees with Hurum and Sabath (2003) that the posterior part of the skull and posterior face of the braincase of the Asian species is markedly narrower than that of the North American counterpart. However, Molnar (1991, 2000) and Hurum and Sabath (2003) restored the posterior temporal boxes of *T. rex* and *T. bataar*, respectively, as too triangular in dorsal view, judging from a number of skulls both articulated and restored from uncrushed elements (Fig. 18.3D, E; Paul 1988a, Carr 1999). Hurum and Sabath (2003) restored the rostrum of *T. bataar* much too narrow compared with the least crushed articulated skulls (cf. Hurum and Sabath 2003, figs. 1B and 2A2 to fig. 15A). The medial palatal shelves of the maxillae and the anterior part of the vomer of *T. bataar* are just as broad as those of other tyrannosaurids. Because the rostrum remained fairly consistently broad in tyrannosaurids, the greater disparity between the breadths of the anterior and posterior halves of the skull present in some tyrannosaurids was due mainly to the broadening of the aft section. The increasing breadth of the posterior portion of the skull does not appear to have been size related. Other gigantic avepods lack the feature. *Daspletosaurus* had a broader skull than similar-sized *A. libratus*, and the even bigger *T. bataar* and juvenile *T. rex* (=*Nanotyrannus*) skulls had exceptionally broad temporal boxes (Carr 1999), although apparently not as broad as in the large specimens. Variations in the breadth of the posterior part of the skull and the braincase plate probably reflect corresponding relative differences in the temporal jaw adductors and anterior cervical muscles. These were least developed in *A. libratus* and most powerful in *T. rex*. It is possible that only the latter could crush bones (Erickson and Olson 1996; Erickson et al. 1996; Varricchio 2001; Meers 2002; Rayfield 2004), a hypothesis supported by macerated bones in a coprolite (Chin et al. 1998).

Hurum and Sabath (2003) detailed differences in the skulls of tyrannosaurids, with emphasis on *T. bataar* and *T. rex*. Different articulations between the nasals, lacrimals, and frontals indicate differing abilities to handle prey with divergent combat abilities (the suture pattern restored for the type *T. bataar* in Fig. 18.3D differs somewhat from that for ZPAL MgD-1/4 in Hurum and Sabath 2003). As usual *T. rex* appears to have the skull best adapted to absorb high-stress loads, even when its large size is taken into account. In accord with this pattern, teeth generally became fewer in number, larger, and more conical progressing from *Albertosaurus*, to *Daspletosaurus*, to *T. bataar*, to *T. rex*.

Bakker et al. (1988) illustrated *A. libratus* with a standard avepodian ventral braincase in which the basitubera was positioned well posterior to the basipterygoid process. However, examination of specimens has failed to confirm that the configuration greatly differs from those of other tyrannosaurids (cf. Fig. 18.6C to fig. 7 in Bakker 1988). The basitubera-basipterygoid complex was somewhat more anteroposteriorily telescoped in *T. bataar* and *T. rex*, and formed a nearly continuous, subvertical plate (Fig. 18.6F, G; Bakker et al. 1988). But the difference relative to *Albertosaurus* and *Daspletosaurus* (Fig. 18.6C–E) is not extensive, being more a matter of degree. The occiput of *A. libratus* may have been less ventrally oriented than those of other tyrannosaurids and that of *T. rex* more so (Fig. 18.6C–G), reflecting possible differences in ventral flexion of the head. The differences seem modest, and a larger sample is required for verification and to confirm a consistent pattern in the group.

The basic features of the head that distinguished tyrannosaurids from other giant avepods became increasingly extreme starting with *Albertosaurus* and progressing through *Daspletosaurus* and *T. bataar*, and they reach their extreme in *T. rex* (Fig. 18.3). *T. rex* also has the largest mouth and is capable of a bite 1 m long (Paul 1988a). Postcranially, differences among the tyrannosaurids are few (Fig. 18.4), with *Albertosaurus* being the most gracile, *Daspletosaurus* and *T. bataar* intermediate, and *T. rex* the most massively constructed. The slenderness of the albertosaurine femur is especially notable, but it is not known whether this was unique among megaavepods. Limb bone strengths were lower in albertosaurines and *T. rex* than in *Daspletosaurus* and *T. bataar* (Christiansen 2000). *Albertosaurus* was somewhat more compactly built than *Daspletosaurus* (Fig. 18.1A–C), and should have been a little more agile. There is some variation in the relative size of the arms in tyrannosaurid specimens, but they are never large, and they do not appear to have either progressively diminished or greatly enlarged during the evolution of the group (Currie 2003).

Direct Fossil Evidence for Tyrannosaurid Predation and Other Interactions

Healed bite marks on gigantic hadrosaurs (Carpenter 2000; Wegweiser et al. 2004) and horned ceratopsids (Happ this volume) have been attributed to *T. rex*. Hadrosaur tail bite marks suggest an attack from the rear. The bite marks on the horn of the *Triceratops* indicate head-to-head combat, much like that portrayed in classic children's books (Geis 1960; Watson and Zallinger 1960), as well as Knight's famed portrayal of *T. rex* facing off with *Triceratops* (Glut this volume).

Pertinent to the discussion is fossil evidence for predation by other gigantic theropods. Carpenter et al. (2005) document evidence of healed wounds that indicate combat between allosaurs and stegosaurs. Thomas and Farlow (1997) concluded that the famous Texas theropod-sauropod trackway records an act of predation. Research of the trackway's unusual footfall pattern during the construction of full-sized sculptures representing the incident (Maryland Science Center) caused sculptor Hall Train and me to agree that the predator probably bit the base of the tail of the sauropod. If the slower, weaker-skulled allosauroids involved in these cases were active predators, it is unlikely that the more powerfully armed, faster tyrannosaurids were not.

Tyrannosaurid skeletons typically bear the marks of numerous injuries, some apparently caused by other dinosaurs (Hanna 2000; Rothschild et al. 2001; Tanke and Currie 2000; Larson and Donnan 2002; Brochu 2003; Rothschild and Tanke 2005; Rothschild and Molnar this volume). Some are not attributable to specific taxa and may represent damage experienced during predation, conflicts over possession of carrion, or during intraspecific disputes, while others are identifiable as having been inflicted by other tyrannosaurids.

To date, surprisingly few footprints or trackways have been attributed to tyrannosaurids (Lockley and Hunt 1994; Manning this volume). This lack of data hinders attempts to reconstruct the habits of these dinosaurs.

Lifestyle Reconstruction

Predation and Scavenging

Currently at least one large terrestrial carnivore scavenges little or not at all (see Holtz this volume). Cheetahs are too delicately constructed to compete over carcasses with other big cats, canids, and hyenas (Caro 1994). There is nothing in the fossil record that establishes that tyrannosaurids scavenged, so it is possible that all carcasses they fed upon were killed during active predation. However, because tyrannosaurids were the top predators in their habitats, there is no reason to conclude that they would have turned down a free meal. Scavenging is so probable that it should be assumed to have occurred unless compelling evidence indicates otherwise.

No living land carnivores are limited to pure scavenging, although all gain a significant portion of their energy and nutrition from scavenging (Houston 1979; Nowak 1999). Pure scavengers have not been positively identified among Cenozoic fossil mammals or ground birds. The absence of a corollary in itself casts doubt on the concept of scavenging-only tyrannosaurids. Walking scavengers may not be able to find sufficient carcasses to survive without hunting, and only soaring scavengers have the visual range and speed and energy efficiency to scout out and reach distant carcasses to make it energetically workable (Houston 1979). However, Ruxton and Houston (2002) concluded that terrestrial scavenging is not inherently energy ineffective, unless competing aerial scavengers are present. The only known large fliers of the latest Cretaceous, the giant azdarchid pterosaurs, were ill suited for scavenging, judging from their slender, nonhooked beaks and light skeletons (Kellner and Langston 1996). Marabou storks with stouter, tapering beaks scavenge, but they rely on hook-beaked vultures to open the carcasses and mostly intimidate the smaller vultures into dropping what they have cut out of the carcass (Hoyo et al. 1992). Azdarchids would have enjoyed none of these advantages. Instead, the delicately built pterosaurs would have been vulnerable to predation by competing tyrannosaurids unless they used a hidden deterrent, such as unpalatable flesh. It is therefore concluded that tyrannosaurids did not suffer from competition from scavenging pterosaurs. Even so, the concept of land creatures living by pure scavenging remains speculative in the absence of known examples.

Tyrannosaurids lived amid a dense population of potential victims (Farlow 1976, 1994; Ruxton and Houston 2002). If the adults had the ability to hunt prey, then it is improbable that they would have failed to kill what they needed when carcasses were not available. Mature tyrannosaurids would only have scavenged if they lacked the adaptations needed to hunt. This raises another problem concerning the lack of known extant obligate scavengers. Without such examples, it is difficult to determine the anatomical and other characteristics expected in such forms. This is especially true because adaptations for predation and scavenging in modern hunters are highly variable, and adaptations used for scavenging can also be used for predation and vice versa. One attribute that can be projected for a gigantic obligatory scavenger would be a low top speed, comparable to that of elephants, and low maneuverability because of the lack of need to pursue prey and the lack of a need to escape other predators. An obligate scavenging tyrannosaurid could do without a selective evolutionary need to move fast or be agile, while enhanced safety factors would select for a low top speed. As a result, there would be no concentration of mass away from the ends of the body and into the legs, or elongation of the legs, as discussed above. That the tyrannosaurids' well-developed running and turning adaptations were better developed than those of giant avepod dinosaurs is not only compatible with their being active pursuit predators, but also suggests that they scavenged less than other mega-avepods.

Carpenter and Smith (2001) and Lipkin and Carpenter (this volume) have suggested that the forelimbs of tyrannosaurids were used to hold prey during the attack. The extreme smallness of the arms makes it difficult to see how they could have been used to reach and engage prey of any size. Although the arms were stronger and more powerfully muscled than human arms, relative to the size of their owners and potential prey, they appear to have been too weak to be of important practical use. The absence of powerful, raptoral forelimbs has been cited as evidence against active predation (Horner and Lessem 1993). In this view, grappling limbs are a necessary adjunct for catching and dispatching prey in the manner of cats and raptoral birds. This hypothesis may be discarded on the basis of the absence of grappling appendages in canids, hyaenids, and thylacines, as well as a large number of extinct mammalian carnivores (see Holtz this volume).

Well-developed olfaction has been cited as a specialization for scavenging (Horner and Lessem 1993; Horner 1994; Horner and Dobb 1997). The ability to smell well is compatible with and useful for discovering carcasses that are not visually detectable. However, the hypothesis that high-level olfaction is limited to scavengers cannot be true: oras, canids, and hyenas possess this feature (Auffenberg 1981; Kruuk 1972; Schaller 1972; Nowak 1999). As documented by Holtz (this volume), tyrannosaurids did not have deficient vision, which would hinder predation. In any case, scavengers need and have visual acuity to spot carcasses at a distance. Farlow (1994) noted that the 4- to 6-m height of a rearing adult tyrannosaurid would have improved its ability to spot carrion, especially in a habitat lacking aerial scavengers whose descent indicated the location of carrion.

The bone-crushing ability of adult tyrannosaurids, especially of *T. rex*, would have been a useful adaptation for scavengers, and is a feature shared by hyenas. The adaptation is also advantageous for predators, both in providing extreme biting power for killing prey and subsequently consuming it efficiently, as proven by hunting hyenas.

The abbreviated life spans of tyrannosaurids indicate that they lived dangerous lives. Attacking similarly large herbivores would have provided the requisite level of danger. Feeding on herbivores after they no longer posed a threat explains the short lives of the tyrant avepods. The anatomy and life history of all tyrannosaurids, from the smallest to the largest, are in no regard incompatible with active predation, and in all respects are either compatible with or indicative of active predation. In any case, the scavenging versus predation debate is effectively moot in the face of healed bite wounds inflicted by the biggest known tyrannosaurid on the bones of adult hadrosaurs and ceratopsids. The marks could have occurred only if the theropods engaged in combat with the herbivores when the prey were living (Carpenter 1998; Wegweiser et al. 2004; Happ this volume). The question, therefore, is not whether giant tyrannosaurids, as well as nontyrannosaurid mega-avepods, were active predators. Rather, the question concerns the detail of their predatory activities. These activities were potentially complex because different tyrannosaurid taxa and populations may have exhibited dramatically different predatory habits depending on anatomical adaptations, age, sex, behavioral genetics, prey availability, and habitat structure.

The information provided by the healed bite marks goes beyond the fundamental conclusion that active predation occurred. That the herbivores survived and lived for a long time after the attacks shows that those individuals were healthy enough to successfully cope with the assault. It is even possible that the horned dinosaur killed its attacker. Therefore, adult

Figure 18.7. Reconstruction of a successful attack procedure by *Tyrannosaurus* to hunt *Triceratops*. Surprised from behind, the horned dinosaurs have been forced to attempt to flee, exposing their vulnerable rears and flanks to the tyrannosaurids. As a *T. rex* runs alongside its chosen target, it begins to deliver a deep, punch-pull, cookie-cutter wound intended to cripple the locomotor thigh muscles of the ceratopsid. The projected results of the wound are shown in the inset. The disabled herbivore can then be dispatched more safely. Whether the giant, small-brained predators hunted gregariously is speculative. For a restoration of an actual failed encounter between *T. rex* and *Triceratops*, see Happ (this volume).

giant tyrannosaurids were active big-game predators that repeatedly attempted to kill similarly massive, and in some cases potentially dangerous, prey. Giant tyrannosaurids did not consistently target individuals that were too small, or too weak from illness, malnutrition, injury or age, to successfully flee or defend themselves.

Because the healed wounds documented by Carpenter (1998), Wegweiser et al. 2004), and Happ (this volume) were delivered in failed attacks, they may not represent normal, successful tactics. It is especially unlikely that tyrannosaurids regularly engaged horned ceratopsids head-to-head because doing so would maximize danger while minimizing the chance of a successful kill, as the survival of the *Triceratops* shows. Biting a horn is not a productive mode of killing, and the act suggests that the *T. rex* was trying to protect itself from being wounded in an attack that had gone bad for the predator. Charging horned ceratopsids of 1.5 to perhaps 9 tonnes (all mass estimates in this section based on Paul 1997) were probably the most challenging and dangerous prey for tyrannosaurids. The prey vision appears to have been modest, with eyes set low and possibly unable to see forward. Very large nasal cavities suggest that the sense of smell was well developed. The example of the *Triceratops* documented by Happ (this volume) indicates that ceratopsids used their horns for active defense, so these structures were not exclusively used for interspecific identification and intraspecific display or conflicts (partly contra Sampson et al. 1997; Goodwin et al. 2006).

The use of horns for defense is not surprising. It occurs among modern horned ungulates, especially cattle-type bovids and rhinos. These animals are hefty enough to put up a stout fight, and they lack the speed needed to readily flee (Schaller 1972; Owen-Smith 1988; Nowak 1999). Asian rhinos also use their incisors to fend off predators (Owen-Smith 1988), and ceratopsids could have used their parrotlike beaks to bite attackers. Armed defense was probably necessary, considering the slow speed of these ornithischian. The means by which the wide variety of ceratopsids utilized their highly divergent horn arrays to defend against predators is little studied, perhaps as a result of an overemphasis on investigating the evolution of the structures for intraspecific activities. Centrosaurs with long, erect nasal horns may have used upward jabbing head motions intended to damage the chest or belly of a predator from below. The tyrannosaurids's gastralia may have provided some protection against ventral thrusts. Chasmosaurines with long, anteriorly directed brow horns may have preferred a straight-in charge to the flanks of the attacker. Even within *Triceratops*, horn use may have differed between the 2 morphs (long snout, short nasal horn, big brow horn versus deeper snout, longer nasal horn, shorter brow horn). Boss-nosed *Pachyrhinosaurus* and *Achelosaurus* (assuming their bone bosses did not anchor keratinous horns), hook-horned *Einosaurus*, and the short-horned *Chasmosaurus* may have been the most prone to biting, but the ramming effect of their blunt weapons, backed by tonnes of body mass, would have been formidable as well. The poorly developed horns of juveniles (Sampson et al. 1997; Goodwin et al. 2006) suggest that they too used their beaks, in addition to fleeing or relying on adults for protection. Although cranial frills and the horns and hornlets that ceratop-

sids often bore may have evolved mainly for interspecific and intraspecific identification and display functions, the structures also served to help protect the neck. The posterior ribs of ceratopsids were tightly packed together and articulated with the prepubis (Paul and Christiansen 2000), forming a cuirass that provided protection to the abdomen. The low-slung, quadrupedal ceratopsids may have been able to out-turn the longer limbed, 2-legged tyrannosaurids. Considering their substantial cranial weaponry, the optimal defensive maneuver was to face attacking tyrannosaurids, which may have deterred and aborted an attack. Conversely, tyrannosaurids needed to avoid the potentially dangerous horns and beaks and attack from the rear, where they could disable the ceratopsid by biting the thigh or the caudofemoralis (Fig. 18.7). A rear assault could be achieved either by ambushing the horned dinosaurs or by intimidating the horned dinosaurs into fleeing. Paul (1988a) and Meers (2002) concluded that the biting force of a single *T. rex* was sufficient to kill a *Triceratops*.

Sauropods were also dangerous prey as a result of their combination of enormous bulk (about 10 to 20 tonnes for the titanosaurs present in tyrannosaurid habitats), towering height (which made it difficult to reach the vulnerable neck and head), elongated, strongly muscled tail, and long, club-footed legs. For example, the tail of a 15-tonne titanosaur weighed about 1 metric tonne (Paul 1997) and was capable of delivering a powerful blow. Vision would have been enhanced by the great height of the head. It is difficult to determine the best strategy for attacking sauropods. Although a sauropod is much too slow to outrun a tyrannosaurid, it could try to move away while using its tail to cover its rear. Avoiding the tail would appear to be prudent, but the Texas trackway discussed above suggests an attack on the tail base during a rather slow (due to deep mud) pursuit, probably to disable the caudofemoralis hind limb retractor. An attack on the flank risked kicks from the clawed hind feet, and frontal attack may involve kicks from the forelimbs.

The ground sloth–like therizinosaurs, which weighed up to 3 tonnes, were too slow to escape tyrannosaurids. Instead, they probably stood their ground and fought with their long arms bearing enormous claws. Large sclerotic rings show the eyes were large, so vision should have been good. Tyrannosaurids should have preferred to approach from the rear. It is more difficult to assess the defense of the poorly known deinocherians, which may have reached 6 tonnes. It is not known whether they were fast enough to attempt to flee, or whether they had to face tyrannosaurids and fight them off with their long, big-clawed arms. Again, a rear attack was most advantageous.

In ankylosaurs, vision does not appear to have been well developed, and vision was hindered by the low position of the head. Olfaction appears to have been good, judging from the large, often complex nasal passages. Their massive armor is additional evidence that mega-avepods were predators because such extensive protection is unlikely to evolve unless it is needed to protect against powerful enemies. Weighing in at 2 to 6 tonnes, the low-slung ankylosaurids were armed with a large tail club that could damage the lower legs of tyrannosaurids (Coombs 1995). These armored dinosaurs, which were too slow to outrun the avepods, lacked large defensive spines protecting the body and neck; armor formed a passive defense but was well developed. If tyrannosaurids chose to attack ankylosaurids, then a frontal approach to avoid the tail club was called for. The ankylosaurid probably tried to present its posterior to the attacker, either spinning around to do so, or fleeing while the club swept back and forth to clear the tyrannosaurid from its rear. Fleeing into heavy brush when available could have been an effective defensive strategy for these low-profile, heavy-bodied herbivores. Nodosaurids, which typically weighed about 2 tonnes and which were too slow to readily escape pursuit, lacked a tail club or other posterior defensive weapons. Therefore, presenting the rear to the enemy was not beneficial. Large lateral spikes protected the neck and shoulders. One defensive tactic may have been to lie on the ground and rely on passive defense, with the anterior spikes protecting the neck. An alternative was to present the front to the predator, perhaps assisted by short rushes with the shoulder spikes. The need to emphasize a frontal defense explains the extreme armoring of the nodosaurid head, which includes ossification of the cheek tissues.

Hadrosaurs, whose weight ranged from 2 to 10 tonnes or more, lacked major defensive weapons or armor. Large sclerotic rings imply good vision that could be boosted by rearing to see in the distance. The size and complexity of the nasal passages imply that olfaction may have been well developed. Although they lacked even the thumb spikes of their iguanodont predecessors, their only significant physical defensive action was to kick with their powerful, heavy-footed hind limbs, or to swing the tail. The modest length, lateral flattening, and stiffening of a rhombic lattice of ossified tendons suggest that the hitting reach and power of the tail was limited. Duckbills could bite, but the defensive effectiveness of the broad beaks is problematic. Because ornithopods had a large lumbar region that lacked long ribs, the abdomen was vulnerable. The neck was lightly constructed and set low on the animal. A few hadrosaurs were so gigantic that sheer size should have conferred some degree of protection. Otherwise, the only apparent option available to hadrosaurs was to flee at maximum speed while attempting to land any kicks that they could, even though they were probably not able to easily outrun the even more powerfully and longer limbed tyrannosaurids. Hadrosaurs, which were able to use their forelimbs for locomotion, may have been able to out-turn the bipedal tyrannosaurids. Hadrosaurs appear to have been gregarious (Horner and Dobb 1997), and losing oneself in the herd would have been a priority. Fleeing into dense brush might have been advantageous, because hadrosaurs were somewhat lower slung than their attackers and could use their forelimbs to help push the body through heavy vegetation. Heavy vegetation would have also made it more difficult for the predator to position itself to deliver effective bites. Primary targets for attacking tyrannosaurids would have been the caudofemoralis on the tail and the thigh muscles in order to disable to locomotor system, the vulnerable abdomen to eviscerate the victim, and the neck, which could have resulted in the most rapid death by cutting the trachea and major blood vessels (Molnar 2000).

Among smaller prey, ornithomimids of 50 to 500 kg and nonhadrosaur ornithopods of 50 to 100 kg, lacked substantial weapons. They could kick with their legs, and small ornithopods could bite with their unhooked beaks. Their best defense was flight. Excellent vision, indicated by large sclerotic rings and tall height, gave ornithomimids exceptional early warning. They may have been the only nonpredators able to outpace pursuing tyrannosaurids. Small ornithopods may have been a little slower than the even more gracile limbed tyrannosaurids, so escaping into dense brush would have been a good tactic. Oviraptorosaurs were fairly well armed with hand claws, but they probably did not pose an extreme danger. They probably tried to flee like small ornithopods and fought only when cornered or captured. Escaping into dense vegetation could have been effective for the low-slung, heavily built pachycephalosaurs. The domeheads may have also adopted an offensive defense by using their heavily constructed skull roofs, backed by their bulky bodies, to ram the flanks of tyrannosaurids. The latter therefore may have preferred a rear attack. Even more dangerous prey were protoceratopsids of 50 to 200 kg. They could have used their large parrot beaks to bite attackers (as recorded by the famous fighting *Protoceratops* and *Velociraptor*; Carpenter 2000; Holtz 2002). Again, a rear attack would have been advantageous for attacking tyrannosaurids.

Whether herding ceratopsids and sauropods used organized defensive tactics is unknown; their reptilian brains cast some doubt on the possibility and suggest any such cooperative behavior was limited in sophistication. The production of large numbers of offspring (Paul 1994) also suggests that any parental defense was less intense than among slower-breeding mammals of similar size. The possibility that horned dinosaurs formed a protective ring around the young, a rare practice even among big-brained, horned ungulates (Owen-Smith 1988; Nowak 1999), must be rated as low. The herds of giant dinosaurs may have been unorganized collections of juveniles and adults who randomly associated with one another in order to enjoy the passive protection provided by large numbers (Paul 1998). It was to the benefit of tyrannosaurids to either attack an isolated individual, or to cull one away from the herd. One defense tactic that was not likely to succeed was retreating to water. Tyrannosaurids were probably as good or better swimmers than their prey (Bakker 1986; Paul 1988a), and ungulates that flee into water are usually killed by hyaenas and canids (Kruuk 1972; Nelson and Mech 1984).

The means by which tyrannosaurids initiated attacks is difficult to restore because the tactics they used depended on numerous, often uncertain, variables (Paul 1988a). Presumably prey was detected and tracked by a combination of excellent vision and olfaction, supplemented by good hearing. Rearing would have helped locate distant prey, but had to be utilized carefully to minimize exposure. The exceptional olfactory performance was especially useful for sensing low-profile prey in heavily vegetated areas. Potential herbivore prey countered with varying arrays of the same senses to perceive attackers. The degree to which the predators and their prey were camouflaged in order to minimize their detectability is not known. Smell to detect prey would caused tyrannosaurids to approach prey from downwind, thus minimizing the chance of being detected too early. Whether reptilian-brained predators are knowledgeable enough to deliberately avoid being upwind of their targets is not certain (the ability of oras to circumvent the detection of their odor is not clear; Auffenberg 1981). The tendency of herbivores to vocalize when in herds or while breeding, defending territory, and the like would have aided long-distance detection by predators. The great size and long, birdlike legs of adult tyrannosaurids do not appear to have been conducive to catlike stalking tactics. However, adopting a slow, plantigrade gait would have reduced the height profile. Plantigrade avepod trackways are surprisingly common (Kuban 1986), and birds sometimes stalk flat footed (Paul 1988a). A stealthy, close approach would have been most feasible if vegetation was sufficiently dense. Alternately, tyrannosaurids approached their prey with minimal stalking, as do canids and hyenids. Big cats can switch between stalking prey and running at it, depending on circumstances (Schaller 1972); the same may have been true of tyrannosaurids. Certainly they did not stop and roar at their victims, as commonly portrayed in cable documentaries these days.

Tachyenergetic tyrannosaurids could have had the high aerobic exercise capacity needed to chase prey over long distances, but whether they actually did so depended on myological cellular factors that are difficult to assess without soft tissue. If giant adults required high power density, short-burst leg muscle fibers in order to run, then they were all short-distance ambush predators. If the adults did not need short-burst muscles to move fast, then a number of options are possible: they could have been long-range pursuit predators powered by highly aerobic, sustained power leg muscles, or they may have been configured for short-range pursuit, or some taxa may have been extended-range runners and others may have been optimized for brief dashes. If the latter mixture of pursuit types was true, then gracile *Albertosaurus* may qualify as a longer-range runner than stouter *Daspletosaurus*, which may have emphasized a burst of speed over shorter distances. Heavily built *T. bataar* and *T. rex* may have been short-range runners, but long-range pursuit performance cannot yet be ruled out.

Tyrannosaurids used their speed both as a means to catch prey and to avoid being injured by the victim. Mega-avepods probably did not contact prey while running toward it at high-collision speeds because doing so would have resulted in serious injury to the predator. Predators attempt to match the speed and cadence of their victim in order to minimize the speed differential at the moment of attack, maximizing the predator's safety and its ability to apply its weapons optimally (Thomas and Farlow 1997). A tyrannosaurid would continue to run fast while assaulting its victim only if the latter were also moving fast and parallel to the predator. If the conflict turned into a turning dogfight, the speed would drop because inertial forces prevent large, fast-moving objects from turning quickly, and because of the dangers of accidental collision or tripping. The predator would especially desire to avoid an accident during what is a normal feeding event, whereas the prey would be more pressed to take risks due to the potentially terminal nature of the situation.

High aerobic exercise capacity should have given tyrannosaurids the ability to engage in lengthy combat when necessary. However, because of the dangers associated with combat and with falling (Paul 1987a, 1988a; Farlow et al. 1995), it may have been to the advantage of tyrannosaurids to minimize physical contact and extended, close-in struggles with similarly sized prey, especially the well-armed ceratopsids and sauropods. Hit-and-run tactics in which the mega-avepods dashed in to deliver crippling punch-and-pull or toxic bites, followed by a quick retreat, would have best utilized the speed of tyrannosaurids (Fig. 18.7; Paul 1987a, 1988a). Some sharks use hit-and-run attack modes (Tricas and McCosker 1984; Paul 1988a; Klimley 1994), as do oras when assaulting large ungulates (Auffenberg 1981). Tyrannosaurid attacks could have been quickly repeated until the desired effect was achieved. Minimizing the high risk of direct-contact struggles is in line with the rather low strength factor of tyrannosaurid legs, especially albertosaurines and *T. rex*, particularly the gracile morph. Such a modus operandi is also logical considering the exceptional power of the bite of tyrannosaurids, which may have been an adaptation for minimizing the contact time required to critically injure prey.

Overlapping fields of vision may have further optimized the precision and rapidity of the bites. Attacks on limb muscles would have been effective at slowing the prey down or even crippling it, rendering it easier to reapproach and safer to reattack (Fig. 18.7). Alternatively, the attacker may have waited for the long-term effect of wounds to set in. The latter tactic is used by oras, whose bites are highly septic (Auffenberg 1981). If tyrannosaurids had lips able to harbor copious amounts of bacteria-laden salvia, then they may have been able to deliver virulently infectious bites. Abler (1992) suggested that the dinosaur's tooth serrations harbored decaying, bacteria-breeding flesh. The septic bite hypothesis is not verifiable. It would be undermined if fossil soft tissues establish that tyrannosaurids lacked moisture-retaining lips. The evidence for infection of healed bite wounds (Carpenter 2000; Wegweiser et al. 2004) may be compatible with the septic bite hypothesis, but it does not verify it because the infections were not virulent enough to kill the prey, and because any wound would become infected to some degree. Septic bites can take days to sicken the prey. This would be acceptable for predators as big as tyrannosaurids, who would normally not feed for many days between meals (Paul 1988a). Infectious bites would have been especially efficacious in dealing with the armored dinosaurs because a small wound could be lethal.

Venomous bites could have also been advantageous to tyrannosaurids. Fry et al. (2006) reported that varanids have fast-acting toxins in their saliva. This adaptation has arisen within the squamate clade. There is no osteological evidence that theropods evolved oral toxins, but the same is true of venomous varanids. If these dinosaurs lacked lips, then oral venom would appear unlikely. If they possessed saliva-retaining lips, then toxic bites are plausible but speculative.

In any case, waiting for wounded prey to weaken increases the risk of other tyrannosaurids taking over the operation. It is therefore possible that big tyrannosaurids quickly killed prey in one bout of combat, and then consumed it before other tyrannosaurids could get in on the proceedings. The massive-headed, strong-legged daspletosaurs appear best suited for this tactic. When targeted prey was not large enough and was too weakly armed to pose a serious danger, tyrannosaurids probably utilized their ability to bite, hold, and manipulate prey with the jaws to quickly and severely injure and kill it. Rapid lethality is especially likely when attacking the relatively weakly protected hadrosaurs, which could have been hastily dispatched with a bite to the neck. The tyrannosaurids' superior ability to grab and manage prey with the head may have compensated for the lack of the large arms other mega-avepods used for handling prey.

In addition to availability, prey selection would have varied depending on tyrannosaurid taxon, age, and perhaps sex and population. Russell (1970, 1989), Lehman (2001), Sampson et al. (2003), and Sampson and Loewen (2005) observed that contemporary *Albertosaurus* and *Daspletosaurus* were fairly diverse and provincial, suggesting a degree of specialization. The robust *Daspletosaurus* may have preyed on ceratopsids and ankylosaurs more often than the more lightly constructed *Albertosaurus*, which could have specialized in the more vulnerable hadrosaurs (Russell 1970, 1989; Paul 1988a). Russell (1970, 1989) further noted that *Albertosaurus* and hadrosaurs were both common relative to *Daspletosaurus* and ceratopsids, respectively. On the basis of bone bite marks in an *Albertosaurus-Daspletosaurus* habitat, Jacobson (1998) concluded that tyrannosaurids fed on hadrosaurs 3 times more often than on ceratopsids.

T. bataar did not experience competition as top predator, nor did this moderately heavily constructed dinosaur have to cope with ceratopsids. Its primary prey was probably hadrosaurs, although *Saurolophus* and *Shantungosaurus* were exceptionally large, with prey supplemented by ankylosaurs and sauropods (Hurum and Sabath 2003). Like *T. bataar*, but unlike earlier North American tyrannosaurids, *T. rex* lacked competition in its adult size class. The wide range of habitats and herbivore faunas it lived among suggest that it was a generalist predator despite its highly specialized anatomy (Lehman 2001; Sampson and Loewen 2005). Because *Triceratops* was common, it was probably an important portion of the diet, but exactly how important relative to the other common prey item, the more vulnerable edmontosaurs, has not been documented by a bone mark survey. Gigantic *Ankylosaurus* and smaller *Edmontonia* were too rare in *T. rex* habitats to be an important part if its food intake. The major differences in limb bone strength suggested from the 2 morphs of *T. rex* suggests divergent predation habits. This is true if the 2 morphs (or taxa) represent different weight classes, and it is even more probable if the 2 types were similar in total mass but differed in bone strength. In either case, the robust morph appears better suited for battling with ceratopsids than its gracile counterpart. The specialized adaptations of *T. rex* for hunting the horned giants is most compatible with it being an active predator.

Although the evidence shows that they did assault healthy giants, large tyrannosaurids probably followed the standard predator tactic of preferentially targeting individuals that were not in prime condition. Tyrannosaurids were most likely opportunists that attacked and consumed any prey item within their abilities to capture and kill. Large extant predators often feed on small and juvenile tetrapods, including predators, and cannibalism of young occurs (Schaller 1972; Auffenberg 1981; Nowak 1999). Bone bite marks, gut contents, and coprolite contents demonstrate that adult tyrannosaurids often fed on small and young dinosaurs (Chin et al. 1998; Jacobson 1998; Varricchio 2001; Larson and Donnan 2002). Tyrannosaurids may have been cannibals, as has been documented for another mega-avepod (Rogers et al. 2003), even if they practiced parenting (as do lions; Schaller 1972), but any cannibalism was relatively rare according to the results in Jacobson (1998).

If the gracile juvenile tyrannosaurids, as well as any small-bodied adults, were independent hunters, then their hunting practices can be expected to have differed dramatically from those of the much larger, more massively constructed gigantic adults (Russell 1970; Paul 1988a; Farlow 1976; Farlow and Holtz 2002, 2004). The more delicately built and toothed small-bodied tyrannosaurids, whether adults or juveniles, were less adapted for combat with powerfully armed prey than were the grown-ups. Suitable prey for the swift, small-bodied tyrannosaurids would be the similarly fast ornithomimids, as well as oviraptorosaurs and small ornithopods. Pachycephalosaurs and protoceratopsids were probably targeted on occasion, as well as juvenile hadrosaurs and ceratopsids. Again, prey segregation may have occurred between small and juvenile tyrannosaurid taxa. For example, if juvenile *Albertosaurus* was more gracile than *Daspletosaurus* of equivalent growth size, then the former may have been more prone to hunt lighter, faster prey than the latter. It is not possible to directly assess the predatory habits of the smallest, post-hatchling tyrannosaurids because remains are lacking. Because they weighed only a few kilograms, they presumably the fed on invertebrates and small vertebrates, from insects to birds.

There is no evidence that tyrannosaurids, which lacked the long, slender snouts of spinosaurs, of any age were adapted for or particularly prone to feeding on aquatic organisms, although they may have done so on occasion. Adult tyrannosaurids may have preyed on smaller crocodilians along shorelines. However, the gigantic North American deinosuchians were the only predators that posed a potential threat to adult tyrannosaurids by ambushing and pulling the dinosaurs into the water and drowning them (Schwimmer 2002). Hence, tyrannosaurids may have been careful when near large bodies of water, at least where giant crocodilians were common enough to pose a serious threat. Considering that footprints are usually preserved on shorelines, fear of supercrocs may help explain the perplexing paucity of tyrannosaurid trackways. Alternatively, I have elsewhere speculated (Paul 1988b) that the tachyenergetic tyrannosaurids could turn the tables on the bradyenergetic deinosuchians at high latitudes when winter left the crocodilians torpid, which may explain the absence of crocodilians from arctic watercourses at that time.

Assuming that *"Nanotyrannus"* is a juvenile *T. rex*, then the morphological changes during growth, including alteration of the teeth (Carr 1999), may reflect changes in the prey selected during ontogeny. The prey of young *T. rex* presumably consisted largely of fairly safe ostrich mimics, small ornithopods, and juvenile herbivores. By adulthood, *T. rex* engaged in combat with elephant-sized, heavily armed, rhino-speed ceratopsids. Even while growing, *T. bataar* did not experience an equivalent prey transformation in its ceratopsid-free habitat, so its form did not need to undergo the same degree of modification. An extraordinary anatomical ontogenetic alteration in *T. rex* forced by its ontogenetic shift in victims may be obscuring recognition of the juvenile status of tyrannosaurid specimens from the late Maastrichtian of western North America.

Mental Performance and the Question of Organization during Hunting

The modest enlargement of tyrannosaurid brains implies that their hunting behavior was more complex than that of smaller brained mega-avepods. The problem of hunting the fast and dangerous ceratopsids may have posed an exceptional mental challenge. It is difficult to comparatively assess and to restore hunting in the 2 groups because there are so many uncertainties. The sophistication of tyrannosaurid hunting tactics was probably limited. Their neural networks were still reptilian in organization, and much smaller than those of predaceous birds and mammals. The dinosaurs' predatory behavior should have been correspondingly restricted in scope, as well as stereotyped compared with raptors and mammalian carnivores. (Contrary to the common impression, predators are not generally bigger brained than their prey. Indeed, the opposite may be true, as in phorusrhacids compared with their mammalian victims; Paul 1988a.)

The question of mental capacity affects another problem: whether tyrannosaurids normally hunted singly, in packs, or both. Among modern predators, closely related taxa or even populations within a species can exhibit dramatically different hunting modes. Tigers are solitary hunters, whereas lions usually live in prides that deploy cooperative prey capture tactics (Schaller 1972; Nowak 1999). However, the sophistication of lion group hunting tactics should not be exaggerated (Schaller 1972). Coyotes usually solitary, but they use group hunting tactics on occasion, especially when wolves are absent in the area (Bekoff 1978). Similar inconsistency may have occurred among mega-avepods.

Monospecific bone beds of tyrannosaurid and nontyrannosaurid mega-avepod of differing age classes, and gregarious trackways of the latter have been presented as evidence of groups that hunted as packs or even family groups (Larson 1997; Larson and Donnan 2002; Currie 2000; Farlow and Holtz 2002, 2004; Currie et al. 2005; Coria and Currie 2006). Such group tactics would have multiplied the offensive power of predatory avepods, allowing them to better overcome the defensive power of prey taxa. Currie (2000) further suggested that tyrannosaurids of differing ages played different roles during group hunting: the faster juveniles are postulated as herding prey toward the slower, more powerful adults. However, it is not certain whether such a marked speed differential between juveniles and adults was present. Nor is it clear that small, lightly built juveniles could have intimidated large herbivores into fleeing in the desired direction. If group hunting was a simple matter of the pack or family attacking en masse, then the disparate fighting power of the juveniles and the adults could pose problems if the prey were large, powerful adults. The exceptional killing power of individual tyrannosaurids, as suggested by Meers (2002), may have reduced their need to work in groups.

It is problematic whether the small-brained tyrannosaurids and other mega-avepods were mentally capable of using truly cooperative hunting tactics. Even if they did have the mental acuity to form organized packs, they may not have done so. After all, even big-brained predators can be solitary hunters, as are most cats and raptors. The tyrannosaurids would seem capable of more complex social and parenting behavior than other, smaller brain mega-avepods, but this does not guarantee that they did so.

Also casting doubt on the depth or existence of parent-offspring interactions is the unusual ontogenetic deepening of the tyrannosaurid skulls. This is the opposite of the pedomorphic pattern expected in animals that parent their young. Instead, the long-snouted, many-toothed juveniles appear well suited for fending for themselves (Paul 1988a). Therrien et al. (2005) concluded that the juveniles' jaws were as well adapted for feeding as those of the adults. Juvenile tyrannosaurids were probably able to survive on their own even if they normally received some degree of parenting.

Reproduction and Parenting Potential

The same modest levels of mental capacity that probably placed constraints on tyrannosaurid predation should have limited the extent and sophistication of any social organization, including parenting, that they may have practiced, if they practiced any at all. It is questionable whether such small-brained animals would have cared for their progeny over the many years that they appear to have remained juveniles. In some mammals (Nowak 1999) and especially in reptiles, sexual maturity is reached well before final adult size. For example, oras are sexually mature when just a few years old and at a small fraction of adult mass (Auffenberg 1981). According to the data in Schweitzer et al. (2005, this volume), a *T. rex* that was about half of its maximum mass, in its late teens, and about 5 years short of reaching final mass (Erickson et al. 2004) was sexually mature (elephants often become sexually mature in their teens; Nowak 1999). Whether tyrannosaurids were sexually mature considerably earlier than the specimen reported by Schweitzer et al. is not yet known. A larger survey of tyrannosaurid specimens will, I hope, answer this important question. Among the short-lived tyrannosaurids, it was selectively advantageous for reproduction to begin well before growth was completed in order to maximize the reproductive span, but even then, the reproductive period would have been unusually brief for such large animals.

It would have also been selectively advantageous to compensate for the short reproductive span by investing resources in rapid reproduction. Although the mode of reproduction is currently not documented by direct fossil evidence in this particular group, it was probably by modest-sized, hard-shelled eggs (Paul 1994, 2002; Larson 2000). Because it was able to deposit large numbers of eggs each year, and because it lived in danger-filled habitats, it follows that tyrannosaurids were probably fast-breeding r-strategists (Paul 1994). If so, then their population should have been dominated by juveniles, and the energy-inefficient tachyenergetic adults should have been rarer than adult mammals of similar size. The probable production of large numbers of eggs and the limited breeding period of the short-lived adults further suggest that parenting was limited or absent. It is therefore possible that monospecific adult-juvenile bone beds do not record parent-offspring interactions. Juvenile tyrannosaurids may have been independent hunters from hatching on. Some juveniles may have associated with whatever adults were available in order to benefit from regular scavenging of the kills of the latter. A partial modern analog is the arctic fox. The little canid often follows polar bears and wolves in order to obtain carrion (Nowak 1999); at other times, it hunts on its own. The remora is another example of an association of a small predator with a much larger one. The *T. rex* juveniles may have enjoyed a degree of passive protection by being near adults. However, whether they were related or not, juvenile tyrannosaurs may have been in danger of being consumed by adults they associated with. If juveniles were markedly faster than the adults, then this speed may have provided their means of escape. The superior agility of small individuals would have provided them with some protection. Of all avepod dinosaurs, young tyrannosaurids were, because of their short arms, the least suited for escaping adults and other predators by climbing into brush and trees, like juvenile oras (Paul 2002).

It is possible that parent-offspring relationships varied among tyrannosaurids, ranging from none in some to extensive in others. Some or all tyrannosaurids may simply have deposited their eggs in the ground, then abandoned them. Or they built more elaborate nests that they then left, or that they maintained and guarded. There is no evidence that tyrannosaurids created the ring-shaped nests common to more birdlike dinosaurs (Larson and Donnan 2002; Paul 2002). No data are currently available on the nesting behavior of tyrannosaurids. Direct incubation of partly exposed eggs, as inferred in smaller avepod dinosaurs (Paul 2002), is improbable because eggs would have been crushed, and because adults lacked the extensive feather pelage needed to insulate them (contra Larson and Donnan 2002). Instead, incubation of buried eggs would have been by indirect solar heating of the soil, and perhaps fermentation of plant materials. The great size disparity between adults and hatchlings would have posed serious problems for parenting tyrannosaurids. The young would have been in serious danger of being stepped on by adults, who would have had trouble keeping track of a brood of such small charges, and juveniles weighing just a few kilograms would have been hard-pressed to keep up with multitonne adults. It is possible that some or all posthatchling tyrannosaurids immediately left the nest and led solitary lives. Or some or all posthatchlings may have formed juvenile pods that lived separately from larger juveniles and adults, who in turn may have hunted the small juveniles when the opportunity arose. Again, there are no available fossil data on the matter.

The extensive, apparently intraspecific injuries observed in tyrannosaurids indicate that they were often violent among themselves. This is not surprising because predators are commonly aggressive toward one another (Kruuk 1972; Schaller 1972; Auffenberg 1981; Nowak 1999). It is not currently possible to determine whether the injuries occurred during social disputes within organized groups, over territorial and breeding contests, or while contending for carcasses. Some combination of the above is probable.

Conclusion

Much is now known about tyrannosaurid biology. There is no reason to doubt that they fed on carcasses on a regular basis, but they were not specialized for scavenging. No other similar-sized predaceous dinosaurs concentrated so much mass in the head or legs, and were as well adapted for running. Healed bite marks on adult hadrosaurs and ceratopsids show that tyrannosaurids actively hunted giant herbivores healthy, fast, and powerful enough to either flee or fight off their attacker. Tyrannosaurids were tachyenergetic, experienced a period of rapid growth, and they lived only about 3 decades at most.

If the oviparous tyrannosaurids were r-strategists, then they were species whose high rates of reproduction, rapid growth, and ability as juveniles to survive on their own gave them a greater population recovery potential than giant mammals, which are big-brained, slow-breeding K-strategists with extended parental care (Paul 1994). Rapid reproduction and growth would have allowed tyrannosaurids to replace the high rate of losses that resulted from dying so young. Unlike slow-breeding, giant mammals that acquire large knowledge bases and invest in long lives, the presumably fast-reproducing tyrannosaurids whose reptilian brains stored limited information appear adapted for short lives of extreme danger in which safety was not a primary selective factor. Gigantic tyrannosaurids can therefore be confidently assessed as fast-growing predators who lived in the energetic and locomotor fast lanes, engaging in high-risk combat with equally large prey at ungulatelike speeds, and consequently dying young. This extreme lifestyle was neither reptilian nor fully mammalian. A partial modern analog is the mountain goat, which risks death with virtually every step taken on towering cliffs. Another analog is the cheetah, which are exceptionally fast predators. The speed is achieved at the cost of a lightly built body that is vulnerable to damage, and even a slight injury can disable the cats' speed to the degree that they cannot successfully hunt. Wild cheetahs do not normally live as long as other big cats (Caro 1994).

The probability that tyrannosaurids lived high-risk lives with low safety factors further complicates attempts to assess their athletic abilities by using bone dimension and strength factors relative to body mass. Tyrannosaurid mass estimates extrapolated from correlations of bone diameters in living bipeds may be too low if the tyrannosaurids were anatomically adapted to accept higher loss rates during locomotion and combat. Conversely, estimates of speed that are based on limb bone strength and risks of falling (Alexander 1989, 1996; Farlow et al. 1995, 2000; Christiansen 2000) may be too low for the same reason (Alexander 1996; Paul 2000). Albertosaurine tyrannosaurids should have been, if anything, faster than allosaurids, yet the latter have the more robust and apparently stronger femora (Christiansen 2000). Likewise, speed estimates that are based on the risks associated with falling when running may be too low. Furthermore, although fast running was itself dangerous for giants, the use of a high-speed attack reduced the danger that the prey would injure the predator, and it also reduced the risk that the prey would escape or survive the attack. The short-lived, combative tyrannosaurids may have been under strong selective pressures to reduce safety factors to a minimum despite their colossal size. This trend appears to have been taken to an extreme in the slender-legged albertosaurines and huge *T. rex*, in which the limb strength factors are the lowest. If so, the small-brained but gigantic-bodied tyrannosaurids might be viewed as readily replaced, throwaway predators.

As much as we now know about tyrannosaurids, much else remains uncertain. It is not known whether their bites were septic, toxic, or neither. The presence or degree of cranial kinesis is unclear. The oft-cited overlapping fields of vision may or may not have provided true stereoscopic vision. Nor is it known whether the forward-facing eyes evolved under direct selective pressures, or whether they were a beneficial or incidental secondary effect of lateral expansion of the jaw muscle accommodating the temporal box. It is not known whether the superpredators hunted by day, at night, or both. Modern predators often hunt at all hours of the day (Kruuk 1972;

Schaller 1972; Nowak 1999), and Molnar (1991) suggested that the tyrannosaurids' overlapping vision fields may have aided nocturnal predation.

Limited mental capacity and size disparity problems constrained the complexity of whatever gregarious interactions that may have occurred and may have prevented complex social activities. Virtually nothing is known about nesting or the care, if any, of eggs, hatchlings, and small juveniles. The tyrannosaurids' unusual ontogenetic skull allometry suggests that parenting was not well developed or was absent. Single-species skeletal accumulations may record true sociality between age classes, or alternatively, it may indicate that juveniles tagged along with adults, hoping to picking up leftovers.

If juvenile tyrannosaurids were largely or entirely independent hunters, they would have competed with adult deinonychosaurs, and they may have suppressed the deinonychosaurs' population sizes and influenced their evolution (Paul 1988a). There was a general size decrease in flightless dromaeosaurs in the late Cretaceous (Paul 1988a, 2002). Juvenile tyrannosaurids, with larger brains, overlapping fields of vision, and better running performance, may have posed more serious competition than the young of other mega-avepods. However, because they were ultimately configured to become enormous adults, the anatomy of juvenile tyrannosaurids could not be optimized for their size, so their body form was a compromise between the needs of small predators and the need to grow into giants. Deinonychosaurs were selectively optimized for their size class. In addition, because they descended from ancestors with a history of climbing and flight (Paul 1988a, 2002; Mayr et al. 2005), the birdlike, larger-brained, strong-armed, large-clawed, and agile deinonychosaurs, which may have been more social and parental than other dinosaurs (Paul 2002), were adapted for hunting in a manner dramatically different from young tyrannosaurids. Dromaeosaurs may have even been able to kill proportionally larger prey (Paul 1988a; Therrien et al. 2005). Deinonychosaurs may therefore have enjoyed selective advantages over equal-sized tyrannosaurids in many circumstances. If juvenile tyrannosaurids did have a strong effect on deinonychosaurs, it may have been because adult tyrannosaurids flooded the habitats with so many young that they could not help but adversely affect the deinonychosaurs.

In comparison to tyrannosaurids, other mega-avepods were not as fast, as agile, or as powerful at a given size. Their sense of smell was also less developed, perhaps because it is not difficult to find large sauropods in the drier, more open habitats. Nontyrannosaurid mega-avepod heads were more suited for slashing soft tissues with long rows of bladed teeth powered by weaker jaw muscles. Less overlap of vision fields suggests bites were applied less precisely. In these examples, part of the attack was delivered by powerful, large clawed arms. The ability to inflict crippling or killing wounds appears to have been less than that of tyrannosaurids, so attacks may have taken longer to be effective. Nontyrannosaurid mega-avepods appear to be better candidates than the more powerfully jawed, conical-toothed tyrannosaurids for having septic or venomous bites. Nontyrannosaurid mega-avepods commonly lived among sauropods, and they probably used medium-speed, slashing attacks on the much larger herbivores

(Thomas and Farlow 1997; Paul 1987a, 1988a; Novas et al. 2005). The extreme size of carcharodontosaurs may have evolved to facilitate preying on especially enormous titanosaurid sauropods. Some tyrannosaurids lived in regions that lacked prey significantly larger than themselves. The less extreme bulk of tyrannosaurids, higher speeds, and punch-pull attack mode may have evolved to cope with these less massive, faster herbivores. This hypothesis is complicated by the consistent presence of sauropods in the habitats of tarbosaurs, and later in some *T. rex* habitats (e.g., the North Horn Formation of Utah).

Another explanation for the different adaptations of nontyrannosaurid mega-avepods and tyrannosaurids is their different evolutionary heritages. In this scenario, the tyrannosaurids descended from coelurosaurs more advanced than the basal ancestors of other mega-avepods. This led the 2 groups to evolve different suites of adaptations that accomplished the same need to kill large prey. It is possible that differing evolutionary histories and different prey characteristics combined to distinguish tyrannosaurids from other mega-avepods. It is interesting that the only other mega-avepods that rivaled tyrannosaurids in arm reduction and speed enhancement—certain abelisaurs—appeared at about the same time. Allosaurids show that at least some other mega-avepods also led brief, danger-filled lives. Whether any nontyrannosaurid mega-avepods had skeletal safety factors as low as that seen in albertosaurines and *T. rex* is not yet certain.

Some aspects of tyrannosaurid hunting techniques can be restored with substantial confidence. Other aspects remain highly speculative, but the tactics probably varied widely depending on factors such as tyrannosaurid taxon, age, sex, habitat, and prey type. The unusual speed of tyrannosaurids, ceratopsids, and hadrosaurs, the upgrading of weaponry in tyrannosaurids and ceratopsids, and the upscaling of the size of the last tyrannosaurids, ceratopsids, and ankylosaurids may represent a Red Queen arms race, aided by the expansion of the resource base as the interior seaway retreated from North America (Sampson et al. 2003). Osborn's (1916) view that *T. rex* combined a unique level of size, killing power, and running speed among known terrestrial predators is verified by the results of this study. How the descendants of *T. rex* and its prey could have evolved with additional time is unknown because the K-T crisis aborted the evolutionary experiment.

Acknowledgments

I acknowledge the assistance and discussion over many years of Philip Currie, Kenneth Carpenter, Thomas Holtz, Peter Larson, Jorn Hurum, James Farlow, Blaire Van Valkenburgh, William Sellers, Per Christiansen, John Happ, Mary Schweitzer, Hall Train, Oliver Wings, and numerous others.

References Cited

Abler, W. L. 1992. The serrated teeth of tyrannosaurid dinosaurs and biting structures in other animals. *Paleobiology* 18: 161–183.

———. 1999. The teeth of the *Tyrannosaurus*. *Scientific American* 281: 40–41.

Abourachid, A., and Renous, S. 2000. Bipedal locomotion in ratites (Paleognatiform): example of cursorial birds. *Ibis* 142: 538–549.

Alexander, R. M. 1989. *Dynamics of Dinosaurs and Other Extinct Giant.* Columbia University Press, New York.

———. 1996. *Tyrannosaurus* on the run. *Nature* 379: 121.

Alexander, R. M., Langman, V. A., and Jayes, A. S. 1977. Fast locomotion of some African ungulates. *Journal of Zoology* 183: 291–300.

Alexander, R. M., Maloiy, G. M., Njau, R., and Jayes, A. S. 1979. Mechanics of running of the ostrich. *Journal of Zoology* 187: 169–178.

Anderson, J. F., Hall-Martin, A., and Russell, D. A. 1985. Long bone circumference and weight in mammals, birds and dinosaurs. *Journal of Zoology* 207: 53–61.

Auffenberg, W. 1981. *The Behavioral Ecology of the Komodo Monitor.* University of Florida Press, Gainesville.

Bakker, R. T. 1986. *The Dinosaur Heresies.* William Morrow, New York.

Bakker, R. T., Williams, M., and Currie, P. J. 1988. *Nanotyrannus*, a new genus of pygmy tyrannosaur, from the latest Cretaceous of Montana. *Hunteria* 1: 1–30.

Barsbold, R. 1983. Carnivorous dinosaurs of from the Cretaceous of Mongolia (in Russian). *Sovmestma Sovetsko-Mongolskad Paleontologiceskad Ekspediticad, Trudy* 19: 1–120.

Bekoff, M. 1978. *Coyotes.* Academic Press, New York.

Blanco, R. E., and Mazzetta, G. 2001. A new approach to evaluate the cursorial ability of the giant theropod *Giganotosaurus carolinii*. *Acta Palaeontologica Polonica* 46: 193–202.

Brochu, C. A. 2000. A digitally-rendered endocast for *Tyrannosaurus rex*. *Journal of Vertebrate Paleontology* 20: 1–6.

———. 2003. Osteology of *Tyrannosaurus rex*: insights from a nearly complete skeleton and high-resolution computed tomographic analysis of the skull. *Journal of Vertebrate Paleontology* 22(Suppl. 4): 1–138.

Bybee, P. J., Lee, A. H., and Lamm, E. T. 2006. Sizing the Jurassic theropod dinosaur *Allosaurus*: assessing growth strategy and evolution of ontogenetic scaling of limbs. *Journal of Morphology* 267: 347–59.

Caro, M. 1994. *Cheetahs of the Serengeti Plains.* University of Chicago Press, Chicago.

Carpenter, K. 2000. Evidence of predatory behavior by carnivorous dinosaurs. P. 135–144 in Pérez-Moreno, B. P., Holtz, T. J., Sanz, J. L., and Moratalla, J. (eds.). *Aspects of Theropod Paleobiology. Gaia: Revista de Geociencias, Museu Nacional de Historia Natural, Lisbon*, 15.

Carpenter, K., Sanders, F., McWhinney, L. A., and Wood, L. 2005. Evidence for predator-prey relationships: examples for *Allosaurus* and *Stegosaurus*. P. 325–350 in Carpenter, K. (ed.). *The Carnivorous Dinosaurs.* Indiana University Press, Bloomington.

Carpenter, K., and Smith, M. 2001. Forelimb osteology and biomechanics of *Tyrannosaurus rex*. P. 90–116 in Tanke, D. H., and Carpenter, K. *Mesozoic Vertebrate Life.* Indiana University Press, Bloomington.

Carr, T. D. 1999. Craniofacial ontogeny in Tyrannosauridae. *Journal of Vertebrate Paleontology* 19: 497–520.

Case, T. J. 1978a. On the evolution and adaptive significance of postnatal growth in the terrestrial vertebrates. *Quarterly Review of Biology* 53: 243–282.

———. 1978b. Speculations on the growth rate and reproduction of some dinosaurs. *Paleobiology* 4: 320–328.

Chin, K. et al. 1998. A king-sized theropod coprolite. *Nature* 393: 680–682.

Christiansen, P. 1997. Locomotion in sauropod dinosaurs. *Gaia: Revista de Geociencias, Museu Nacional de Historia Natural, Lisbon*, 14: 45–75.

———. 2000. Strength indicator values of theropod long bones, with comments on limb proportions and cursorial potential. P. 241–255 in Pérez-Moreno, B. P., Holtz, T. J., Sanz, J. L., and Moratalla, J. (eds.). *Aspects of Theropod Paleobiology. Gaia: Revista de Geociencias, Museu Nacional de Historia Natural, Lisbon*, 15.

Christiansen, P., and Paul, G. S. 2001. Limb bone scaling, limb proportions, and bone strength in neoceratopsian dinosaurs. *Gaia: Revista de Geociencias, Museu Nacional de Historia Natural, Lisbon*, 16: 13–29.

Colinvaux, P. 1978. *Why Big Fierce Animals Are Rare*. Princeton University Press, Princeton.

Coombs, W. P. 1978. Theoretical aspects of cursorial adaptations in dinosaurs. *Quarterly Review of Biology* 53: 393–418.

———. 1995. Ankylosaurian tails clubs of middle Campanian to early Maastrichtian age from western North America. *Canadian Journal of Earth Sciences* 32: 902–912.

Coria, R. A., and Currie, P. J. 2002. The braincase of *Giganotosaurus carolinni* from the Upper Cretaceous of Argentina. *Journal of Vertebrate Paleontology* 22: 802–811.

———. 2006. A new carcharodontosaurid from the Upper Cretaceous of Argentina. *Geodiveritas* 28: 71–118.

Currie, P. 2000. Possible evidence of gregarious behavior in tyrannosaurids. P. 271–277 in Pérez-Moreno, B. P., Holtz, T. J., Sanz, J. L., and Moratalla, J. (eds.). *Aspects of Theropod Paleobiology. Gaia: Revista de Geociencias, Museu Nacional de Historia Natural, Lisbon*, 15.

———. 2003. Allometric growth in tyrannosaurids from the Upper Cretaceous of North America and Asia. *Canadian Journal of Earth Sciences* 40: 651–665.

Currie, P., Trexler, D. Koppelhus, E. B., Wicks, K., and Murphy, N. 2005. An unusual multi-individual tyrannosaurid bonebed in the Two Medicine Formation. P. 313–324 in Carpenter, K. (ed.). *The Carnivorous Dinosaurs*. Indiana University Press, Bloomington.

Currie, P., and Zhao X. 1993. A new carnosaur from the Jurassic of Xinjiang, People's Republic of China. *Canadian Journal of Earth Sciences* 30: 2037–2081.

Erickson, G. M., and Brochu, C. A. 1999. How the "terror crocodile" grew so big. *Nature* 398: 205–206.

Erickson, G. M., Makovicky, P. J., Currie, P. J., Norell, M. A., Yerby, S. A., and Brochu, C. A. 2004. Gigantism and comparative life-history parameters of tyrannosaurid dinosaurs. *Nature* 430: 772–775.

Erickson, G. M., and Olson, K. H. 1996. Bite marks attributable to *Tyrannosaurus rex*: preliminary description and implications. *Journal of Vertebrate Paleontology* 16: 175–178.

Erickson, G. M., Rogers, K. C., and Yerby, S. A. 2001. Dinosaurian growth patterns and rapid avian growth rates. *Nature* 412: 429–432.

Erickson, G. M., Van Kirk, S. D., Su, J., Levenston, M. E., Caler, W. E., and Carter, D. R. 1996. Bite-force estimation for *Tyrannosaurus rex* from tooth-marked bones. *Nature* 382: 706–708.

Farlow, J. O. 1976. Speculations about the diet and foraging behavior of large carnivorous dinosaurs. *American Midland Naturalist* 95: 186–191.

———. 1994. Speculations about the carrion-locating ability of tyrannosaurs. *Historical Biology* 7: 159–165.

Farlow, J. O., Brinkman, D. L., Abler, W. L., and Currie, P. J. 1991. Size, shape and serration density of theropod lateral teeth. *Modern Geology* 16: 161–198.

Farlow, J. O., Gatesy, S. M., Holtz, T. R., Jr., Hutchinson, J. R., and Robinson, J. M. 2000. Theropod locomotion. *American Zoologist* 40: 640–663.

Farlow, J. O., and Holtz, T. R. 2002. The fossil record of predation in dinosaurs. *Paleontological Society Papers* 8: 251–265.

Farlow, J. O., Smith, M. B., and Robinson, J. M. 1995. Body mass, bone "strength indicator," and cursorial potential of *Tyrannosaurus rex*. *Journal of Vertebrate Paleontology* 15: 713–725.

Franzosa, J., and Rowe, T. 2005. Cranial endocast of the Cretaceous theropod dinosaur *Acrocanthosaurus atokensis*. *Journal of Vertebrate Paleontology* 25: 859–864.

Fry, B. G., Vidal, N., Norman, J. A., Vonk, F. J., Scheib, H., Ramjan, S. F. R., Kuruppu, S., Fung, K., Hedges, S. B., Richardson, M. K., Hodgson, W. C., Ignjatovic, V., Summerhayes, R., and Kochva, E. 2006. Early evolution of the venom system in lizards and snakes. *Nature* 439: 584–588.

Gatesy, S., Baker, M. and Hutchinson, J. 2006. How *Tyrannosaurus* didn't move: constraint-based exclusion of limb poses for reconstructing dinosaur locomotion. *Journal of Vertebrate Paleontology* 26(3): 66A.

Geis, D. 1960. *The How and Why Wonder Book of Dinosaurs*. Wonder Books, New York.

Goodwin, M. B., Clemens, W. A., Horner, J. R., and Padian, K. 2006. The smallest known *Triceratops* skull: new observations on ceratopsid cranial anatomy and ontogeny. *Journal of Vertebrate Paleontology* 26: 103–112.

Halstead, L. B., and Halstead, J. 1981. *Dinosaurs*. Blandford Press, Poole, U.K.

Hanna, R. R. 2000. Dinosaurs got hurt too. P. 119–126 in Paul, G. S. (ed.). *The Scientific American Book of the Dinosaur*. St. Martin's Press, New York.

Henderson, D. M. 1999. Estimating the masses and centers of mass of extinct animals by 3-D mathematical slicing. *Paleobiology* 25: 88–106.

Henderson, D. M., and Snively, E. 2004. Tyrannosaurus en pointe: allometry minimized rotational inertia of large carnivorous dinosaurs. *Proceedings of the Royal Society of London* 271: 557–560.

Herd, R. M., and Dawson, T. J. 1984. Fiber digestion in the emu, *Dromaius novaehollandiae*, a large bird with a simple gut and high rates of passage. *Physiological Zoology* 57: 70–84.

Holtz, T. R. 1994. The arctometatarsalian pes, an unusual structure of the metatarsus of Cretaceous theropods. *Journal of Vertebrate Paleontology* 14: 480–519.

———. 2002. Theropod predation: evidence and ecomorphology. *Topics in Geobiology* 17: 325–340.

———. 2004. Tyrannosauroidea. P. 111–136 in Weishampel, D., Dodson, P., and Osmólska, H. (eds.). *The Dinosauria*. University of California Press, Berkeley.

Horner, J. R. 1994. Steak knives, beady eyes, and tiny little arms (a portrait of *T. rex* as a scavenger). P. 157–164 in Rosenberg, G. D., and Wolberg, D. L. (eds.). *DinoFest*. Paleontological Society Special Publication 7.

Horner, J. R., and Dobb, E. 1997. *Dinosaur Lives: Unearthing an Evolutionary Saga*. Harcourt Brace, San Diego.

Horner, J. R., and Lessem, D. 1993. *The Complete T. rex*. Simon & Schuster, New York.

Horner, J. R., and Padian, K. 2004. Age and growth dynamics of *Tyrannosaurus rex*. *Proceedings of the Royal Society of London B* 271: 1875–1880.

Houston, D. C. 1979. The adaptations of scavengers. P. 263–286 in Sinclair, A.

R. E., and Norton-Griffiths, M. (eds.). *Serengeti: Dynamics of an Ecosystem.* University of Chicago Press, Chicago.

Hoyo, J., Elliott, A., and Sargatal, J. 1992. *Handbook of the Birds of the World.* Vol. 1. Lynx Edicians, Barcelona.

Hurum, J., and Currie, P. 2000. The crushing bite of tyrannosaurids. *Journal of Vertebrate Paleontology* 20: 619–621.

Hurum, J., and Sabath, K. 2003. Giant theropod dinosaurs from Asia and North America: Skulls of *Tarbosaurus bataar* and *Tyrannosaurus rex* compared. *Acta Palaeontologica Polonica* 48: 161–190.

Hutchinson, J. R. 2004a. Biomechanical modeling and sensitivity analysis of bipedal running ability. 1. Extant taxa. *Journal of Morphology* 262: 421–440.

———. 2004b. Biomechanical modeling and sensitivity analysis of bipedal running ability. 2. Extinct taxa. *Journal of Morphology* 262: 441–461.

Hutchinson, J. R., Anderson, F. C., Blemker, S. S., and Delp, S. L. 2005. Analysis of hindlimb muscle moment arms in *Tyrannosaurus rex* using a three dimensional musculoskeletal computer model: implications for stance, gait, and speed. *Paleobiology* 31: 676–701.

Hutchinson, J. R., Famini, D., Lair, R., and Kram, R. 2003. Are fast-moving elephants really running? *Nature* 422: 493–494.

Hutchinson, J. R., and Garcia, M. 2002. *Tyrannosaurus* was not a fast runner. *Nature* 415: 1018–1021.

Hutchinson, J. R., and Gatesy, S. M. 2006. Beyond the bones. *Nature* 440: 292–294.

Jacobson, A. R. 1998. Feeding behavior of carnivorous dinosaurs as determined by tooth marks on dinosaur bones. *Historical Biology* 13: 17–26.

Kellner, A. W. A., and Langston, W. 1996. Cranial remains of *Quetzalcoatlus* from Late Cretaceous sediments of Big Bend National Park, Texas. *Journal of Vertebrate Paleontology* 16: 222–231.

Klimley, A. P. 1994. The predatory behavior of the white shark. *American Scientist* 82: 122–133.

Kruuk, H. 1972. *The Spotted Hyena.* University of Chicago Press, Chicago.

Kuban, G. J. 1986. A summary of the Taylor site evidence. *Creation/Evolution* XVII 6(1): 10–18.

Lambe, L. M. 1917. The Cretaceous theropodous dinosaur *Gorgosaurus*. *Geological Survey of Canada Memoir* 100: 1–84.

Larson, P. L. 1997. The king's new clothes: a fresh look at *Tyrannosaurus rex*. P. 65–72 in Wolberg, D. L., Stump, E., and Rosenberg, G. D. *Dinofest International Proceedings.* Academy of Natural Sciences, Philadelphia.

———. 2000. The theropod reproductive system. P. 389–397 in Pérez-Moreno, B. P., Holtz, T. J., Sanz, J. L., and Moratalla, J. (eds.). *Aspects of Theropod Paleobiology. Gaia: Revista de Geociencias, Museu Nacional de Historia Natural, Lisbon,* 15.

Larson, P. L., and Donnan, K. 2002. *Rex Appeal.* Invisible Cities Press, Montpelier, VA.

Larsson, H. C. E., Sereno, P., and Wilson, J. A. 2000. Forebrain enlargement among nonavian theropod dinosaurs. *Journal of Vertebrate Paleontology* 20: 615–618.

Leahy, G. D. 2000. Noses, lungs and guts. P. 52–63 in Paul, G. S. (ed.). *The Scientific American Book of the Dinosaur.* St. Martin's Press, New York.

Lehman, T. M. 2001. Late Cretaceous dinosaur provinciality. P. 310–330 in Tanke, D. H., and Carpenter, K. (eds.). *Mesozoic Vertebrate Life.* Indiana University Press, Bloomington.

Lockley, M., and Hunt, A. P. 1994. A track of the giant theropod dinosaur *Tyrannosaurus* from close to the Cretaceous/Tertiary boundary, northern New Mexico. *Ichnos* 3: 213–218.

Madsen, J. H. 1976. *Allosaurus fraglis: A Revised Osteology*. Utah Geological Survey Bulletin 109.

Martin, G. R., and Katzir, G. 1995. Visual fields in ostriches. *Nature* 374: 19–20.

Matthews, P., and McWhirter, N. 1992. *The Guinness Book of Records*. Guinness Publishing, Middlesex, U.K.

Mayr, G., Pohl, B., and Peters, D. S. 2005. A well-preserved *Archaeopteryx* specimen with theropod features. *Nature* 310: 1483–1486.

McMahon, T. A., and Bonner, J. T. 1983. *On Size and Life*. W. H. Freeman, New York.

Meers, M. R. 2002. Maximum bite force and prey size of *Tyrannosaurus rex* and their relationships to the inference of feeding behavior. *Historical Biology* 16: 1–12.

Molnar, R. E. 1991. The cranial morphology of *Tyrannosaurus rex*. *Palaeontographica* A 217: 137–176.

———. 2000. Mechanical factors in the design of the skull of *Tyrannosaurus rex*. P. 193–218 in Pérez-Moreno, B. P., Holtz, T. J., Sanz, J. L., and Moratalla, J. (eds.). *Aspects of Theropod Paleobiology. Gaia: Revista de Geociencias, Museu Nacional de Historia Natural, Lisbon*, 15.

Nelson, M. E., and Mech, L. D. 1984. Observation of a swimming wolf killing a swimming deer. *Journal of Mammalogy* 65: 143–144.

Novas, F. E., Valais, S. Vickers-Rich, P., and Rich, T. 2005. A large Cretaceous theropod from Patagonia, Argentina and the evolution of carcharodontosaurids. *Naturwissenschaften* 92: 226–230.

Nowak, R. M. 1999. *Walker's Mammals of the World*. Johns Hopkins University Press, Baltimore, MD.

O'Connor, P. M., and Claessens, P. A. M. 2005. Basic avian pulmonary design and flow-through ventilation in non-avian theropod dinosaurs. *Nature* 436: 253–256.

Owen-Smith, R. N. 1988. *Megaherbivores*. Cambridge University Press, Cambridge.

Osborn, H. F. 1912. Crania of *Tyrannosaurus* and *Allosaurus*. *Memoirs of the American Museum of Natural History* 1: 1–30.

———. 1916. Skeletal adaptations of *Ornitholestes, Struthiomimus, Tyrannosaurus*. *Bulletin of the American Museum of Natural History* 35: 733–771.

Padian, K., Ricqles, A. J., and Horner, J. R. 2001. Dinosaurian growth rates and bird origins. *Nature* 412: 405–408.

Paul, G. S. 1987a. Predation in the meat-eating dinosaurs. P. 171–176 in *Fourth Symposium on Mesozoic Terrestrial Ecosystems*. Royal Tyrrell Museum of Paleontology, Drumheller, Alberta, Canada.

———. 1987b. The science and art of restoring the life appearance of dinosaurs and their relatives. P. 4–49 in Czerkas, S. J., and Olson, E. C. (eds.). *Dinosaurs Past and Present*. Vol. 2. Natural History Museum of Los Angeles County, Los Angeles.

———. 1988a. *Predatory Dinosaurs of the World*. Simon & Schuster, New York.

———. 1988b. Physiological, migratorial, climatological, geophysical, survival, and evolutionary implications of Cretaceous polar dinosaurs. *Journal of Paleontology* 62(4): 640–652.

———. 1994. Dinosaur reproduction in the fast lane: implications for size, success and extinction. P. 244–255 in Carpenter, K., Hirsch, K. F., and Horner,

J. R. (eds.). *Dinosaur Eggs and Babies*. Cambridge University Press, Cambridge.

———. 1997. Dinosaur models: the good, the bad, and using them to estimate the mass of dinosaurs. P. 129–154 in Wolberg, D. L., Stump, E., and Rosenberg, G. D. *Dinofest International Proceedings*. Academy of Natural Sciences, Philadelphia.

———. 1998. Terramegathermy and Cope's rule in the land of titans. *Modern Geology* 23: 179–217.

———. 2000. Limb design, function and running performance in ostrich-mimics and tyrannosaurs. P. 257–270 in Pérez-Moreno, B. P., Holtz, T. J., Sanz, J. L., and Moratalla, J. (eds.). *Aspects of Theropod Paleobiology. Gaia: Revista de Geociencias, Museu Nacional de Historia Natural, Lisbon*, 15.

———. 2001. Were the respiratory complexes of predatory dinosaurs like crocodilians and birds? P. 463–482 in Gauthier, J., and Gall, L. F. (eds.). *New Perspectives on the Origin and Early Evolution of Birds*. Peabody Museum of Natural History, New Haven, CT.

———. 2002. *Dinosaurs of the Air*. Johns Hopkins University Press, Baltimore, MD.

———. 2005. Body and tail posture in theropod dinosaurs. P. 238–246 in Carpenter, K. (ed.). *The Carnivorous Dinosaurs*. Indiana University Press, Bloomington.

Paul, G. S., and Christiansen, P. 2000. Forelimb posture in neoceratopsian dinosaurs: implications for gait and locomotion. *Paleobiology* 26: 450–465.

Paul, G. S., and Leahy, G. D. 1994. Terramegathermy in the time of titans: Restoring the metabolics of colossal dinosaurs. P. 177–198 in Rosenberg, G. D., and Wolberg. D. L. (eds.). *DinoFest*. Paleontological Society Special Publication 7.

Patak, A., and Baldwin, J. 1993. Structural and metabolic characteristization of the muscles used to power running in the emu, a giant flightless bird. *Journal of Experimental Biology* 175: 233–249.

Perry, S. F. 2001. Functional morphology of the reptilian and avian respiratory systems and its implications for theropod dinosaurs. P. 429–442 in Gauthier, J., and Gall, L. F. (eds.). *New Perspectives on the Origin and Early Evolution of Birds*. Peabody Museum of Natural History, New Haven, CT.

Rayfield, E. J. 2004. Cranial mechanics and feeding in *Tyrannosaurus rex*. *Proceedings of the Royal Society of London* B 271: 1451–1459.

Rayfield, E. J., Norman, D. B., and Upchurch, P. 2002. Cranial design and function in a large theropod dinosaur. *Nature* 409: 1033–1037.

Robertson-Bullock, W. 1962. The weight of the African elephant. *Proceedings of the Zoological Society* 138: 133–135.

Rogers, R. R., Krause, D. W., and Rogers, K. C. 2003. Cannibalism in the Madagascan dinosaur *Majungatholus atopus*. Nature 422: 515–518.

Rothschild, B., and Tanke, D. 2005. Theropod paleopathology: state of the art review. P. 351–365 in Carpenter, K. (ed.). *The Carnivorous Dinosaurs*. Indiana University Press, Bloomington.

Rothschild, B., Tanke, D. H., and Ford, T. L. 2001. Theropod stress fractures and tendon avulsions as a clue to activity. P. 331–336 in Tanke, D., and Carpenter, K. (eds.). *Mesozoic Vertebrate Life*. Indiana University Press, Bloomington.

Rozhdestvensky, A. K. 1965. Growth changes in Asian dinosaurs and some problems of their taxonomy (in Russian). *Palaeontological Zhurnal* 13: 95–105.

Russell, D. A. 1970. Tyrannosaurs from the Late Cretaceous of Western Canada. *National Museum of Natural Sciences Publications in Paleontology* 1: 1–34.

———. 1977. *A Vanished World: The Dinosaurs of Western Canada*. National Museums of Canada, Ottawa.

———. 1989. *An Odyssey in Time: The Dinosaurs of North America*. University of Toronto Press, Toronto.

Ruxton, G. D., and Houston, D. C. 2002. Could *Tyrannosaurus rex* have been a scavenger rather than a predator? An energetics approach. *Proceedings of the Royal Society of London* B 270: 731–733.

Sampson, S. D., and Farlow, J. O., and Carrano, M. T. 2003. Ecological and evolutionary implications of gigantism in theropod dinosaurs. *Journal of Vertebrate Paleontology* 23: 92A.

Sampson, S. D., and Loewen, M. A. 2005. *Tyrannosaurus rex* from the Upper Cretaceous North Horn Formation of Utah: Biogeographic and paleoecological implications. *Journal of Vertebrate Paleontology* 25: 469–472.

Sampson, S. D., Ryan, M. J., and Tanke, D. H. 1997. Craniofacial ontogeny in centrosaurine dinosaurs: taxonomic and behavioral implications. *Zoological Journal of the Linnean Society* 121: 293–337.

Schaller, G. B. 1972. *The Serengeti Lion*. University of Chicago Press, Chicago.

Schweitzer, M. H., Wittmeyer, J. L., and Horner, J. R. 2005. Gender-specific reproductive tissue in ratites and *Tyrannosaurus rex*. *Science* 308: 1456–2005.

Schwimmer, D. R. 2002. *King of the Crocodylians: The Paleobiology of Deinosuchus*. Indiana University Press, Bloomington.

Seebacher, F. 2001. A new method to calculate allometric length-mass relationships of dinosaurs. *Journal of Vertebrate Paleontology* 21: 51–60.

Sellers, W. I., Dennis, L. A., and Crompton, R. H. 2003. Predicting the metabolic energy costs of bipedalism using evolutionary robotics. *Journal of Experimental Biology* 206: 1127–36.

Sellers, W. I, and Paul, G. S. 2004. Speed in giant tyrannosaurs: evolutionary computer simulation. *Journal of Vertebrate Paleontology* 24: 111–112A.

Short, H. L. 1963. Rumen fermentations and energy relationships in white-tailed deer. *Journal of Wildlife Management* 27: 184–195.

Snively, E., and Russell, A. P. 2003. Kinematic model of tyrannosaurid arctometatarsus function. *Journal of Morphology* 255: 215–227.

Stevens, K. In press. Binocular vision in theropod dinosaurs. *Journal of Vertebrate Paleontology*.

Tanke, D. H., and Currie, P. J. 2000. Head-biting behavior in theropod dinosaurs: paleopathological evidence. P. 167–184 in Pérez-Moreno, B. P., Holtz, T. J., Sanz, J. L., and Moratalla, J. (eds.). *Aspects of Theropod Paleobiology. Gaia: Revista de Geociencias, Museu Nacional de Historia Natural, Lisbon*, 15.

Therrien, F., Henderson, D. M., and Ruff, C. B. 2005. Bite me: Biomechanical models of theropod mandibles and implications for feeding behavior. P.179–237 in Carpenter, K. (ed.). *The Carnivorous Dinosaurs*. Indiana University Press, Bloomington.

Thomas, D. A., and Farlow, J. O. 1997. Tracking a dinosaur attack. *Scientific American* 277: 74–79.

Thulborn, R. A. 1982. Speeds and gaits of dinosaurs. *Paleogeography, Palaeoclimatology, Palaeoecology* 38: 227–256.

Tricas, T. C., and McCosker, J. E. 1984. Predatory behavior of the white shark, with notes on its biology. *Proceedings of the California Academy of Sciences* 43: 221–238.

Van Valkenburgh, B., and Molnar, R. E. 2002. Dinosaur and mammalian predators compared. *Paleobiology* 28: 527–543.

Varricchio, D. J. 2001. Gut contents from a Cretaceous tyrannosaurid: implications for theropod dinosaur digestive tracts. *Journal of Paleontology* 75: 410–406.

Watson, J. W., and Zallinger, R. F. 1960. *Dinosaurs*. Golden Press, New York.

Wegweiser, M., Breithaupt, B., and Chapman, R. 2004. Attack behavior of tyrannosaurid dinosaur(s): Cretaceous crime scenes, really old evidence, & "smoking guns." *Journal of Vertebrate Paleontology* 24: 127A.

Wings, O. 2004. Identification, Distribution, and Function of Gastrolithes in Dinosaurs and Extant Birds with Emphasis on Ostriches. Ph.D. thesis. University of Bonn, Bonn.

Xu, X, Norell, M. A., Kuang, X., Wang, X., Zhao, Q., and Jia, C. 2004. Basal tyrannosauroids from China and evidence for protofeathers in tyrannosauroids. *Nature* 431: 680–884.

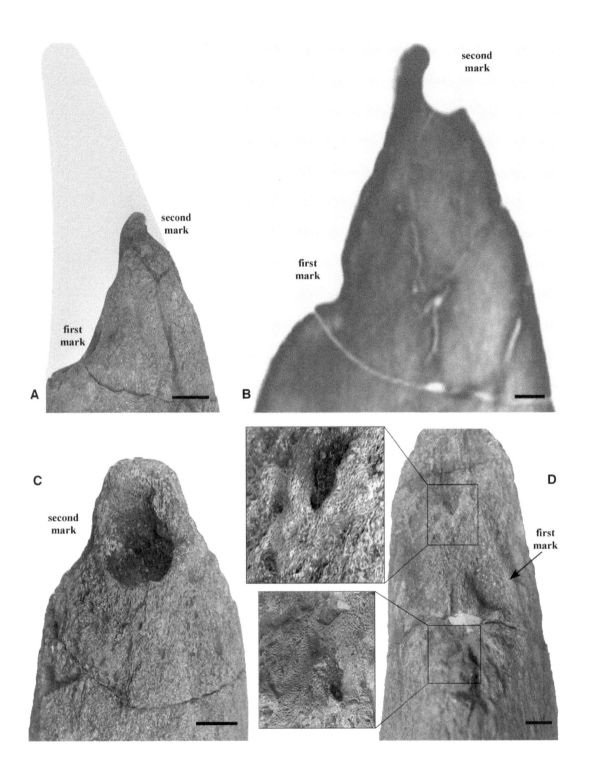

Figure 19.1. Rostral portion of left supraorbital horn core of **Triceratops** SUP 9713. (A) Left lateral view. Shadow is right supraorbital horn core. (B) CT image from left lateral view showing sagittal plane passing through midline of horn core. (C) Dorsal view of second mark. (D) Ventral view of horn break showing first mark. Areas of pronounced osteoproliferative changes are enlarged. Scale bars = 2 cm (A), 1 cm (B, C, D).

AN ANALYSIS OF PREDATOR-PREY BEHAVIOR IN A HEAD-TO-HEAD ENCOUNTER BETWEEN *TYRANNOSAURUS REX* AND *TRICERATOPS*

19

John Happ

Introduction

Evidence from ancient examples of predator-prey relationships can provide insight into the ecology and behavior of ancient animal populations (Bishop 1975; Erickson and Olson 1996; Happ and Morrow 1997; Jacobsen 1998; Farlow and Holtz 2002). When predators attack their prey, they can potentially leave enduring evidence of their behavior in the fossil record. The pattern of predator damage that is seen in fossil animal bones correlates with the type of carnivore (Haynes 1980, 1981). For example, theropods can sometimes be identified to the family or genus level by matching spacing between parallel tooth marks on bone with intertooth distances of specific theropod skulls. The size and shape of a tooth mark provide additional diagnostic information (Currie and Jacobsen 1995; Jacobsen 1995; Erickson and Olson 1996).

Predation marks are inflicted by predators before or at the time of death; scavenging marks are inflicted only after death. The distinction between the 2 feeding behaviors is often impossible to discern from the fossil record (Brain 1981; Horner and Lessem 1993; Currie and Jacobsen 1995; Jacobsen 1995, 2001). Evidence of predation is more clear when tooth marks are accompanied by healing (Rogers 1990; Williamson 1996). For example, bone regrowth is observed in tooth marks of *Tyrannosaurus rex* left in caudal vertebrae of *Edmontosaurus annectens* (Carpenter 1998). In addition, healing of tooth-marked bone requires that the bitten specimen survive the encounter for at least the time required for new bone to form. This is illustrated in rehealed tooth-strike trauma in cranial material of *Herrerasaurus ischigualastensis* (Sereno and Novas 1993), *Sinraptor dongi*, *Gorgosaurus libratus*, and *Daspletosaurus torosus* (Tanke and Currie 1998), where the bite marks provide evidence of intra- or interspecific biting among theropods. Here, I analyze bite marks on the skull of *Triceratops* that are attributed to predation by *Tyrannosaurus rex* (Happ 2003) to estimate the position of their heads at the time of the encounter and to consider behavioral consequences inferred from their position.

Description of the *Triceratops* Skull

A partial skull of *Triceratops* sp., SUP 9713, was collected in 1997 from the upper portion of the Hell Creek Formation (upper Maastrichtian, Upper

Cretaceous) near Jordan, Garfield County, MT (exact locality information on file at Shenandoah University). The skull lay in a gray siltstone in which the dominant clay minerals were mixed-layer smectite-illite, as determined by Fastovsky (1987), who interpreted these facies as floodplain deposits. The *Triceratops* skull was buried during a flood in fine-grained sediment that preserved details of carnivore damage. Burial occurred soon after death, but before the horn sheath and its underlying complex network of blood vessels could decay (Happ and Morrow 2000). As a result, there was minimal weathering of the skull surface. The skull, which was approximately 70% complete, was found inverted with both supraorbital horn cores detached. The right horn core was lying against the right side of the skull, and the left horn core was within 8 cm of the maxilla. Additional cranial elements included, the rostrum, premaxillary, maxillary, nasal horn core, nasal, left jugal, right squamosal, fragmented parietal, fragmented left squamosal, and the region surrounding the brain cavity. The large relative size and degree of coossification of skull elements indicate that the specimen was an adult at the time of death. For example, the nasal horn core measures 30 cm from the ventral surface of the nasal to apex of the horn core and represents the largest of 41 *Triceratops* nasal horn cores measured. No postcranial material of SUP 9713 was found at the quarry site.

Description of Injuries

The skull shows evidence of injuries to both the left supraorbital horn core and left squamosal. The right supraorbital horn core is complete and is approximately 600 mm long. The left horn core is damaged, and the remaining portion is about 360 mm long. If both horns were of comparable size before the damage, then the left horn core would be missing one-third (240 mm) of its original length. The horn was broken diagonally for 120 mm in the rostral direction to the dorsal side (Fig. 19.1A). On the edges of the damaged area on opposite sides of the horn are 2 symmetrically opposing conical depressions. Both lie in a sagittal plane that passes through the midline of the horn (Fig. 19.1B). The first depression (Fig. 19.1B, first mark) is 33 mm inside the break on the ventral side, and is 15 mm wide and 12 mm deep. The opposing depression (Fig. 19.1B and 19.1C, second mark) occurs 11 mm outside the break on the dorsal surface, and is 18 mm wide and 16 mm deep. Both depressions preserve smooth edges, presumably from osteoproliferative changes. On the surface of the break, new reactive bone growth has produced smooth lamellar bone with numerous confluent, small perforations resulting in a fine filigree texture (Fig. 19.1D). The surface of the break is also characterized by additional rugose overgrowths of bone (Fig. 19.1D). This pattern is not observed in the broken portion of 7 additional supraorbital horn cores in the collection at Shenandoah University that show no signs of healing.

Three surface lacerations that are symmetrically linear and parallel appear on the rostral portion of the left squamosal close to the depressions previously described (Fig. 19.2A). The first is 60 mm long, 22 mm wide, and prominent and is 65 mm from the second mark, which is 95 mm long, 6 mm wide, and more superficial. The second mark is 63 mm from the

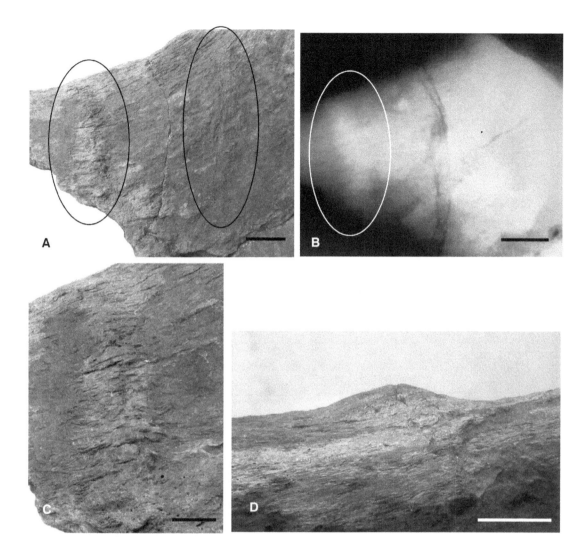

Figure 19.2. Rostral portion of left squamosal of ***Triceratops*** SUP 9713. (A) Dorsal view with parallel score marks circled. (B) Radiograph of same view showing higher radiodensity at first score mark. (C) Enlargement of first score mark. (D) Lateral view of periosteal new bone or involucrum at first score mark. Scale bars = 2 cm (A, B), 1 cm (C, D).

third, which is 90 mm long, 6 mm wide, and faint. At the first mark, bone fibers are expanded and interlace the length of the mark, producing a disfigurement (Fig. 19.2C). Rugose overgrowth of bone extends 3 mm above its normal surface (Fig. 19.2D). The remaining 2 laceration marks are elongate, gently curving in cross section, and have ragged margins.

The internal structure of both left horn and squamosal was analyzed by computed tomography (CT) with a CTi-GE scanner. Areas of increased radiodensity that reflect accumulation of higher density bone extend subperiosteally in CT scans at both injuries. The first scar on the squamosal is detected in CT images to a depth of 12 mm, and the second scar is detected to a depth of 8 mm into the bone. The third scar is too faint to be seen in CT scans.

Source of Damage

The size and configuration of the depressions on the supraorbital horn suggest that they were caused by blows from blunt objects, such as teeth. The depressions are typical of large theropod tooth-strike trauma (Jacob-

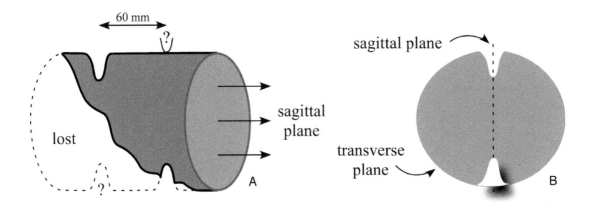

Figure 19.3. Left supraorbital horn of *Triceratops*. (A) Left lateral view, caudal direction to the right. Questions marks signify possible tooth marks. (B) Caudal view of cross section of horn showing displacement of tooth marks in different transverse planes.

sen 1995), and their morphology is consistent with the peglike shape of stout tyrannosaurid teeth (Farlow et al. 1991; Abler 1992). In Figure 19.1B, the second depression compares to the apical portion of the crown of a maxillary lateral tooth of an adult *T. rex* (Erickson and Olsen 1996, fig. 3). The symmetrically opposing positions of the depressions on the dorsal and ventral surfaces of the horn that pass through a common sagittal plane through the midline of the horn and disparate directions of tooth penetration (Fig. 19.1B) strongly imply that the punctures result from a single bite. Because the horn core is composed of a thick outer layer of compact bone at the points of penetration, the damage requires a bite of unusual force. Analysis of cranial design and function in tyrannosaurids indicates that the jaws are capable of generating crushing bites (Hurum and Currie 2000; Meers 2002; Rayfield 2004; Rayfield et al. 2001). In fact, Erickson et al. (1996) estimated that *T. rex* had a bite force that rivals the largest bite force measured for any modern taxon. Analysis of bone fragments in a coprolite attributed to *T. rex* provides additional evidence of its bone-crushing ability (Chin et al. 1998). Therefore, the depressions and break in the horn of *Triceratops* are attributed to a bone-crushing bite of *T. rex*.

Irregular pits on the short, bosslike supraorbital horn cores or exhibit irregular depressions above the orbits rather than well-formed supraorbital horns occur in certain chasmosaurine taxa (*Chasmosaurus belli*, *C. russelli*, *C. irvenensis*) and centrosaurine taxa (*Styracosaurus*, *Pachyrhinosaurus*, *Einiosaurus*) (Sampson et al. 1997; Holmes et al. 2001; Dodson et al. 2004). The process responsible for the horn core erosion remains unknown (Sampson et al. 1997). In *Triceratops*, *Chasmosaurus mariscalensis*, *Pentaceratops*, and other chasmosaurines that are characterized by long and well-formed supraorbital horns, these depressions are absent (Holmes et al. 2001). In fact, the presence of these depressions in *Chasmosaurus irvinensis* is one of the defining characters that distinguishes *C. irvinensis* from *Triceratops* and other chasmosaurines. In addition, pits in other ceratopsian horn cores are irregular and asymmetric (Holmes et al. 2001; Dodson et al. 2004), whereas the ones in *Triceratops* described above show a pattern of symmetry. They are symmetrically situated in opposing positions on the dorsal and ventral surfaces of the horn and lie in a common sagittal plane that passes through the midline of the horn. Therefore, because the erosional pits observed in cen-

Table 19.1. Intertooth Distances of *Tyrannosaurus rex*

Speciman	Maximum Intertooth Distance (mm)
CM 9380	67
AMNH 5027	66
RTMP 81.6.1	57
SDSM 12047	73
LACM 23844	66
BHI 3033	78
FMNH PR2081	70
SUP 9713	Distance between 3 squamosal marks 63, 65

trosaurine and certain other ceratopsids appear in an irregular pattern and do not occur in *Triceratops*, the puncture marks on the SUP 9713 horn core are more easily explained as resulting from a tyrannosaurid bite.

The laceration marks on the squamosal resemble large theropod tooth scrapes over bone, and their morphology is also consistent with the shape of tyrannosaurid teeth (Jacobsen 1995). Because the squamosal was lacerated by teeth from just a single jaw, the wounds do not show the degree of bone crushing produced by both jaws in the horn bite. The distance between linear and parallel marks (63 and 65 mm) corresponds to the maximum intertooth distance of 8 specimens of *T. rex* (Table 19.1). *T. rex* is the only theropod from this formation with an intertooth distance of this size. The next largest intertooth distance recorded for a theropod from this formation is for *Ricardoestesia* sp. and is less than 10 mm (Jacobsen 1995). Lacking a taxon of the size of *T. rex* in collections or in the literature, the laceration marks on the squamosal as well as the closely associated puncture marks on the horn core are attributed to the same source: *T. rex*.

Evidence of Healing and Infection

Reactive bone formation is detected in the injuries to the horn and squamosal. Tooth punctures in unhealed bone typically have sharp, ragged edges (Tanke and Currie 1998, fig. 5; Binford 1981, figs. 3.01 and 3.02). But for SUP 9713, rough edges of the tooth marks have been rounded over by apparent osseous tissue development. Smoothing of the puncture marks by weathering or other taphonomic causes is unlikely because of the well-preserved fine structure of the skull that was buried soon after death. Additional fine structure related to the original traumatic injury, such as traces of tooth carinae, have been obscured by osteoproliferative changes.

The rugose growth and disfigurement of new bone in both the horn break and squamosal laceration marks are an indication that bone infection, or osteomyelitis, accompanied the open bone wounds. The smooth lamellar bone with numerous confluent small perforations that appears in the supraorbital horn break matches the filigree type of bone reaction observed in the broken supraorbital horn core of the chasmosaurine *Anchiceratops* described by Rothschild and Tanke (1992, fig. 6). Figure 19.2D shows a mound of reactive bone tissue or involucrum at the wound site on the squamosal. The presence of the involucrum indicates that osteomyelitis had progressed to subperiosteal abscess formation and is an additional diagnostic feature of bone infection (Huether and McCance 2000). Radiographs (Fig. 19.2B) and CT images further highlight subperiosteal regions of increased radiodensity at wound sites. The wound repair process can lay down new compact lamellar bone of higher density than surrounding bone, causing reduced radiolucence. Areas of increased radiodensity are not the result of postmortem mineral percolation of heavier elements such as iron or manganese. A clear postmortem breakage area elsewhere on the squamosal does not show this effect in radiographs or CT images. Thus, the increased radiodensity is helpful in examining the subsurface extent of bone infection.

Fungi, parasites, viruses, and bacteria can all cause bone infection. One source of infection occurs when an animal inoculates buccal bacterial flora into skin or bone during a septic bite. As an example, the bite of an extant *Varanus komodoensis* (Komodo monitor) is notoriously infectious (Auffenberg 1981). Because of similarities in structure between tyrannosaurid teeth and those of the living Komodo monitor, Abler (1992) proposed that the bite of tyrannosaurs was likely to have been extremely infectious as well. The trapping of meat debris, especially from carrion, in the tooth rows of tyrannosaurids may have acted to promote the growth of bacteria, as in the Komodo monitor. Tanke and Currie (1998) provided evidence that the bite of the tyrannosaurid *Gorgosaurus libratus* was septic. Therefore, a septic bite by *T. rex* is a probable cause of the osteomyelitis in the wounds on SUP 9713.

Orientation of the Heads

An interpretation of the orientation of the heads during the bite to the supraorbital horn should conform to the position of tooth punctures. Both of the opposing teeth marks lie in the same sagittal plane passing through the midline of the horn: one occurring on the dorsal surface and the other on the ventral surface (Fig. 19.3). However, they lie in different transverse planes and are displaced by 60 mm in the sagittal plane. The position of punctures in the horn should also conform to arrangement of teeth in the jaws of *T. rex*. When its jaws were closed, teeth of the lower tooth row nested inside those of the upper jaw (Farlow and Brinkman 1994; Meers 2002). Although the lower jaw was rather rigid, with little movement at the intramandibular joint (Hurum and Currie 2000), the upper jaw may have been slightly mobile at the maxilla-jugal contact (Rayfield 2004). In any case, a tooth mark by the upper jaw should be displaced from one by the lower jaw, and they should not directly occlude tooth to tooth, as in mammal teeth. Postmortem bites may not necessarily require maximum bite force because the carnivore can concentrate on soft tissue. But here, the bite marks were produced before death on tough bone of a large, mobile prey. So the bite is likely to have required near to the maximal force of which the predator was capable. Because of the potential to exert higher forces from teeth situated closer to the region of muscle attachment to the jaws (Molnar 1998, this volume), a bite from a more caudal portion of the jaw would be more capable of crushing the compact bone of the horn.

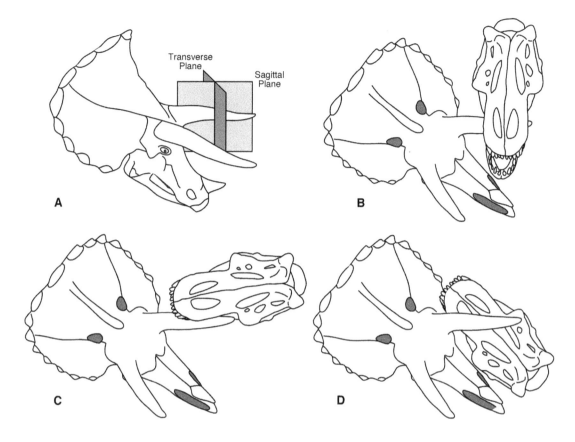

Figure 19.4. Position of heads of **Triceratops** and **T. rex** during bites. (A) Sagittal and transverse planes of left supraorbital horn. (B) Perpendicular sagittal planes of horn and skull of **T. rex**. (C) Parallel sagittal planes of horn and skull of **T. rex**. (D) Bite of squamosal.

Two possible configurations of the head of *T. rex* producing different arrangements of teeth marks in the horn of Triceratops are illustrated in Figure 19.4. I assume that the *Triceratops* head was upright because the puncture marks were centered on the dorsal and ventral sides of the horn. It is also assumed that *T. rex* used its upper jaw to make the dorsal mark and its lower jaw to make the ventral mark. The first orientation of the head of *T. rex* is one in which the sagittal plane of its head is perpendicular to that of the *Triceratops* horn (Fig. 19.4B). In this orientation, marks from the right maxillary and dentary teeth would form on the dorsal and ventral surfaces of the horn. Punctures made by the left jaws would not be preserved. The mark on the horn's dorsal surface should be more caudally oriented than the one on the opposite side because of the displacement of jaws. But this is not observed. Therefore, the punctures could not have been made with the heads in this configuration.

If the head of *T. rex* were turned 90° to the right in a second orientation, then the bite would be made by the left maxillary and lateral row of dentary teeth (Fig. 19.4C). Because teeth of *T. rex* were continually being replaced and varied greatly in height along the jaw, some teeth may not have made contact with bone. If the first puncture mark on the ventral side of the horn is from a larger tooth in the dentary and the second mark on the ventral side is from a larger one in the maxilla, then the bite could result in the maxillary tooth mark occurring more rostrally and the dentary tooth mark more caudally than observed. The 2 marks on the horn were

Figure 19.5. Encounter between ***T. rex*** and ***Triceratops*** as reconstructed from cranial injuries preserved in SUP 9713. Illustration by Gregory S. Paul.

separated in the sagittal plane by 60 mm. That distance is consistent with the intertooth distances of 63 and 65 mm observed between score marks on the squamosal. The second position of the heads is consistent with the location of puncture marks in the horn, as well as arrangement of teeth in the *T. rex* jaws. In this orientation, the sagittal plane of the head of *T. rex* would be approximately parallel to that of the supraorbital horn, and the 2 animals would be almost face to face (Fig. 19.5).

The score marks on the dorsal surface of the left squamosal were made as *T. rex* scraped teeth from the upper jaw along the bone. Marks from the lower jaw did not appear on the opposite side of the squamosal. Most likely, the raking marks were made by either larger caniniform teeth of the maxilla or smaller incisiform teeth of the anterior portion of the maxilla and premaxilla. The other, more medial, premaxillary teeth were too short to make contact with bone. Therefore, these marks were made when the sagittal planes of the heads of both animals were either parallel when heads were face to face, or counterparallel when the heads were facing in the same direction (Fig. 19.4D).

Inferred Behavior

Antipredator Behavior

When confronted with a modern mammal predator, there are 3 primary defensive responses by antlered or horned prey: flee, remain in place, or approach the predator and fight (Mech and Peterson 2003; Estes 1991). Speed and agility are a defense for smaller prey able to run from a predator. Larger prey may also choose flight, but it more often chooses to face the predator and stand its ground, remaining motionless and staring constantly at the predator. Just the appearance of aggressiveness can often deter attack. A third option for prey possessing antlers or horns as weapons is to approach the predator and fight (Mech and Peterson 2003; Estes 1991).

The antipredator behavior of modern mammals may provide clues to the defensive strategies of *Triceratops*. Because its flank would be exposed to potential danger while running (Paul 1988), flight may have been a less successful option. In addition, the running ability of *Triceratops* is debatable (Bakker 1986). An alternative defense would have been to stand its ground with the appearance of aggressiveness. Ostrom and Wellnhofer (1986) suggest that a defensive pose for *Triceratops* is one in which the head is lowered so that both the nasal horn and brow horns point forward toward the aggressor and the neck frill is raised for a more impressive upright display. Such an aggressive display may have potentially deterred a predator attack (Dodson 1996).

Another possible option available to *Triceratops* is to fight using its horns as weapons for stabbing. The horns have long been described as defensive weapons (Hatcher et al. 1907; Lull 1933). However, this function has been questioned more recently by Alexander (1989). On the basis of comparisons of horn cross-sectional area versus body mass in *Triceratops* and horned bovids, Alexander concluded that the *Triceratops* brow horn was relatively short, thin, and weak for its body mass when engaging the base of its horn in intraspecific horn wrestling. When using the tip in stabbing, however, the brow horn of *Triceratops* does not need to be as strong because impact forces are transmitted directly through the horn's longitudinal axis rather than at an oblique angle. Farlow (1990) points out that the aggressive pose causes the nasal horn to make stabbing contact at the same time, thereby reducing stresses to the 2 supraorbital horns. This seems even more reasonable considering the large size of the nasal horn of SUP 9713. In addition, the 3 horns would have penetrated soft tissue rather than engaging other horns in intraspecific horn wrestling. At any rate, when faced with a life-threatening confrontation, it seems reasonable that *Triceratops* would use its horns for defense.

Predator Behavior

A live and aggressive *Triceratops* may not have been the preferred food source for *T. rex*. A study of tooth-marked dinosaur bones from the Dinosaur Park Formation shows a relatively high percentage (14%) of tooth-marked hadrosaurid bones relative to 5% of tooth-marked ceratopsid bones (Jacobsen 1998). Although this suggests a prey preference by Judithian tyrannosaurs for hadrosaurs, it may also reflect a preservational bias. Jacobsen suggests that the horned ceratopsians were more capable of protecting themselves against tyrannosaurs by using their frills and horns and were eaten less frequently as a result. Although unconfirmed, a similar feeding preference is expected by Maastrichtian tyrannosaurids in the Hell Creek Formation.

Big cats may strangle prey by holding closed the nostrils and mouth or trachea of the prey until the animal suffocates (Schaller 1972). If *T. rex* approached *Triceratops* from the front and used this technique of suffocation, then evidence should be present in the premaxilla or nasal regions. Because the premaxilla and nasal of SUP 9713 were free of pathological marks, attempts at suffocation by obstructing the muzzle did not occur. In addition, a frontal approach would place *T. rex* precariously close to the horned weapons of *Triceratops*. In any vertebrate, the neck is a vulnerable point. Tanke and Currie (1998) observe unhealed tyrannosaurid tooth marks in hadrosaur cranial elements, indicating that tyrannosaurs grabbed their victims by the head or neck region with crushing bites. Attack from above to the neck of *Triceratops* would seem to have been less effective because of the large protective frill (Molnar 1998). Both from a logistic standpoint and potential peril to the predator, frontal attack seems unlikely.

Modern predators often attack horned prey from the rear and rarely from the front (Mech 1970). In a rear approach, they are out of view of the prey and are further from the horns, which are potential weapons. There is evidence of this mode of attack by *Tyrannosaurus* toward *Edmontosaurus annectens*. Carpenter (1998) attributes damage to 4 neural spines in the tail of this hadrosaur to a quick bite by *Tyrannosaurus* in an approach from the rear. Bone regrowth after the event indicates that the hadrosaur survived the bite. Usually, damage to weight-bearing bones of modern mammals is fatal (Brandwood et al. 1986; Bulstrode et al. 1986). Molnar (2001) finds that the lack of survivable pathologies in rear weight-bearing elements such as the femur, tibia, and sacrum of bipedal dinosaur prey indicates that injuries to these areas were also seldom survivable.

Finding evidence of rear attack is difficult because of a bias against preservation of postcranial elements of *Triceratops*. No postcranial material of SUP 9713 was found at the quarry site. The normal course of disarticulation of modern large mammal carcasses often begins with separation of the skull from the body because of the relative mobility of the skull on the atlas (Toots 1965; Dodson 1971; Haynes 1981). Dismemberment of the carcass by predators and scavengers concentrates on postcranial portions of the body, where soft tissue is more abundant. The cranial elements of *Triceratops* were more difficult to dismember and swallow because of their size and hardness. Therefore, it is not unusual that only the skull was found at the quarry site (Kruuk 1972; Jacobsen 1995, 1998). Unfortunately, any evidence of pathologies to postcranial material is lost in the case of SUP 9713, and an attack by *T. rex* to the flank or rear cannot be confirmed or denied.

Aggressive Prey Behavior

There are various reasons why *Triceratops* might respond aggressively toward a threatening approach of *Tyrannosaurus*. If the actions of *Triceratops* resembled modern mammalian herbivores (Estes 1991), they would be expected to actively defend their territory or protect their young, family, harem, herd, or nesting sites. On the other hand, it is also possible that a passive *T. rex* was reacting to a direct assault by an aggressive *Triceratops*. Accordingly, the data presented do not preclude the possibility that *T. rex* was not the aggressor in the encounter.

Conclusions

This chapter is an attempt to make reasonable inferences about aspects of the encounter on the basis of available data. It seems reasonable and appropriate antipredator behavior for *Triceratops* to face its aggressor, and the data indicate that *Triceratops* was facing *T. rex* when its left horn was bitten. On the other hand, this places *T. rex* in the unexpected and unfortunate position of facing the remaining horns of *Triceratops*. This is not an advantageous position for a predator. If *T. rex* used long-distance pursuit rather than ambush as a hunting strategy, as some have suggested (Paul 1988; Van Valkenburgh and Molnar 2002), then *Triceratops* may have had sufficient warning of a rearward approach to turn and face the predator. When *T. rex* damaged the left horn, it disabled a major weapon of *Triceratops* and left *Triceratops* in a more vulnerable state. At that same moment, the right supraorbital horn and nasal horn of *Triceratops* were near to the underside of *T. rex*. It is likely that *Triceratops* was capable of a powerful thrust (Bakker 1986), and it may have attempted to use its remaining horns for defense. Whatever happened next, *Triceratops* was not fatally injured. Tyrannosaurids were successful carnivores that were not in the habit of making bad decisions. Therefore, *T. rex* may have decided to abandon the fight because its position was not favorable.

Acknowledgments

I thank Christopher Morrow, Jason Kelley, Stanley Snyder, Randi Clifton, and Joanne Happ for their help in fieldwork and preparation. I appreciate the help of Marty Monroe and Winchester Medical Center staff for providing computed tomography and Winchester Orthopaedic Associates for X-ray photography. I am grateful to John and Sylvia Trumbo, and the Bureau of Land Management (permit M79223) for permitting access to the land and its fossils. Financial support for this research was provided by grants from the Ohrstrom Foundation, the Little River Foundation, and Shenandoah University.

References Cited

Abler, W. L. 1992. The serrated teeth of tyrannosaurid dinosaurs, and biting structures in other animals. *Paleobiology* 18: 161–183.

Alexander, R. M. 1989. *Dynamics of Dinosaurs and Other Extinct Giants*. Columbia University Press, New York.

Auffenberg, W. 1981. *The Behavioral Ecology of the Komodo Monitor*. University Presses of Florida, Gainesville.

Bakker, R. T. 1986. *The Dinosaur Heresies*. Kensington Publishing, New York.

Binford, L. R. 1981. *Bones, Ancient Men and Modern Myths*. Academic Press, New York.

Bishop, G. A. 1975. Traces of predation. P. 261–281 in Frey, R. W. (ed.). *The Study of Trace Fossils*. Springer-Verlag, New York.

Brain, C. K. 1981. *The Hunters or the Hunted? An Introduction to African Cave Taphonomy*. University of Chicago Press, Chicago.

Brandwood, A., Jayes, A. S., and Alexander, R. M. 1986. Incidence of healed fracture in the skeleton of birds, mollusks, and primates. *Journal of Zoology* 208: 55–62.

Bulstrode, C., King, J., and Roper, B. 1986. What happens to wild animals with broken bones? *Lancet* January 4: 29–31.

Carpenter, K. 1998. Evidence of predatory behavior by carnivorous dinosaurs. P. 135–144 in Pérez-Moreno, B. P., Holtz, T. J., Sanz, J. L., and Moratalla, J.

(eds.). *Aspects of Theropod Paleobiology. Gaia: Revista de Geociencias, Museu Nacional de Historia Natural, Lisbon,* 15.

Chin, K., Tokaryk, T. T., Erickson, G. M., and Calk, L. C. 1998. A king-sized theropod coprolite. *Nature* 393: 680–682.

Currie, P. J., and Jacobsen, A. R. 1995. An azhdarchid pterosaur eaten by a velociraptorine theropod. *Canadian Journal of Earth Science* 32: 922–925.

Dodson, P. 1971. Sedimentology and taphonomy of the Oldman Formation (Campanian), Dinosaur Provincial Park, Alberta (Canada). *Palaeogeography, Palaeoclimatology, Palaeoecology* 10: 21–74.

———. 1996. *The Horned Dinosaurs.* Princeton University Press, Princeton, NJ.

Dodson, P., Forster, C. A., and Sampson, S. D. 2004. Ceratopsidae. P. 494–513 in Weishampel, D., Dodson, P., and Osmólska, H. (eds.). *The Dinosauria.* University of California Press, Berkeley.

Erickson, G. M., and Olson, K. H. 1996. Bite marks attributable to *Tyrannosaurus rex*: preliminary description and implications. *Journal of Vertebrate Paleontology* 16: 175–178.

Erickson, G. M., Van Kirk, S. D., Su, J., Levenston, M. E., Caler, W. E., and Carter, D. R. 1996. Bite-force estimation for *Tyrannosaurus rex* from tooth-marked bones. *Nature* 382: 706–708.

Estes, R. D. 1991. *The Behavior Guide to African Mammals.* University of California Press, Berkeley.

Farlow, J. O. 1990. Dynamic dinosaurs. *Paleobiology* 16: 234–241.

Farlow, J. O., and Brinkman, D. L. 1994. Wear surfaces on the teeth of tyrannosaurs. P. 165–175 in Rosenberg, G. D., and Wolberg, D. L. (eds.). *Dino Fest.* Paleontological Society Special Publication 7.

Farlow, J. O., Brinkman, D. L., Abler, W. L., Currie, P. J. 1991. Size, shape, and serration density of theropod dinosaur lateral teeth. *Modern Geology* 16: 161–198.

Farlow, J. O., and Holtz, T. R., Jr. 2002. The fossil record of predation in dinosaurs. P. 251–265 in Kowalewski, M., and Kelley, P. H. (eds.). *The Fossil Record of Predation.* Paleontological Society Papers 8.

Fastovsky, D. 1987. Paleoenvironments of vertebrate-bearing strata during the Cretaceous-Paleogene transition, Eastern Montana and Western North Dakota. *Palaios* 2: 282–295.

Happ, J. W. 2003. Periosteal reaction to injuries of the supraorbital horn and squamosal of an adult *Triceratops* (Dinosauria: Ceratopsidae). *Journal of Vertebrate Paleontology* 23(3, Supplement): 59A.

Happ, J. W., and Morrow, C. M. 1997. Bone modification of subadult *Triceratops* (Dinosauria: Ceratopsidae) by crocodylian and theropod dining. *Journal of Vertebrate Paleontology* 17(Suppl. 3): 51A.

———. 2000. Evidence of soft tissue associated with nasal and supraorbital horn cores, rostral and exoccipital of *Triceratops. Journal of Vertebrate Paleontology* 20(Suppl. 3): 47A.

Hatcher, J. B., Marsh, O. C., and Lull, R. S. 1907. *The Ceratopsia.* United States Geological Survey Monograph 49.

Haynes, G. 1980. Evidence of carnivore gnawing on Pleistocene and recent mammalian bones. *Paleobiology* 6: 341–351.

———. 1981. Prey bones and predators: potential ecologic information from analysis of bone sites. *Ossa* 7: 75–97.

Holmes, R. B., Forster, C., Ryan, M., and Shepherd, K. M. 2001. A new species of *Chasmosaurus* (Dinosauria: Ceratopsia) from the Dinosaur Park Formation of southern Alberta. *Canadian Journal of Earth Science* 38: 1423–1438.

Horner, J. R., and Lessem, D. 1993. *The Complete T. rex.* Simon & Schuster, New York.

Huether, S. E., and McCance, K. L. 2000. *Understanding Pathophysiology.* Mosby, St. Louis.

Hurum, J. H., and Currie, P. J. 2000. The crushing bite of tyrannosaurids. *Journal of Vertebrate Paleontology* 20: 619–621.

Jacobsen, A. R. 1995. Predatory Behavior of Carnivorous Dinosaurs: Ecological Interpretations Based on Tooth Marked Dinosaur Bones and Wear Patterns of Theropod Teeth. M.Sc. thesis, University of Copenhagen.

———. 1998. Feeding behavior of carnivorous dinosaurs as determined by tooth marks on dinosaur bones. *Historical Biology* 13: 17–26.

———. 2001. Tooth-marked small theropod bone: an extremely rare trace. P. 58–63 in Tanke, D., and Carpenter, K. (eds.). *Mesozoic Vertebrate Life.* Indiana University Press, Bloomington

Kruuk, H. 1972. *The Spotted Hyena: A Study of Predation and Social Behavior.* University of Chicago Press, Chicago.

Lull, R. S. 1933. *A Revision of the Ceratopsia, or Horned Dinosaurs.* Memoirs of the Peabody Museum of Natural History 3.

Mech, L. D. 1970. *The Wolf: The Ecology and Behavior of an Endangered Species.* Natural History Press, Garden City, NY.

Mech, L. D., and Peterson, R. O. 2003. Wolf-prey relations. P. 131–160 in Mech, D., and Boitani, L. (eds.). *Wolves: Behavior, Ecology, and Conservation.* University of Chicago Press, Chicago.

Meers, M. B. 2002. Maximum bite force and prey size of *Tyrannosaurus rex* and their relationships to the inference of feeding behavior. *Historical Biology* 16: 1–12.

Molnar, R. E. 1998. Mechanical factors in the design of the skull of *Tyrannosaurus rex* (Osborn, 1905). P. 193–218 in Pérez-Moreno, B. P., Holtz, T. J., Sanz, J. L., and Moratalla, J. (eds.). *Aspects of Theropod Paleobiology. Gaia: Revista de Geociencias, Museu Nacional de Historia Natural, Lisbon,* 15.

———. 2001. Theropod pathology: a literature survey. P. 337–363 in Tanke, D., and Carpenter, K. (eds.). *Mesozoic Vertebrate Life.* Indiana University Press, Bloomington

Ostrom, J. H., and Wellnhofer, P. 1986. The Munich specimen of *Triceratops* with a revision of the genus. *Zitteliana* 14: 111–158.

Paul, G. S. 1988. *Predatory Dinosaurs of the World.* Simon & Schuster, New York.

Rayfield, E. J. 2004. Cranial mechanics and feeding in *Tyrannosaurus rex. Proceedings of the Royal Society of London B* 271: 1451–1459.

Rayfield, E. J., Norman, D. B., Horner, C. C., Horner, J. R., Smith, P. M., Thomason, J. J., and Upchurch, P. 2001. Cranial design and function in a large theropod dinosaur. *Nature* 409: 1033–1037.

Rogers, R. R. 1990. Taphonomy of three dinosaur bone beds in the Upper Cretaceous Two Medicine Formation of Northwestern Montana: evidence for drought-related mortality. *Palaios* 5: 394–413.

Rothschild, B. M., and Tanke, D. H. 1992. Paleoscene 13. Paleopathology of vertebrates: insights to lifestyle and health in the geological record. *Geoscience Canada* 19: 73–82.

Sampson, S. D., Ryan, M. J., and Tanke, D. H. 1997. Craniofacial ontogeny in centrosaurine dinosaurs (Ornithischia: Ceratopsidae): taxonomic and behavioral implications. *Zoological Journal of the Linnean Society* 121: 293–337.

Schaller, G. B. 1972. *The Serengeti Lion: A Study of Predator-Prey Relations.* University of Chicago Press, Chicago.

Sereno, P., and Novas, F. E. 1993. The skull and neck of the basal theropod *Herrerasaurus ischigualastensis. Journal of Vertebrate Paleontology* 13: 451–476.

Tanke, D., and Currie, P. 1998. Head-biting behavior in theropod dinosaurs: paleopathological evidence. P. 167–184 in Pérez-Moreno, B. P., Holtz, T. J., Sanz, J. L., and Moratalla, J. (eds.). *Aspects of Theropod Paleobiology. Gaia: Revista de Geociencias, Museu Nacional de Historia Natural, Lisbon,* 15.

Toots, H. 1965. Sequence of disarticulation in mammalian skeletons. *University of Wyoming Contributions to Geology* 4: 37–39.

Van Valkenburgh, B., and Molnar, R. E. 2002. Dinosaurian and mammalian predators compared. *Paleobiology* 28: 527–543.

Williamson, T. E. 1996. ?*Brachychampsa sealeyi,* sp. nov. (Crocodylia, Alligatoroidea), from the Upper Cretaceous (Lower Campanian) Menefee Formation, Northwestern New Mexico. *Journal of Vertebrate Paleontology* 16: 421–443.

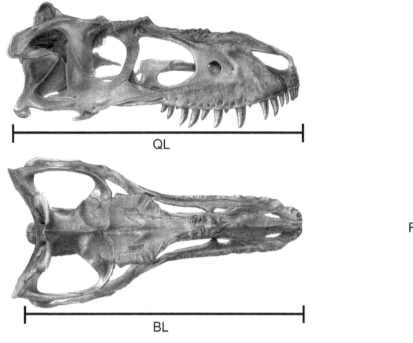
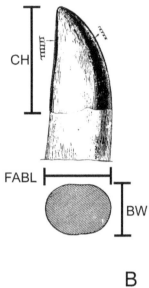

Figure 20.1. Morphometric dimensions measured for this analysis. (A) Skull of juvenile **Gorgosaurus libratus**, after Carr (1999). BL, skull base length; QL, skull quadrate length. (B) Dentary tooth of **Tarbosaurus bataar**, after Maleev (1974). Abbreviations: FABL, fore-aft base length; BW, base width; CH, crown height. See text for discussion.

A CRITICAL REAPPRAISAL OF THE OBLIGATE SCAVENGING HYPOTHESIS FOR *TYRANNOSAURUS REX* AND OTHER TYRANT DINOSAURS

20

Thomas R. Holtz Jr.

Introduction

The biology of the giant latest Cretaceous coelurosaur *Tyrannosaurus rex* and its kin, the Tyrannosauridae, has been of great interest to both paleontologists and the general public. Of particular interest is the ecological behavior of these dinosaurs: specifically, were tyrant dinosaurs predators or scavengers?

Among modern carnivores (that is, animals that derive the majority of their food requirements in the form of flesh), both scavenging (obtaining food from animals already dead by other means) and predation (killing other animals for food) are found. Indeed, large-bodied animals that obtain their food solely from one or the other behavior seem to be vanishingly rare (DeVault et al. 2003). *Crocuta crocuta* (the spotted hyena) was once thought to be primarily a scavenger (e.g., Walker 1964), but direct field observations revealed that they obtain much of their food by predation (Kruuk 1972; Holekamp et al. 1997), although the smaller *Hyaena hyaena* and *H. brunnea* (the striped and brown hyena, respectively) do obtain more food by scavenging than by killing (Kruuk 1976; Owens and Owens 1978). *Panthera leo* (the lion), the archetypal mammalian predator, obtains approximately 10% of its food by scavenging (Mills and Biggs 1993). It is therefore difficult to define a scavenger versus a predator. It might be more accurate to say that there exists an ecological category called *carnivore*, and that carnivores vary in terms of the degree of scavenging and predation behaviors by which they obtain food.

Determining the relative frequency of scavenging versus predation is extraordinarily difficult even for modern predators. Of several different field techniques (stomach analysis, fecal analysis, tracking spoor, opportunistic encounter, radio location, and direct observation), the least biased method, and the one in which such factors as prey selection, kill frequencies, and consumption rates can be measured, is direct observation (Mills 1996; Radloff and DuToit 2004). Of course, direct field observation of tyrannosaurid food acquisition behavior is impossible for paleontologists.

Furthermore, recognizing a scavenging event from a successful predation event from fossil remains is difficult conceptually: in both, the animal would be dead by the end of the feeding episode. Consequently, successful kills and scavenged carrion will be identical in that the food item will not be able to generate a healing response that might be recovered from fossil material. Possible clues that the food item was scavenged by a particular predator type (e.g., a tyrannosaurid) rather than predated might include the following: tyrannosaurid tooth marks that cut across previously existing tooth marks of some other carnivore, which would have necessarily been feeding there first (whether the nontyrannosaur was a predator or scavenger would remain a separate issue for investigation); tyrannosaurid tooth marks that cross-cut traces of decay (e.g., invertebrate or fungal trace fossils) or weathering; and evidence of lethal injuries to the food item not produced by a tyrannosaur (e.g., extremely traumatic trample marks).

However, the debate concerning tyrannosaurids has not consisted of the relative frequency of scavenged versus predated meat in their diet, but rather whether tyrant dinosaurs in general, or *Tyrannosaurus rex* in particular, were capable of acquiring prey at all. In other words, the hypothesis offered is that tyrannosaurids were obligate scavengers.

The concept that tyrannosaurids may have been incapable of hunting has been suggested since the 1910s (Lambe 1917; Halstead and Halstead 1981; Barsbold 1983). However, most of these versions of this concept were not framed as testable, scientific hypotheses. Some workers who have organized their hypotheses in a testable framework are Colinvaux (1978) and Horner (1994, 1997; Horner and Lessem 1993; Horner and Dobb 1997). Colinvaux's arguments were concerned primarily with theoretical ecology and have been reexamined elsewhere (Farlow 1993). Horner and his colleagues have been primarily ecomorphological—that is, they concern the anatomical features of tyrannosaurids and their interaction with their environment.

Several different aspects of tyrannosaurid ecomorphology have been offered to suggest that they were incapable of routinely killing other animals (and thus obligating them to a scavenging behavior). These include the apparently small size of the eye socket relative to the size of the skull; the comparatively short length of the tibia to the femur; the extraordinarily reduced forelimb length; and the observation that tyrannosaurid teeth, unlike those of typical theropods, are not flat and bladelike, but instead have a much wider cross section (Horner and Lessem 1993; Horner 1994, 1997; Horner and Dobb 1997). These observations and their implications are examined in this study. Larson and Donnan (2002) have previously independently examined several of these same ideas.

Methods and Materials

Each of the claims (concerning orbit size, hind limb proportions, arm length, and tooth structure) will be examined separately. In each case, 2 important questions will be examined: is the size of the structure involved unexpectedly small in tyrannosaurids, relative to other forms of comparable size? And would the particular state of that feature in fact prohibit tyrannosaurs from acquiring prey?

The first aspect can be answered by morphometric means: measuring the observed dimensions for the structure at hand, in a number of tyrannosaurid and nontyrannosaurid specimens, and plotting these dimensions versus body size (or some proxy thereof). In the case of the orbit, the smallest diameter across the upper portion of the orbit was used as the largest effective diameter of the eye (see Chure 2000 for a discussion of the position of the eyeball in the orbit of theropods with a noncircular orbital foramen). Orbit diameter was compared with 2 different proxies of body size in theropods: the skull base length, defined as the linear measurement from the anteriormost tip of the premaxilla to the posteriormost point of the occipital condyle; and the skull quadrate length, defined as the linear measurement parallel to the skull base length from the anteriormost tip of the premaxilla to the posteriormost point of the mandibular articulation of the quadrate (Fig. 20.1A). (Skull length would make a poor proxy for body size across diverse dinosaur clades. For example, *Pachycephalosaurus* and *Diplodocus* might have comparable skull lengths, but they have body sizes different by orders of magnitude. However, the theropods examined here have comparable body shapes, so skull length might serve as an approximate estimate of total size.) In the case of most coelophysoids, ceratosaurs, basal tetanurines, and carnosaurs, the skull quadrate length is greater than the skull base length; in the case of coelurosaurs and some coelophysoids, the reverse is true (Holtz 2000).

The data examined were collected directly from specimens as well as from the literature. Theropods of various sizes, from *Scipionyx* to *Tyrannosaurus* and *Giganotosaurus*, were examined.

For limb proportions, the maximum linear dimension parallel to the main shaft of the femur, tibia, or metatarsal III in anterior view was measured for a variety of theropod taxa. These data were derived primarily from Holtz (1995), Gatesy and Middleton (1997), and Farlow et al. (2000), but also include several corrections and additional specimens. Femur length was used as a proxy of body size (as in Holtz 1995). From the total database, only theropods with a femoral length of 200 mm or greater were examined in this study. Additionally, the same measurements were taken from ceratopsians, hadrosaurids, and other bipedal ornithischians (*Thescelosaurus* and pachycephalosaurs) of the Late Cretaceous of western North America for comparison with the American tyrannosaur data. These measurements include data from the literature.

For tooth data, the primary measurements taken were those used by Van Valkenburgh and Ruff (1987) and Farlow et al. (1991): fore-aft (mesial-distal) base length; the base (labiolingual) width, and the crown height (Fig. 20.1B). Fore-aft base length and base width were measured at the basal limit of the enamel-covered part of the tooth for isolated specimens, and at the level of the tooth socket for in situ teeth. Tooth crown height was the vertical distance from the base of the tooth crown to the top of the tooth tip measured perpendicular to fore-aft base length (and thus disregards tooth curvature). These data were measured directly from both isolated teeth and from teeth still articulated in dentaries and maxillae. Premaxillary teeth were not examined in this analysis.

Results

Beady Little Eyes?

Horner (1994) observes that tyrannosaurs look as if they had "beady little eyes," unlike the large eyes of forms such as *Velociraptor*, which he considers to be predators. Although Horner admitted that he did not know whether or not this was a significant feature, it serves as a potential case study in examining the relative size of a feature in animals of much different size.

Two questions might be asked concerning the size of tyrannosaur eyes: were they unexpectedly small for their size? And how does their size relative to the skull compare with their size relative to their function? To answer the former, we may use the techniques of allometric analysis. As has long been observed, not all body parts of an organism grow at the same rate (for reviews, see Schmidt-Nielsen 1984; and McGowan 1991). This difference in growth rate, or allometry, can be manifested in 2 different ways: positive allometry, where the body part in question grows faster than the other factor examined; and negative allometry, where the body part in question does not grow as fast as the other factor. When we compare a baby human with an adult, we see that head grows with nega-

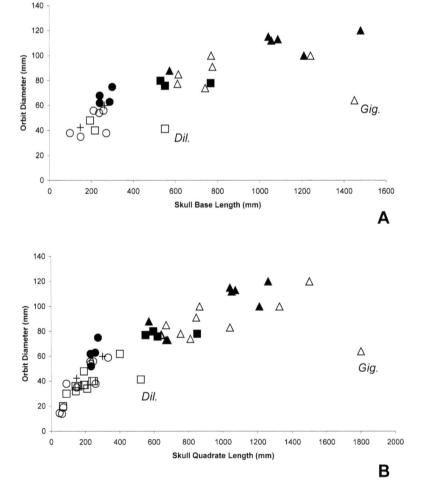

Figure 20.2. *Plot of orbit diameter against skull base length (A) and skull quadrate length (B) for various theropod taxa. Symbols: +, basal theropods (**Eoraptor,** herrerasaurids); open squares, coelophysoids; solid squares, ceratosaurs; open triangles, carnosaurs; open circles, non-ornithomimosaur, nontyrannosaurid coelurosaurs (compsognathids, therizinosauroids, oviraptorosaurs, dromaeosaurs); solid circles, ornithomimosaurs; solid triangles, tyrannosaurids. Abbreviations: Dil.,* **Dilophosaurus wetherilli;** *Gig.,* **Giganotosaurus.**

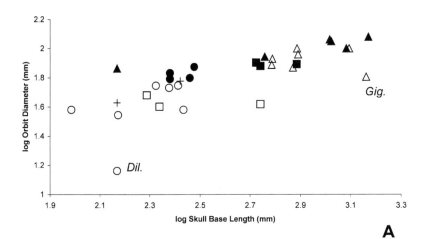

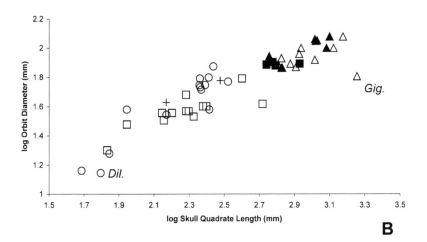

Figure 20.3. *Plot of log orbit diameter against skull base length (A) and skull quadrate length (B). Symbols as in Figure 20.2.*

tive allometry relative to total body height, while leg length grows at positive allometry.

Plotting the orbit diameter versus the skull base length (Fig. 20.2A) and skull quadrate length (Fig. 20.2B) shows that orbit size increases as skull size increases. Furthermore, it can be seen that the orbit diameter size of tyrannosaurids is not atypically small, but is instead comparable to that of other large theropods (carnosaurs and ceratosaurs) of the same skull lengths. (The immense carnosaur *Giganotosaurus* and the coelophysoid *Dilophosaurus wetherilli*, however, do have orbit diameters that plot well below those of other theropods of comparable skull length, and these taxa might well be characterized as beady-eyed.)

Plotting the base 10 logarithms of orbit diameter and skull lengths (Fig. 20.3) allows for the calculation of the allometric equation for these data. If the slope of the regression line of the plotted data is less than 1.0, the feature in question demonstrates negative allometry relative to the feature against which it is plotted. The reduced major axis regression lines for theropod orbits have slopes of 0.44 (when plotted against skull base length) and 0.48 (when plotted against skull quadrate length). Thus, orbit size has a fairly large negative allometry compared with skull size—or, to put it a different way, the rest of the skull grows faster than the orbit size. Similar patterns can be seen in the growth of individual species of various dinosaurs (e.g., Carpenter et al. 1994; Horner and Currie 1994; Long and McNamara 1995, 1997; Carr and Williamson 2004). Thus, the orbits of tyrannosaurids are not unexpectedly small, but their smaller relative size compared with those of *Velociraptor* (for example) is a product of allometry.

Additionally, the eye functions as a photon-catching device. Although the orbit of a tyrannosaurid is smaller in relative terms compared with skull length than in smaller theropods, it is still a much larger opening in absolute terms. Indeed, the orbit diameter of the largest measured specimen of a tyrannosaurid (*Tyrannosaurus rex*, FMNH PR2081) is 120 mm. Even though the actual aperture (pupil) of their eyes would be smaller than this 120-mm diameter, tyrannosaurid eyes potentially had large light-catching surfaces.

Hind Limb Proportions

Although Horner (1994) could not find a particular ecological significance to the small eyes of *Tyrannosaurus*, he is clear with his functional interpretation of the hind limb proportions of tyrant dinosaurs. Horner and colleagues (Horner 1994, 1997; Horner and Lessem 1993; Horner and Dobb 1997) observe that tyrannosaurids, unlike small theropods such as dromaeosaurids, have tibiae that are only as long or shorter than their femora. In modern animals, a tibia/femur ratio that is greater than 1 is often associated with animals adapted to running (Hildebrand 1974; Coombs 1978), whereas lower values are associated with animals incapable of running. Horner and colleagues (Horner 1994, 1997; Horner and Lessem 1993; Horner and Dobb 1997) argue that if tyrannosaurs were incapable of running, they would be incapable of chasing down live prey, and thus would have been restricted to being scavengers.

Before examining the tyrannosaurid condition in particular, it should be pointed out that although there is a general trend for elongation of distal elements in more cursorial animals, the absolute value of the ratio or the metatarsus/femur ratio does not scale directly with speed across clades (Garland and Janis 1993; Carrano 1999). For example, modern species of *Equus* typically have tibiae that are as short or shorter than their femora (values ranging from 0.84 to 1.00, average 0.92, for 18 individuals; Holtz 1995), yet they are undeniably cursorial animals.

In the present analysis, 2 aspects of tyrannosaurid limb proportions will be examined. First, how do tyrannosaurid hind limb proportions compare with those of other theropods, particularly with regards to allometry? Second, how do tyrant dinosaur hind limb proportions compare with those of their potential prey items, large-bodied ornithischians such as hadrosaurids and ceratopsids?

Theropod limb proportions have been subject to a number of previous studies (Coombs 1978; Gatesy 1991; Holtz 1995; Gatesy and Middleton 1997; Christiansen 1997, 2000; Carrano 1999; Farlow et al. 2000). Of par-

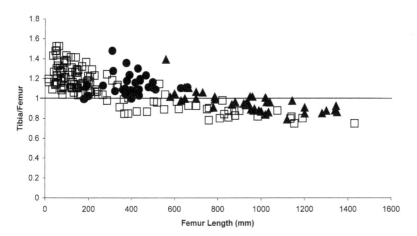

Figure 20.4. *Plot of tibia/femur ratio against femur length for various theropod dinosaurs. Symbols: solid triangles with apex upward, tyrannosaurids; solid circles, ornithomimosaurs; open squares, other theropods.*

ticular interest are the studies of Holtz (1995), Gatesy and Middleton (1997), Carrano (1999), and Farlow et al. (2000), in which the hind limb proportions of different clades of theropods were compared with each other. The present analysis updates aspects of these works.

Figure 20.4 plots the ratio of the tibia length to femur length (T/F) against femur length (used as a proxy for body size) for various theropod dinosaurs. Among nonavian theropods, there is a decrease in T/F as femur length increases (as noted by Gatesy 1991; Holtz 1995; and Carrano 1999). The hypothesis that tyrannosaurids have tibiae that are only as long or shorter than their femora is not supported by the data: this is true for larger specimens, but not for smaller individuals of tyrant dinosaurs. Indeed, the smallest tyrannosaurs had T/F values as high as those of ornithomimosaurs (generally regarded as among the swiftest of dinosaurs; Barsbold and Osmólska 1990) of the same femoral length and higher than those of other nontyrannosaur, nonornithomimosaur theropods of the same body size. In fact, even the largest individual of *Tyrannosaurus rex* examined (FMNH PR2081) had a T/F value (0.86) equal to that of a much smaller *Herrerasaurus ischigualastensis* (PVL 2566). If T/F values less than 1.0 independent of other features of the anatomy were sufficient to exclude tyrannosaurs from the possibility of predation, then many other theropods (including allosauroids, *Ceratosaurus,* and even herrerasaurids with femora as small as 243 mm long) would also be excluded from predation.

In examining absolute limb lengths in theropods, it is found that both the tibia (Fig. 20.5A) and metatarsus (Fig. 20.5B) increase as femur length increases. These elements both grow with negative allometry (allometric slopes of 0.91 and 0.93 when the log of tibia length or log metatarsal III length are plotted against log femur length, respectively), so that as femur size increases, the relative size of the tibia and metatarsus decreases. This phenomenon is common to many groups of animals, including ungulate, carniverous, and marsupial mammals, and many groups of flightless birds (Holtz 1995). For a given femur length, however, tyrannosaurids and ornithomimids have a longer absolute (and thus relative) tibia and metatarsus length than those of other theropods.

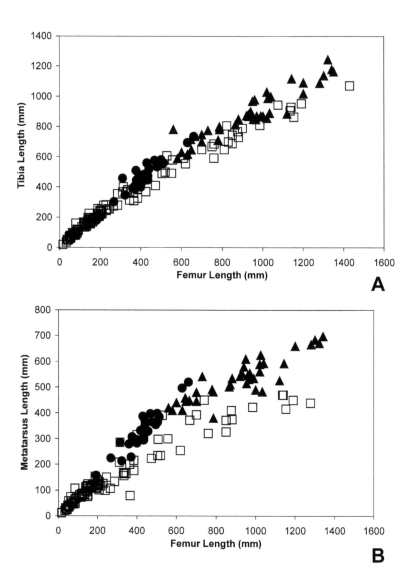

Figure 20.5. *Plot of tibia (A) and metatarsus (B) lengths against femur length for various theropod taxa. Symbols as in Figure 20.4.*

In other words, for a given femur length, tyrannosaurids have a longer distal limb length than those of most other theropods. Similarly, tyrannosaurids and ornithomimids of the same femur length have comparable tibia and metatarsal lengths. Thus, for a given angle of motion of the femur, a tyrannosaur could cover more distance than an allosauroid, ceratosaur, or other large-bodied theropod of the same femur length. Because distance covered per unit of time is the definition of speed, all other things being equal, tyrannosaurids should have been faster than any other comparably sized theropod (see also Carrano 1999). (Note that this does not consider, or even require, a fully suspended phase during this femoral motion; Farlow et al. 1995, 2000; Hutchinson and Garcia 2002; Hutchinson 2004) These data are consistent with a model in which tyrannosaurs were swifter than other potential competitors.

Of more important concern, however, is how the limb proportions of tyrannosaurids compare with those of their potential prey. Examination of

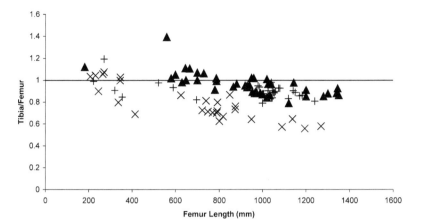

Figure 20.6. *Plot of tibia/femur ratio against femur length for various western North American Late Cretaceous dinosaur taxa. Symbols: solid triangles with apex upward, tyrannosaurids; +, hadrosaurids and other bipedal ornithischians (pachycephalosaurs, **Thescelosaurus**); x, ceratopsians.*

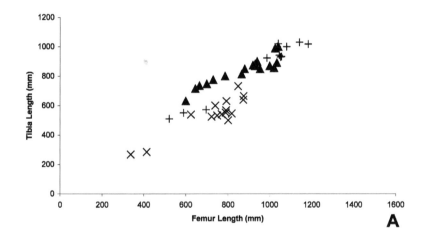

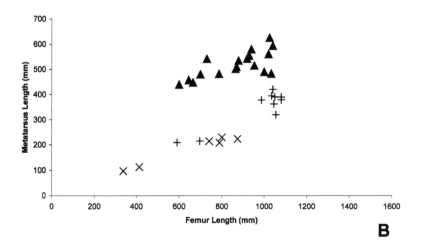

Figure 20.7. *Plot of tibia (A) and metatarsus (B) lengths against femur for various western North American mid- to upper Campanian dinosaur taxa. Symbols as in Figure 20.6.*

Figure 20.8. *Plot of tibia (A) and metatarsus (B) lengths against femur for various western North American Maastrichtian dinosaur taxa. Symbols as in Figure 20.6.*

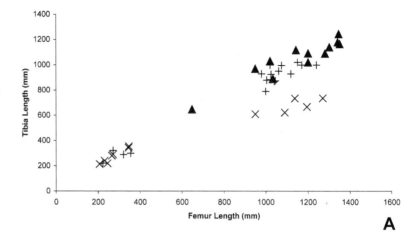

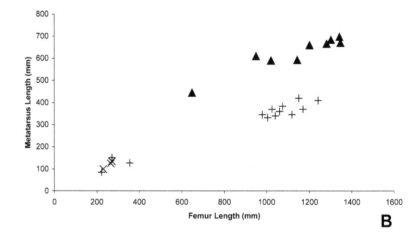

the T/F ratios of western North American Late Cretaceous ceratopsids, hadrosaurids, and other nonankylosaurian ornithischians (Fig. 20.6) shows that these herbivores have as low or lower T/F scores as sympatric tyrannosaurids of the same femur length. Indeed, the values for large hadrosaurids tend to cluster within the cluster of large tyrant dinosaur specimens, while ceratopsians have considerably shorter T/F ratios.

Figure 20.7 plots the tibia and metatarsus against femur length of ornithischians and tyrannosaurids from the mid- to upper Campanian stage Judith River Group and its southwestern U.S. stratigraphic equivalents. Figure 20.8 shows the same for the younger Maastrichtian stage faunas of western North America (e.g., Horseshoe Canyon, Hell Creek, and Lance Formations); and Figure 20.9 combines these data sets. In each of these communities, tyrannosaurids have tibiae at least as long as contemporary hadrosaurids and longer than contemporary ceratopsids of the same femoral length. Furthermore, the metatarsi of tyrannosaurids are much longer than those of ornithischians of the same body size. Thus, the total distal limb length of tyrannosaurids is at least as long, and typically much longer, than contemporary ornithischians.

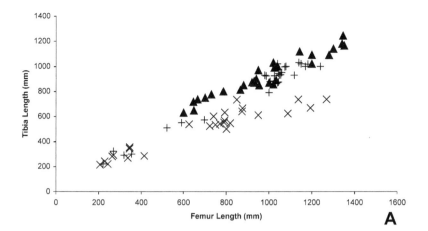
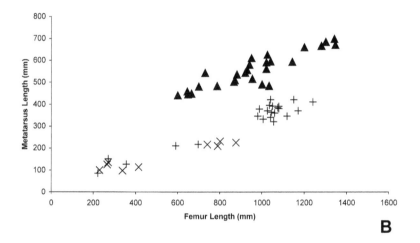

Figure 20.9. *Plot of tibia (A) and metatarsus (B) lengths against femur for various western North American Late Cretaceous dinosaur taxa (combination of data from Figures 20.7 and 20.8). Symbols as in Figure 20.6.*

From this evidence, it is clear that tyrannosaurids would cover more ground for the same angle of femur motion than hadrosaurids and ceratopsids of the same body size. In other words, they would travel further per unit of time (i.e., would be faster) than their potential prey. Again, as before, this does not require a suspended phase on the part of the tyrannosaurid. Limb proportions do not dismiss the possibility that tyrannosaurids could overtake contemporary herbivores, and indeed they are consistent with a model in which tyrant dinosaurs were faster than their potential prey. (Note that the above discussion does not take into account the great disparity between forelimb and hind limb length in ceratopsians. This might indicate that ceratopsids were necessarily slower than a fully bipedal dinosaur of similar hind limb proportions, unless the forelimbs moved with faster steps than the hind limbs in order to keep pace.) To put it another way, if (as Horner argues) the low T/F values of *Tyrannosaurus* and its kin would hinder them from running to catch their prey, the equally low or even lower T/F values of hadrosaurids and ceratopsids would even more greatly hinder the ability of these ornithischians in running *from* a pursuing tyrannosaurid.

Additionally, tyrannosaurids possessed an arctometatarsus, a modification of the foot that biomechanical analysis suggests was more effective at

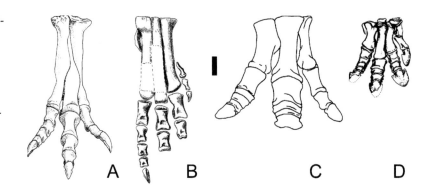

Figure 20.10. *Tyrannosaurid* arctometatarsus (A) compared with more primitive metatarsi of other theropods (B) and ornithischians (C, D). (A) Right pes of the tyrannosaurid **Tarbosaurus**, showing gracile proportions and pinched third metatarsal condition. (B) Right partial pes of the carnosaur **Acrocanthosaurus**, showing more primitive unpinched condition and broader foot. (C) Right pes of the hadrosaurid **Edmontosaurus**. (D) Right pes of the ceratopsid **Centrosaurus**. Scale bar = 100 mm.

the distribution of forces of locomotion (Holtz 1995) (Fig. 20.10). Recent analyses by Snively and Russell (2002, 2003) and Snively et al. (2004) have demonstrated that this adaptation might additionally serve to resist torsional forces, allowing them to turn more rapidly than they might otherwise without risking mechanical failure of their narrow metatarsi. Ceratopsids and hadrosaurids lack this adaptation, or other morphological correlates with more cursorial function (Coombs 1978; Carrano 1999).

Short Arms

Tyrant dinosaurs are characterized by greatly reduced arms (Russell 1970; Molnar et al. 1990; Carpenter 1992; Holtz 2001, 2004; Lipkin and Carpenter this volume). Indeed, tyrannosaur arms are greatly reduced in 2 different senses. First, all known tyrannosaurids are functionally didactyl: they possessed only 2 fingers. More importantly, the overall arm length is quite small compared with the body size. For example, the humerus of tyrannosaurs is only 0.26 to 0.30 times as long as the femur, compared with 0.39 in *Allosaurus* (Gilmore 1920) and *Afrovenator* (Sereno et al. 1994), and 0.44 in *Dilophosaurus* (Welles 1984). The giant allosauroid *Acrocanthosaurus* has a humerus/femur ratio of 0.29 (Currie and Carpenter 2000). At present, the forelimb of the largest allosauroids, *Giganotosaurus* and *Carcharodontosaurus*, are unknown or undescribed, so we cannot determine whether tyrannosaurid arms are uncharacteristically short compared with other carnivorous theropods of this size. (The theropods *Deinocheirus* and *Therizinosaurus*, both of the Nemegt Formation of Mongolia, are large theropods with long forelimbs. However, the former is probably an ornithomimosaur and the latter a therizinosauroid, clades of coelurosaur which were most likely herbivorous [Russell and Dong 1993; Barrett 2005; Kirkland et al. 2005].)

Regardless of whether the arms of tyrannosaurids are uncharacteristically short, the arguments of Horner and colleagues (Horner 1994, 1997; Horner and Lessem 1993; Horner and Dobb 1997) remain valid. Given the reduced size of these structures, it is difficult to envision a method in which the forelimb might have been deployed in prey capture. Carpenter and Smith (2001) and Carpenter (2002) demonstrate, on biomechanical grounds, that these limbs were forcefully designed but had a more limited range of motion compared with *Allosaurus* and *Deinonychus*.

Accepting that the forelimbs of *T. rex* or any tyrannosaurid make unlikely implements for seizing prey items, does this support the argument of obligate scavenging in the tyrant dinosaurs? The mechanics of predation can be divided into 2 main components: prey acquisition and prey dispatch. Among modern terrestrial predatory mammals, there are some taxa that use the forearms in prey capture (e.g., felids), while others do not (hyaenids, canids). Felids use claws to acquire the prey and jaws to dispatch them. The latter forms use the jaws as the primary weapons of prey capture as well as prey dispatch; the forelimbs, if used at all, are primarily used in stabilizing the prey item while the jaws inflict the primary wounds. Carpenter and Smith (2001) support this function in the powerful but short forelimbs of *Tyrannosaurus*.

Modern predatory birds do not use their forelimbs (wings) as the primary weapon of prey capture or dispatch. Instead, their hind limb is used as a primary weapon of prey capture and dispatch (except for falcons, which during prey dispatch use their talons to hold the prey, but the beak to sever the vertebral column) (Brown and Amadon 1968; Cade 1982; Hertel 1995). However, the predatory techniques of flying raptorial birds are difficult to compare directly with those of ground-bound animals such as carniverous mammals or theropod dinosaurs.

The extinct flightless carnivorous birds such as *Diatryma* and the larger phorusrhacids, however, might be used as more informative models for tyrannosaurids. In *Diatryma*, the forelimb is extraordinarily reduced (Matthew and Granger 1917); in at least some of the phorusrhacids, the forelimb is short but apparently quite massively constructed and bore a claw (Chandler 1997). If these forms were indeed predatory, it is unlikely that they could have used their forelimbs in prey acquisition, and in *Diatryma*, they would have been unlikely to have been useful in prey dispatch as well. However, as with nonavian theropods, direct field observation of the predatory techniques (if any) of these forms is denied us.

Although tyrannosaurid arms may have been used to stabilize the food item (prey or scavenged) while feeding, they seem to have been too short to be used to capture prey. However, some modern terrestrial predators (hyaenids and canids) are capable of acquiring prey without using their forelimbs. Similarly, if *Diatryma* and phorusrhacids were predators, they most likely used their skulls alone rather than the forelimbs in prey capture. Thus, a forelimb incapable of prey capture does not obligate a carnivore to a scavenging lifestyle, although it does eliminate some strategies for prey acquisition.

Incrassate Teeth

Tyrannosaurid teeth are distinct from those of other types of theropods, contra Feduccia (1996). Most theropod teeth have a narrow, lens-shaped cross section at the base, and the mesial and distal carinae (bearing the serrations) extend up the front edge and down the rear edge of the tooth. This condition is known as the ziphodont (sword-toothed) condition in the paleocrocodilian literature (e.g., Farlow et al. 1991; Busbey 1995). Tyrannosaurids, however, have lateral (maxillary and dentary) teeth that are expanded basally side to side, and their carinae are offset from the front and rear edges of the tooth. Such dentition has been termed *incrassate* (Holtz 2001, 2003, 2004), after the Latin *incrassatus*, "thickened." (One of many species names Cope proposed for isolated tyrannosaurid teeth was *Laelaps incrassatus*.)

Horner (1997) suggested that tyrannosaurids' incrassate tooth form may have been more associated with bone-crunching ability, which he correlated with scavenging, than with slashing flesh, which he associated with predation. Other authors consider bone crunching, or at least bone biting, to have been part of the feeding repertoire of tyrannosaurs but do not dismiss the possibility of a predatory life style for these theropods (Bakker 1986; Farlow and Brinkman 1987, 1994; Bakker et al. 1988; Abler 1992, 2001; Erickson and Olson 1996; Erickson et al. 1996; Meers 1998, 2003; Carpenter 2000; Hurum and Currie 2000; Hurum and Sabath 2003; Schubert and Ungar 2005; Therrien et al. 2005).

Confirmation of the bone-crunching nature of tyrannosaurid jaws is revealed in a coprolite from the latest Maastrichtian Frenchman Formation of Saskatchewan (Chin et al. 1998). This specimen, almost certainly generated by *Tyrannosaurus rex*, contains the broken fragments of bones of a medium-sized ornithischian dinosaur. Furthermore, specimens of *Edmontosaurus annectens* (Carpenter 2000) and *Triceratops* (Erickson and Olson 1996; Erickson et al. 1996; Happ this volume) demonstrate *Tyrannosaurus* bite marks in which a portion of the bone of the herbivore was removed.

Hyaenids are bone-crunching specialists among the large-bodied carnivorous mammals of the modern world, but as discussed above, all modern hyaenids are known to kill prey, and the largest (*Crocuta crocuta*) obtains the majority of its food in this manner. Bone crunching is accomplished by the molars and premolars in hyaenids (Ewer 1998), but such a feeding mode would result in potentially risk to their canines. Indeed, hyaenids do have thicker canine teeth than found in modern non-bone-crunching dogs, but felids have canines of comparable morphometric proportions to hyaenids. Van Valkenburgh and Ruff (1987) interpret the similarity in canine dimensions of hyaenids and felids not to similar feeding behaviors per se (big cats have not been observed to habitually feed on bones to the same degree as hyaenids), but as adaptations to resist contact with bone during prey capture or dispatch as well as during feeding. Furthermore, Van Valkenburgh and Ruff (1987) demonstrate that the bites of hyaenids and felids generate greater forces than those of canids, and thus teeth that are thicker (and, consequently, more resistant to bending) would be less likely to fail as a result of loads in any direction would have a selective advantage.

The recent discovery of theropods of comparable size to tyrannosaurids allows a comparison of giant ziphodont and incrassate teeth. The subconical teeth of spinosaurid theropods (Holtz 2003), which also differ from the ziphodont condition, are considered a third category for this analysis. Morphometric plots confirm that tyrannosaurid and spinosaurid teeth differ from ziphodont theropods in having greater base widths compared with fore-aft base length (Fig. 20.11A) and crown height (Fig. 20.11B).

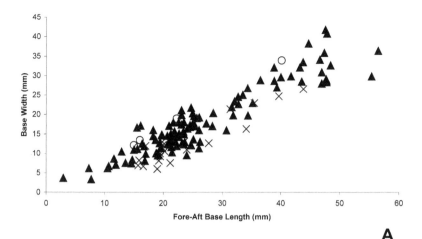

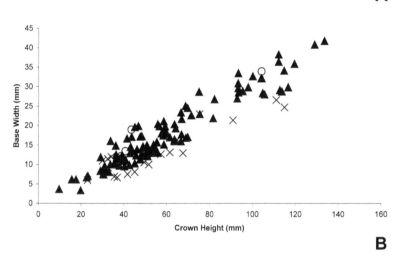

Figure 20.11. *Plot of tooth base width against fore-aft base length (A) and crown height (B) for various theropod taxa. Symbols: solid triangle, tyrannosaurids; x, typical ziphodont theropods; open circle, spinosaurids.*

Van Valkenburgh and Ruff (1987) and Farlow et al. (1991) converted tooth measurements into bending strength indices using beam theory. The present analysis followed the calculations of Farlow et al. and used a rectangular cross-sectional model rather than an oval model for ziphodont and tyrannosaurid teeth. These values do not represent actual strength values, but rather indices comparing the relative resistance of teeth to loads of unit values.

When both the anteroposterior (AP: Fig. 20.12A) and mediolateral (ML: Fig. 20.12B) bending strength index are plotted against crown height, it is found that tyrannosaurid and spinosaurid teeth were more resistant to bending in either directions than ziphodont teeth. This would be consistent with a hyaenid-like bone-crunching habit for tyrant dinosaurs, but would also be consistent with the pattern seen in felids compared with canids: a more powerful bite in tyrannosaurs than in typical theropods, and a better chance of accidental contact between tooth and bone.

That tyrannosaurid teeth contacted bone during feeding is evidenced by various bones with deep furrows or punctures generated by tyrant dinosaur teeth (Horner and Lessem 1993; Erickson and Olson 1996; Erickson

Figure 20.12. *Plot of anteroposterior (A) and mediolateral (B) bending strength indices against crown height for various theropod taxa. Symbols as in Figure 20.11. Tyrannosaurid teeth typically have higher bending strengths than those of other theropods with the same tooth height.*

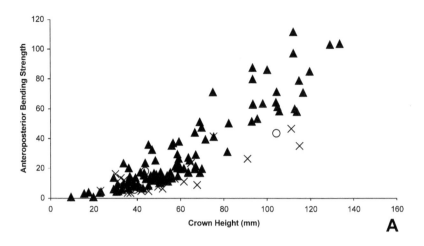

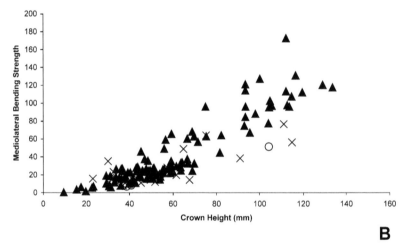

et al. 1996; Carpenter 2000; Happ this volume) and by the Frenchman coprolite (Chin et al. 1998). However, the incrassate teeth of tyrannosaurs may have also functioned during prey capture and dispatch. Horner and colleagues (Horner 1994, 1997; Horner and Lessem 1993; Horner and Dobb 1997) argue that because the forelimbs of tyrannosaurs could not be used to capture prey, the only other likely implement to seize a victim would be the jaws—a view I find reasonable.

However, Horner further argues that the teeth of tyrant dinosaurs would be likely to fail during the stresses generated during prey capture. As shown here, the teeth of tyrannosaurids were mechanically stronger than those of other theropods. Furthermore, Erickson et al. (1996) have demonstrated that tyrannosaurid teeth could withstand considerable stressful contact with bone (see also Meers 1998, 2003; Rayfield et al. 2001; Rayfield 2004). Thus, the data suggest that the teeth of *Tyrannosaurus rex* and its kin would have sufficient strength to absorb the mechanical stresses generated by prey capture.

It is worth noting that the tooth roots of tyrannosaurids and spinosaurids are considerably longer than those of ziphodont theropods. Whereas ziphodonts typically have roots that are subequal to the crown height, tyrannosaurids and spinosaurids have roots typically 150% to 200% of crown height. These may serve to better anchor the teeth and distribute stress against the lateral forces generated during predation and/or feeding that has a greater torsional component than vertical slicing.

Discussion

As seen above, none of the features previously proposed as evidence of an obligate predatory life habit for *Tyrannosaurus* and other tyrant dinosaurs is in fact an indicator of such limitations. The tyrannosaurid eye may appear to be small in relative terms, but its apparent small size is due to the allometrically faster growth of the rest of the skull. In fact, the absolute size of the tyrannosaurid orbit is large. Tyrannosaurid limb proportions do not indicate a necessarily slow speed for these dinosaurs. Instead, tibia/femur ratios in these dinosaurs are higher than those of other large theropods, and as high as or higher than their potential prey. Furthermore, the longer distal limb length of tyrannosaurs compared with duckbills and horned dinosaurs of the same femur length strongly suggests that tyrannosaurs were faster than these herbivores. Regardless of whether tyrannosaurids were capable of a fully suspended running phase or not, the evidence suggests that they could cover more ground per stride than their potential prey, and so could overtake them in a chase. The greatly reduced forelimbs of tyrannosaurids most likely served no function in prey capture, but could have been used to stabilize the prey during dispatch. However, some modern (hyaenids and canids) do not use their limbs to capture their prey, but instead seize them and dispatch them with their jaws alone. The incrassate teeth of tyrannosaurids may indicate a bone-crunching habit for tyrant dinosaurs, but bone crunching and predation are not mutually exclusive behaviors in modern carnivores. Additionally, strong teeth are also found in modern predators as an adaptation to withstand forceful bites and the stresses generated during predation.

Although this reanalysis of the proposed obligate scavenging correlates does not support that hypothesis, it excludes certain predatory behaviors from the tyrannosaurid repertoire, and it is consistent with the findings of others. A catlike model, in which the forelimbs are used in prey capture and a combination of wounds generated by the raking hind limbs and a suffocating bite (Seidensticker and McDougal 1993; Turner and Anton 1997), is excluded because of the improbability that tyrannosaurids could seize prey with their (relatively) tiny arms. Such a behavior would be more consistent with dromaeosaurid dinosaurs, and indeed may be recorded in the famous "fighting dinosaurs" specimens (Jerzykiewicz et al. 1993, fig. 11; Unwin et al. 1995; Carpenter 2000; Holtz 2003). Similarly, a hawklike model of predation, in which the clawed talons were the primary method of killing, is unlikely because both the fore- and hind claws of tyrannosaurids are relatively straight (Holtz 1994, 2004). The hawklike forelimb claws of allosauroids, *Torvosaurus*, *Dryptosaurus*, and some other large theropods indicate that the claws may have served a similar function, although these taxa presumably would have also used their jaws in prey dispatch.

Tyrannosaurid anatomy is consistent with some models of predation, however. The relatively elongate legs of tyrannosaurids suggest that they were faster than their potential prey, although absolute speeds would be difficult to determine without a trackway record. Even though the forelimbs of tyrannosaurids were small, their skulls were large and powerfully built, and (as shown above) their teeth were proportionally stronger and more resistant to bending in various directions than those of other theropods. These data are consistent with canid or hyaenid models for tyrannosaur predation: forms that run down their prey and use the jaws for both prey capture and prey dispatch, and that use the forelimbs only for stabilization during prey dispatch and feeding, if at all. (This analogy describes the behavior of hunting by canids and hyaenids against relatively large prey. While pursuing small items, such as arthropods and small rodents, such predators will use their forelimbs while pouncing on their victims. However, even the most ardent supporter of active tyrannosaurid predation would be unlikely to suggest a coyote-like pounce in the predatory repertoire of *Tyrannosaurus rex!*)

Additional support for the hypothesis that tyrannosaurids were canid- or hyaenid-like jaw-capturing predators can be found in the roof of tyrant dinosaur mouths. Tyrannosaurids differ from other large-bodied theropods, with the exception of spinosaurs (Taquet and Russell 1998; Sereno et al. 1998), in the possession of a substantial ossified secondary palate (Holtz 1998, 2000, 2003, 2004). Busbey (1995) has demonstrated biomechanically that the development of a large bony secondary palate in the skulls of crocodilians resulted in a morphology more resistant to torsional forces than those of the ancestral crocodylomorphs, which had ziphodont teeth and a theropod-like skull form (Russell and Wu 1997). Similarly, a typical theropod skull would be relatively strong in vertical compressive loads, but it would lack solid support to resist strong torsional loads. The solid bony palate of tyrannosaurids, formed by the medial extensions of the maxillae and premaxillae and the greatly expanded diamond-shaped anterior end of the vomer, would allow for greater resistance to torsional loads. As Molnar (2000) argues, however, the vertically oriented maxillae of tyrannosaurids, as opposed to the more horizontally oriented maxillae of crocodilians, indicates that the primary loading in these skulls was still compressive. Mechanical (Erickson et al. 1996), theoretical (Meers 2003), and computer-aided mathematical (Rayfield et al. 2001; Rayfield 2004; Snively et al., personal communication) models of the skull of *Tyrannosaurus* support a powerful bite for this theropod.

Although the data are consistent with this particular model, they do not demonstrate that tyrannosaurids were active predators. As noted previously, such demonstration is difficult even for extant carnivores, except by direct observation. Additionally, as noted above, a successful predation attempt would be difficult, if not impossible, to distinguish from a scavenging event on a carcass, particularly if the traumas that produced the victim's death were not recorded in the hard tissues.

However, there does appear to be direct fossil evidence for at least some unsuccessful predation events by tyrannosaurids. Carpenter (2000) has described a specimen of the hadrosaurid *Edmontosaurus annectens* from the Hell Creek Formation of Montana. This specimen demonstrates a pathological trauma in the caudal region: several consecutive neural spines are damaged and the central one sheared off. The shape of the wound matches the shape of a theropod snout, and pits along this trauma are consistent in size, shape, and position with large theropod teeth. The hadrosaurid survived this trauma, as evidenced by subsequent regrowth of bone in this region. This regrowth indicates that the duckbill was alive at the time of the attack. Happ (this volume) shows similar evidence in the skull of a specimen of *Triceratops horridus*.

At present, *Tyrannosaurus rex* is the only known large theropod from the Hell Creek Formation. It is conceivable that the wound was generated by an as-yet-unknown giant nontyrannosaurid theropod from the latest Maastrichtian of western North America, but at present, an adult *T. rex* is the only likely candidate to have generated this trauma. Although this represents a tiny sample size, these specimens imply that a tyrannosaurid attacked a living hadrosaurid and a living ceratopsid in separate instances.

The data for tyrannosaurids may not support the hypothesis for obligate scavenging, but they do not reject scavenging entirely from the behavior of tyrant dinosaurs. Indeed, as Horner and colleagues (Horner 1994, 1997; Horner and Lessem 1993; Horner and Dobb 1997) and DeVault et al. (2003) correctly point out, carrion represents a food resource that does not require the energy spent in prey capture and dispatch. Furthermore, tyrant dinosaurs were in an excellent position to be effective scavengers—an excellent position literally, in that their great height would allow them a much more extensive view of the landscape in which to find carrion than would smaller carnivores (Farlow 1994); and an excellent position metaphorically, as their much larger body size might allow them to easily chase the smaller dromaeosaurids, troodontids, and other contemporary carnivores away from carcasses.

Additionally, it may be that different growth stages of tyrannosaurids had different life habits (Russell 1970). Perhaps juvenile tyrannosaurids were more active pursuit predators, while adults obtained most of their food as carrion. Furthermore, the relative frequency of predation to scavenging might have varied regionally. Kruuk (1972) observed that 32% of the food of Serengeti populations of *Crocuta crocuta* was from scavenged carcasses, while Ngorongoro populations of the same species scavenged only 7% of their food. There may have been seasonal variation in the frequency of scavenging within species of tyrannosaurids: for example, scavenging might become more important during dry seasons, or during seasons of ornithischian migrations. Finally, it may be that certain species of tyrannosaurid relied more on carrion than prey than did other species, as (for example) the living species of *Hyaena* has a greater fraction of carrion in their diet than does the related *Crocuta*. However, these hypotheses perhaps require direct field observations to test, and thus they remain speculations outside of currently possible scientific inquiry.

Finally, although this short study does not support the hypothesis of obligate scavenging in tyrannosaurids as currently framed, this hypothesis is only provisionally rejected. Additional lines of evidence may indeed support the idea that tyrant dinosaurs were incapable of routinely killing to provide themselves with food. Until such time, however, there is no evidence to suggest that tyrannosaurs were radically different in diet from living large-bodied carnivores, which obtain food both predation and scavenging.

Conclusions

The hypothesis of obligate scavenging in *Tyrannosaurus rex* and other tyrant dinosaurs is provisionally rejected. Previous morphological features suggested as correlates of this hypothetical life habit were not found to reject the possibility of predation in tyrannosaurids. The apparent small size of tyrannosaurid orbits is an artifact of allometry; the hind limbs of tyrannosaurids are (unlike their portrayal in the obligate scavenger model) consistent with a greater locomotor ability in these theropods than in their potential prey; the short arms of tyrannosaurids may have not been used in prey capture, but several living predator groups are known to use their jaws in seizing prey; and the stout incrassate teeth of tyrannosaurs may not have been effective slashers, but were well built to withstand powerful loads.

The anatomy of tyrant dinosaurs is inconsistent with cat- or hawklike predatory behaviors (which necessitate the use of claws in prey capture), but is consistent with canid- or hyaenid-like jaw-capture models. The strong teeth and bony palate of tyrant dinosaurs would allow these dinosaurs to resist stronger twisting loads, and occasional contact with bone, than allosaurids or other typical theropods. Limited fossil evidence documents probable unsuccessful predation attempts of this style by tyrannosaurs. Although obligate scavenging is rejected as a model (pending additional evidence), tyrannosaurids would be effective scavengers. It may be that certain growth stages, or regional or seasonal variations within species, or even whole species, of tyrant dinosaur relied more on scavenged food than killed prey, but barring direct field observation, testing these hypotheses seems impossible.

Acknowledgments

I thank the organizers of the 100 Years of *Tyrannosaurus rex* Symposium for inviting me to participate. I have had (far too many?) discussions on tyrannosaurid paleobiology over the years with various other workers, and I would like to acknowledge, among others, Bob Bakker, Kenneth Carpenter, Philip Currie, Greg Erickson, Jim Farlow, Ralph Molnar, Scott Sampson, and Jack Horner. Although our interpretations of the data may differ, such discussions have helped me to frame my studies of the tyrant dinosaurs. As always, I acknowledge the researchers and other staff at various museums for access to specimens in their collection over the last decade and a half: Ted Daeschler, Academy of Natural Sciences of Philadelphia; Mark Norell and Charlotte Holton, American Museum of Natural History; Peter Larson and Neal L. Larson, Black Hills Museum of Natural History; Michael Henderson and Scott Williams, Burpee Museum of Natural History; Kieran Shephard, Canadian Museum of Nature; Michael Williams,

Cleveland Museum of Natural History; Kenneth Carpenter, Denver Museum of Nature & Science; William Simpson, Field Museum of Natural History; David Whistler and Samuel McLeod, Los Angeles County Museum; Michael Brett-Surman and the late Nicholas Hotton III, National Museum of Natural History; Makoto Manabe, National Science Museum; Angela Milner and Sandra Chapman, the Natural History Museum; Hans-Dieter Sues and Kevin Seymour, Royal Ontario Museum; Don Brinkman, Philip Currie, and Jackie Wilke, Royal Tyrrell Museum of Palaeontology; Sankar Chatterjee, Museum of Texas Tech University; Mary Ann Turner and Christine Chandler, Yale Peabody Museum of Natural History.

References Cited

Abler, W. L. 1992. The serrated teeth of tyrannosaurid dinosaurs, and biting structures of other animals. *Paleobiology* 18: 161–183.

———. 2001. A kerf-and-drill model of tyrannosaurid tooth serrations. P. 84–89 in Tanke, D., and Carpenter, K. (eds.). *Mesozoic Vertebrate Life*. Indiana University Press, Bloomington.

Bakker, R. T. 1986. *The Dinosaur Heresies*. William & Morrow, New York.

Bakker, R. T., Williams, M., and Currie, P. J. 1988. *Nanotyrannus*, a new genus of pygmy tyrannosaur, from the latest Cretaceous of Montana. *Hunteria* 1(5): 1–30.

Barrett, P. M. 2005. The diet of ostrich dinosaurs (Theropoda: Ornithomimosauria). *Palaeontology* 48: 347–358.

Barsbold, R. 1983. Carnivorous dinosaurs from the Cretaceous of Mongolia (in Russian with English summary). *Sovmestnaia Sovestsko-Mongol'skaia Paleontologicheskaia Ekspeditsiia Trudy* 19: 1–117.

Barsbold, R., and Osmólska, H. 1990. Ornithomimosauria. P. 225–244 in Weishampel, D., Dodson, P., and Osmólska, H. (eds.). *The Dinosauria*. University of California Press, Berkeley.

Brown, L., and Amadon, D. 1968. *Eagles, Hawks, and Falcons of the World*. McGraw-Hill, New York.

Busbey, A. B. 1995. The structural consequences of skull flattening in crocodilians. P. 173–192 in Thomason, J. J. (ed.). *Functional Morphology in Vertebrate Paleontology*. Cambridge University Press, Cambridge.

Cade, T. 1982. *The Falcons of the World*. Cornell University Press, Ithaca, NY.

Carpenter, K. 1992. Tyrannosaurids (Dinosauria) of Asia and North America. P. 250–268 in Mateer, N., and Chen, P.-J. (eds.). *Aspects of Nonmarine Cretaceous Geology*. China Ocean Press, Beijing.

———. 2000. Evidence of predatory behavior by carnivorous dinosaurs. P. 135–144 in Pérez-Moreno, B. P., Holtz, T. J., Sanz, J. L., and Moratalla, J. (eds.). *Aspects of Theropod Paleobiology. Gaia: Revista de Geociencias, Museu Nacional de Historia Natural, Lisbon*, 15.

———. 2002. Forelimb biomechanics of nonavian theropod dinosaurs. *Senckenberiana Lethaea* 82: 59–76.

Carpenter, K., and Smith, M. 2001. Forelimb osteology and biomechanics of *Tyrannosaurus*. P. 90–116 in Tanke, D., and Carpenter, K. (eds.). *Mesozoic Vertebrate Life*. Indiana University Press, Bloomington.

Carpenter, K., Hirsch, K. F., and Horner, J. R. 1994. Summary and prospectus. P. 366–370 in Carpenter, K., Hirsch, K. F., and Horner, J. R. (eds.). *Dinosaur Eggs and Babies*. Cambridge University Press, Cambridge.

Carr, T. D. 1999. Craniofacial ontogeny in Tyrannosauridae (Dinosauria, Coelurosauria). *Journal of Vertebrate Paleontology* 19: 497–520.

Carr, T. D., and Williamson, T. E. 2004. Diversity of late Maastrichtian Tyrannosauridae (Dinosauria: Theropoda) of western North America. *Zoological Journal of the Linnean Society* 142: 479–523.

Carrano, M. T. 1999. What, if anything, is a cursor? Categories versus continua for determining locomotor habit in mammals and dinosaurs. *Journal of Zoology* 247: 29–42.

Chandler, R. M. 1997. New discoveries of *Titanis walleri* (Aves: Phorusrhacidae) and a new phylogenetic hypothesis for the phorusrhacids. *Journal of Vertebrate Paleontology* 17(Suppl. 3);36A–37A.

Chin, K., Tokaryk, T. T., Erickson, G. M., and Calk, L. C. 1998. A king-sized coprolite. *Nature* 393: 680–682.

Christiansen, P. 1997. Hind limbs and feet. P. 320–328 in Currie, P. J., and Padian, K. (eds.). *Encyclopedia of Dinosaurs*. Academic Press, San Diego.

———. 2000. Strength indicator values of theropod limb bones, with comments on limb proportions and cursorial potential. P. 241–255 in Pérez-Moreno, B. P., Holtz, T. J., Sanz, J. L., and Moratalla, J. (eds.). *Aspects of Theropod Paleobiology. Gaia: Revista de Geociencias, Museu Nacional de Historia Natural, Lisbon*, 15.

Chure, D. J. 2000. On the orbit of theropod dinosaurs. P. 233–240 in Pérez-Moreno, B. P., Holtz, T. J., Sanz, J. L., and Moratalla, J. (eds.). *Aspects of Theropod Paleobiology. Gaia: Revista de Geociencias, Museu Nacional de Historia Natural, Lisbon*, 15.

Colinvaux, P. 1978. *Why Big Fierce Animals Are Rare: An Ecologist's Perspective*. Princeton University Press, Princeton, NJ.

Coombs, W. P., Jr. 1978. Theoretical aspects of cursorial adaptations in dinosaurs. *Quarterly Review of Biology* 53: 393–418.

Currie, P. J., and Carpenter, K. 2000. A new specimen of *Acrocanthosaurus atokensis* (Dinosauria: Theropoda) from the Lower Cretaceous Antlers Formation (Lower Cretaceous, Aptian) of Oklahoma, USA. *Geodiversitas* 22: 207–246.

DeVault, T. L., Rhodes, O. E., Jr., and Shivik, J. A. 2003. Scavenging by vertebrates: behavioral, ecological, and evolutionary perspectives on an important energy transfer pathway in terrestrial ecosystems. *Oikos* 102: 225–234.

Erickson, G. M., and Olson, K. H. 1996. Bite marks attributable to *Tyrannosaurus rex*: preliminary description and implications. *Journal of Vertebrate Paleontology* 16: 175–178.

Erickson, G. M., van Kirk, S. D., Su, J.-T., Levenston, M. E., Caler, W. E., and Carter, D. R. 1996. Bite force estimation for *Tyrannosaurus rex* from tooth-marked bone. *Nature* 382: 706–708.

Ewer, R. F. 1998. *The Carnivores*. Cornell University Press, Ithaca, NY.

Farlow, J. O. 1993. On the rareness of big, fierce animals: speculations about the body sizes, population densities, and geographic ranges of predatory mammals and large carnivorous dinosaurs. *American Journal of Sciences* 239A: 167–199.

———. 1994. Speculations about the carrion-locating ability of tyrannosaurs. *Modern Geology* 7: 159–165.

Farlow, J. O., and Brinkman, D. L. 1987. Serration coarseness and patterns of wear of theropod dinosaur teeth. *Geological Society of America Abstracts with Programs* 19: 151.

———. 1994. Wear surfaces on the teeth of tyrannosaurs. P. 165–175 in Rosenberg, G., and Wolberg, D. (eds.). *Dino Fest Proceedings*. Paleontological Society Special Publication 7.

Farlow, J. O., Brinkman, D. L., Abler, W. L., and Currie, P. J. 1991. Size, shape, and serration density of theropod dinosaur lateral teeth. *Modern Geology* 16: 161–198.

Farlow, J. O., Gatesy, S. M., Holtz, T. R., Jr., Hutchinson, J. R., and Robinson, J. M. 2000. Theropod locomotion. *American Zoologist* 40: 640–663.

Farlow, J. O., Smith, M. B., and Robinson, J. M. 1995. Body mass, bone "strength indicator," and cursorial potential of *Tyrannosaurus rex*. *Journal of Vertebrate Paleontology* 15: 713–725.

Feduccia, A. 1996. *The Origin and Evolution of Birds*. Yale University Press, New Haven, CT.

Garland, T., Jr., and Janis, C. M. 1993. Does metatarsal/femur ratio predict maximal running speeds in cursorial mammals? *Journal of Zoology* 229: 133–151.

Gatesy, S. M. 1991. Hind limb scaling in birds and other theropods: implications for terrestrial locomotion. *Journal of Morphology* 209: 83–96.

Gatesy, S. M., and Middleton, K. M. 1997. Bipedalism, flight, and the evolution of theropod locomotor diversity. *Journal of Vertebrate Paleontology* 17: 308–329.

Gilmore, C. W. 1920. Osteology of the carnivorous Dinosauria in the United States National Museum, with special reference to the genera *Antrodemus* (*Allosaurus*) and *Ceratosaurus*. *Bulletin of the U.S. National Museum* 110: 1–154.

Halstead, L. B., and Halstead, J. 1981. *Dinosaurs*. Blandford Press, Poole, UK.

Hertel, F. 1995. Ecomorphological indicators of feeding behavior in recent and fossil raptors. *Auk* 112: 890–903.

Hildebrand, M. 1974. *Analysis of Vertebrate Structure*. Wiley and Sons, New York.

Holekamp, K. E., Smale, L., Berg, R., and Cooper, S. M. 1997. Hunting rates and hunting success in the spotted hyena (*Crocuta crocuta*). *Journal of Zoology* 242: 1–15.

Holtz, T. R., Jr. 1994. The phylogenetic position of the Tyrannosauridae: implications for theropod systematics. *Journal of Paleontology* 64: 1100–1117.

———. 1995. The arctometatarsalian pes, an unusual structure of the metatarsus of Cretaceous Theropoda (Dinosauria: Saurischia). *Journal of Vertebrate Paleontology* 14: 480–519.

———. 1998. Large theropod comparative cranial function: a new "twist" for tyrannosaurs. *Journal of Vertebrate Paleontology* 18(Suppl. 3): 51A.

———. 2000. A new phylogeny of the carnivorous dinosaurs. P. 5–61 in Pérez-Moreno, B. P., Holtz, T. J., Sanz, J. L., and Moratalla, J. (eds.). *Aspects of Theropod Paleobiology. Gaia: Revista de Geociencias, Museu Nacional de Historia Natural, Lisbon*, 15.

———. 2001. The phylogeny and taxonomy of the Tyrannosauridae. P. 64–83 in Tanke, D., and Carpenter, K. (eds.). *Mesozoic Vertebrate Life*. Indiana University Press, Bloomington.

———. 2003. Dinosaur predation: evidence and ecomorphology. P. 325–340 Kelley, P. H., Kowalewski, M., and Hansen, T. A. (eds.). *Predator-Prey Interactions in the Fossil Record*. Kluwer Press, New York.

———. 2004. Tyrannosauroidea. *Dinosauria*. 2nd ed. P. 111–136 in Weishampel, D., Dodson, P., and Osmólska, H. (eds.). *The Dinosauria*. University of California Press, Berkeley.

Horner, J. R. 1994. Steak knives, beady eyes, and tiny little arms (a portrait of *T. rex* as a scavenger). P. 157–164 in Rosenberg, G., and Wolberg, D. (eds.). *Dino Fest Proceedings*. Paleontological Society Special Publication 7.

———. 1997. *T. rex* on trial: examining the evidence for meat-eating dinosaurs.

Museum exhibit, Museum of the Rockies, Montana State University, Bozeman, MT. May 24, 1997–April 1999.

Horner, J. R., and Currie, P. J. 1994. Embryonic and neonatal morphology and ontogeny of a new species of *Hypacrosaurus* (Ornithischia, Lambeosauridae) from Montana and Alberta. P. 312–336 in Carpenter, K., Hirsch, K. F., and Horner, J. R. (eds.). *Dinosaur Eggs and Babies.* Cambridge University Press, Cambridge.

Horner, J. R., and Dobb, E. 1997. *Dinosaur Lives: Unearthing an Evolutionary Saga.* Harcourt Brace, San Diego.

Horner, J. R., and Lessem, D. 1993. *The Complete T. rex.* Simon & Schuster, New York.

Hurum, J., and Currie, P. J. 2000. The crushing bites of tyrannosaurids. *Journal of Vertebrate Paleontology* 20: 619–621.

Hurum, J., and Sabath, K. 2003. Giant theropod dinosaurs from Asia and North America: skulls of *Tarbosaurus bataar* and *Tyrannosaurus rex* compared. *Acta Paleontologia Polonica* 48: 161–190.

Hutchinson, J. R. 2004. Biomechanical modeling and sensitivity analysis of bipedal running. II. Extinct taxa. *Journal of Morphology* 262: 441–461.

Hutchinson, J. R., and Garcia, M. 2002. *Tyrannosaurus* was not a fast runner. *Nature* 415: 1018–1021.

Jerzykiewicz, T., Currie, P. J., Eberth, D. A., Johnston, P. A., Koster, E. H., and Zheng J.-J. 1993. Djadokhta Formation correlative strata in Chinese Inner Mongolia: an overview of the stratigraphy, sedimentary geology, and paleontology and comparisons with the type locality in the pre-Altai Gobi. *Canadian Journal of Earth Sciences* 30: 2180–2195.

Kirkland, J. I., Zanno, L. E., Sampson, S. D., Clark, J. M., and DeBlieux, D. 2005. A primitive therizinosauroid dinosaur from the Early Cretaceous of Utah. *Nature* 435: 84–87.

Kruuk, H. 1972. *The Spotted Hyena: A Study of Predation and Social Behavior.* University of Chicago Press, Chicago.

———. 1976. Feeding and social behaviour of the striped hyaena (*Hyaena vulgaris* Desmarest). *East African Wildlife Journal* 14: 91–111.

Lambe, L. B. 1917. *The Cretaceous Theropodous Dinosaur Gorgosaurus.* Memoirs of the Geological Survey of Canada 100.

Larson, P., and Donnan, K. 2002. *Rex Appeal: The Amazing Story of Sue, the Dinosaur that Changed Science, the Law, and My Life.* Invisible Cities Press, Montpelier, VT.

Long, J. A., and McNamara, K. J. 1995. Heterochrony in dinosaur evolution. P. 151–168 in McNamara, K. J. (ed.). *Evolutionary Change and Heterochrony.* Wiley and Sons, New York.

———. 1997. Heterochrony: the key to dinosaur evolution. P. 113–123 in Wolberg, D. L., Stump, E., and Rosenberg, G. D. *Dinofest International Proceedings.* Academy of Natural Sciences, Philadelphia.

Maleev, E. A. 1974. Gigantic carnosaurs of the family Tyrannosauridae (in Russian with English summary). *Sovmestnaia Sovestsko-Mongol'skaia Paleontologicheskaia Ekspeditsiia Trudy* 1: 132–191.

Matthew, W. D., and Granger, W. 1917. The skeleton of *Diatryma*, a gigantic bird from the Lower Eocene of Wyoming. *Bulletin of the American Museum of Natural History* 37: 307–326.

McGowan, C. 1991. *Dinosaurs, Spitfires, and Sea Dragons.* Harvard University Press, Cambridge, MA.

Meers, M. B. 1998. Estimation of maximum bite force in *Tyrannosaurus rex* and its relationship to the inference of feeding behavior. *Journal of Vertebrate Paleontology* 18(Suppl. 3): 63A.

———. 2003. Maximum bite force and prey size of *Tyrannosaurus rex* and their relationships to the inference of feeding behavior. *Historical Biology* 16: 1–12.

Mills, M. G. L. 1996. Methodological advances in capture, census, and food-habit studies of large African carnivores. P. 223–242 in Gittleman, J. L. (ed.). *Carnivore Behavior, Ecology, and Evolution*. Vol. 2. Cornell University Press, Ithaca, NY.

Mills, M. G. L., and Biggs, H. C. 1993. Prey apportionment and related ecological relationships between large carnivores in Kruger National Park. P. 253–268 in Dunstone, N., and Gorman, M. L. (eds.). *Mammals as Predators*. Zoological Society of London Symposium 65. Oxford Science Publications, Oxford.

———. 2000. Mechanical factors in the design of the skull of *Tyrannosaurus rex*. P. 193–218 in Pérez-Moreno, B. P., Holtz, T. J., Sanz, J. L., and Moratalla, J. (eds.). *Aspects of Theropod Paleobiology. Gaia: Revista de Geociencias, Museu Nacional de Historia Natural, Lisbon*, 15.

Molnar, R. E., Kurzanov, S. M., and Dong, Z.-M. 1990. Carnosaurs. P. 169–209 in Weishampel, D., Dodson, P., and Osmólska, H. (eds.). *The Dinosauria*. University of California Press, Berkeley.

Owens, M. J., and Owens, D. D. 1978. Feeding ecology and its influence on social organization in brown hyenas (*Hyaena brunnea*, Thunberg) of the Central Kalahari Desert. *East African Wildlife Journal* 16: 113–135.

Radloff, F. G. T., and DuToit, J. T. 2004. Large predators and their prey in a southern African savanna: a predator's size determines its prey size range. *Journal of Animal Ecology* 73: 410–423.

Rayfield, E. J. 2004. Cranial mechanics and feeding in *Tyrannosaurus rex*. *Proceedings: Biological Sciences* 271: 1451–1459.

Rayfield, E. J., Norman, D. B., Horner, C. C., Horner, J. R., Smith, P. M., Thomason, J. J., and Upchurch, P. 2001. Cranial design and function in a large theropod dinosaur. *Nature* 409: 1033–1037.

Russell, A. P., and Wu, X.-C. 1997. The Crocodylomorpha at and between geological boundaries: the Baden-Powell approach to change? *Zoology* 100: 164–182.

Russell, D. A. 1970. *Tyrannosaurs from the Late Cretaceous of Western Canada*. National Museum Natural Sciences Publications in Palaeontology 1.

Russell, D. A., and Dong, Z.-M. 1993. The affinities of a new theropod from the Alxa Desert, Inner Mongolia, People's Republic of China. *Canadian Journal of Earth Science* 30: 2107–2127.

Schmidt-Nielsen, K. 1984. *Scaling: Why Is Animal Size So Important?* Cambridge University Press, Cambridge.

Schubert, B. W., and Ungar, P. S. 2005. Wear facets and enamel spalling in tyrannosaurid dinosaurs. *Acta Palaeontologica Polonica* 50: 93–99.

Seidensticker, J., and McDougal, C. 1993. Tiger predatory behaviour, ecology and conservation. P. 105–125 in Dunstone, N., and Gorman, M. L. (eds.). *Mammals as Predators*. Zoological Society of London Symposium 65. Oxford Science Publications, Oxford.

Sereno, P. C., Beck, A. L., Dutheil, D. B., Gado, B., Larsson, H. C. E., Lyon, G. H., Marcot, J. D., Rauhut, O. W. M. , Sadlier, R. W., Sidor, C. A., Varricchio, D. D., Wilson, G. P., and Wilson, J. A. 1998. A long-snouted predatory dinosaur from Africa and the evolution of spinosaurids. *Science* 282: 1298–1302.

Sereno, P. C., Wilson, J. A., Larsson, H. C. E., Dutheil, D. B., and Sues, H.-D. 1994. Early Cretaceous dinosaurs from the Sahara. *Science* 255: 845–848.

Snively, E., and Russell, A. P. 2002. The tyrannosaurid metatarsus: bone strain and inferred ligament function. *Senckenbergiana Lethaea* 81: 73–80.

———. 2003. A kinematic model of tyrannosaurid arctometatarsus function (Dinosauria: Theropoda). *Journal of Morphology* 255: 215–227.

Snively, E., Russell, A. P., and Powell, G. L. 2004. Evolutionary morphology of the coelurosaurian arctometatarsus: descriptive, morphometric and phylogenetic approaches. *Zoological Journal of the Linnean Society* 142: 525–553.

Taquet, P., and Russell, D. A. 1998. New data on spinosaurid dinosaurs from the Early Cretaceous of the Sahara. *Comptes Rendus de l'Académie des Sciences Paris, Sciences de la Terre et des Planètes* 327: 347–353.

Therrien, F., Henderson, D. M., and Ruff, C. B. 2005. Bite me: biomechanical models of theropod mandibles and implications for feeding behavior. P. 179–237 in Carpenter, K. (ed.). *The Carnivorous Dinosaurs.* Indiana University Press, Bloomington.

Turner, A., and Anton, M. 1997. *The Big Cats and Their Fossil Relatives: An Illustrated Guide to Their Evolution and Natural History.* Columbia University Press, New York.

Unwin, D. M., Perle A., and Trueman, C. 1995. *Protoceratops* and *Velociraptor* preserved in association: evidence for predatory behavior in dromaeosaurid dinosaurs? *Journal of Vertebrate Paleontology* 15(Suppl. 3): 57A–58A.

Van Valkenburgh, B., and Ruff, C. B. 1987. Canine tooth strength and killing behavior in large carnivores. *Journal of Zoology* 212: 379–397.

Walker, E. P. 1964. *Mammals of the World.* Johns Hopkins University Press, Baltimore, MD.

Welles, S. P. 1984. *Dilophosaurus wetherilli* (Dinosauria, Theropoda): osteology and comparisons. *Palaeontographica Abteilung A* 185: 85–180.

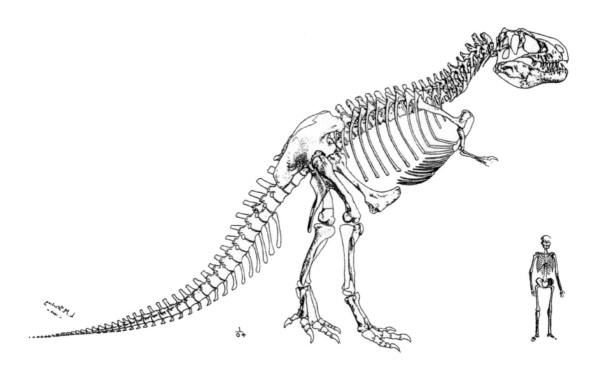

Figure 21.1. *Henry Fairfield Osborn's second and revised skeletal reconstruction of **Tyrannosaurus rex** (reversed for comparison with Fig. 21.2), drawn by L. M. Sterling, based mostly on AMNH 973 (now CM 9380) and including parts of AMNH 5866 (now BMNH R7994). After Osborn (1906).*

TYRANNOSAURUS REX: A CENTURY OF CELEBRITY

21

Donald F. Glut

Today the name *Tyrannosaurus rex* is part of our everyday lexicon, as well known to the public as the names of many human celebrities. Children not yet cognizant of the identity of persons in the public eye often know, and can even spell, the name of this gigantic theropod dinosaur. Given paper and pencil or modeling clay, they can even produce a recognizable approximation of *T. rex*. For well over a decade, one of the most popular and beloved role models among younger children is a purple, singing, dancing *Tyrannosaurus* named Barney, a character created for the Public Broadcasting System. Indeed, the fearsome yet charismatic *T. rex* has evolved into a modern-day multimedia icon.

Over the past century, the familiar image of *Tyrannosaurus rex*—arguably the best known and most popular of dinosaurs among laymen, as well as one of the dinosaurs most studied by paleontologists—has appeared in virtually every popular venue, including prose fiction, poetry, motion pictures, radio and television programs, theatrical presentations, comic books, park exhibits, display posters, toys, games, model kits, food, and advertising. Wherever the image of an animal can be used, *T. rex* has most likely been there (e.g., see Horner and Lessem 1993; Glut 2001). However, before the middle of the first decade of the 20th century, *Tyrannosaurus rex* was unknown.

In 1905, Henry Fairfield Osborn, a professor and founder of the department of vertebrate paleontology at the American Museum of Natural History (AMNH), named the giant, Late Cretaceous carnivorous dinosaur *Tyrannosaurus rex* (Osborn 1905). It was founded on a partial skeleton discovered in 1902 during a field expedition under the direction of the museum's chief fossil collector, Barnum Brown, in the Upper Cretaceous (Maastrichtian) Hell Creek Formation (then called Laramie Formation) of Dawson County, in northern Montana. The same year this initial paper was written and published, additional portions of the specimen were still being excavated (Osborn 1905). Although both jaws and portions of the skull had already been collected, Richard Swann Lull, who was in charge of the specimen's preparation, had not yet completed work on those elements. Consequently, Osborn (1905) did not include any cranial characters in his diagnosis of the new genus and type species, nor did he figure the skull in detail.

The next year, Osborn (1906) published a second paper on *T. rex*, this one providing better figures of the collected material. It was one of these illustrations, as I will show below, that constituted the basis for the first life

Introduction

restoration of *Tyrannosaurus rex* in a stunning and visually powerful painting that would gradually be reproduced in seemingly countless popular publications, thereby shaping the public's conception of *T. rex*. This single work of art would, for almost 3 decades and beyond, constitute *the* accepted image of this dinosaur in the public eye. For a generation of people interested in dinosaurs, this image *was Tyrannosaurus rex*.

Over the decades that followed, there would come a variety of new restorations, some based on earlier artistic interpretations and some more or less accurate than those that came before. All of those images contributed in shaping the public's ideas as to how *T. rex* may have appeared and also how this giant carnivorous dinosaur may have behaved.

T. Rex in the Public Eye

By 1905, the public was already savvy concerning the ongoing discoveries of often spectacular dinosaur specimens, particularly those from the late 19th and early 20th centuries in North America. However, until that year, relatively little information was available to scientist or layman concerning very large carnivorous dinosaurs. Only 2 very large North American theropods had by that time been named and described on the basis of reasonably complete fossil material: *Laelaps* (named by E. D. Cope 1866, but later renamed *Dryptosaurus*) and *Allosaurus* (named by O. C. Marsh 1877). Consequently, the discovery of *Tyrannosaurus*, an animal larger and more spectacular than the geologically older *Allosaurus*, was a newsworthy event. Clearly, Osborn (1905, p. 259) was impressed by *Tyrannosaurus* when he noted that its size "greatly exceeds that of any carnivorous land animal hitherto described."

Osborn offered only a preliminary description of *T. rex*, as well as a brief, and by today's standards barely informative, diagnosis based mostly on size-related features. Two of the features were presumably on observations made by Barnum Brown: humerus (unknown at the time) large and elongate, and absence of dermal plates erroneously thought to be present in *Dynamosaurus imperiosis* (now a junior synonym of *T. rex*). The article also provided the first, somewhat conjectural skeletal reconstruction of *T. rex* (Osborn 1905, fig. 1) drawn by American Museum of Natural History paleontologist William Diller Matthew (see Colbert 1992), a former protégé of Osborn. The figure showed the *T. rex* skeleton upright in left lateral view, along with the skeleton of *Homo sapiens* for size comparison. Although the partially prepared skull could not yet be figured in detail, Osborn did—perhaps significantly, as I will show below—mention the already prepared "supraorbital portion [=lacrimal] of the frontal bone, extremely rugose, constituting a horn above the orbit very similar to that seen in *Allosaurus*" (Osborn 1905, p. 263; cf. Madsen 1976, pl. 1).

The preliminary skull reconstruction includes a number of inaccuracies. Most notably, the skull was drawn triangular in profile, rather than rectangular, as shown by later discoveries. The triangular shape was partly the result of the dentary being drawn too shallow caudally (perhaps patterned after the dentary of *Allosaurus*). The lacrimal was located almost directly above the antorbital fenestra rather than the orbit; and the orbit is drawn as a large, roughly oval-shaped opening approximating the size of the antorbital fenestra. Other inaccuracies in Matthew's drawing included the too-short neck and short cervical neural spines.

Osborn (1906, fig. 1) later presented a revised version of the skull. This illustration corrected several mistakes, including deepening the latter portion of the dentary and repositioning the lacrimal. The orbit was drawn with the vertically narrow shape known today (see also Osborn's 1912 description of the more complete skull of referred specimen AMNH 5027, which was discovered in 1905 and excavated by the American Museum in 1908). Osborn (1906, pl. 39) also included a revised skeletal reconstruction (Fig. 21.1) drawn by L. M. Sterling, based on both AMNH 973 and the referred specimen AMNH 5866 (now BMNH R7994). In this version, the cervical vertebrae were correctly depicted, adding length and bulk to the neck. Assuming, albeit incorrectly, that *Tyrannosaurus* was closely related to the large Late Jurassic carnosaur *Allosaurus*, a fine skeleton (AMNH 4753) of which had already been mounted at the American Museum, Osborn (1905, 1906) based some of the missing parts on that genus in both of his skeletal reconstructions.

Among Osborn's aims, as head of the newly established department of vertebrate paleontology, were to enrich the American Museum's vertebrate fossils collections and to disseminate information about these extinct animals to the public. Because laymen do not, as a rule, read technical articles, Osborn sought another means of making such information generally available. Osborn selected Charles R. Knight, who was regarded as the premier artist of extinct life. Knight was commissioned to recreate in painting and sculpture the various animals whose skeletons were being exhibited in the museum's vertebrate fossils halls (e.g., see Dingus 1996). Knight, who had honed his skills under the guidance of E. D. Cope, was a conscientious, fine artist, and also an amateur paleontologist. After Cope's death, Knight soon found himself under the mentorship of Osborn and working closely with Barnum Brown (Czerkas and Glut 1982).

In an attempt to quickly make his discovery of *Tyrannosaurus* known to the public, Osborn had the pelvis and legs of AMNH 973 mounted for display (Anonymous 1910). To accompany this exhibit, Osborn had Knight prepare a painting (Fig. 21.2) that would convey to museum visitors how this great animal looked in life. However, the skeletal reconstructions available to Knight published by Osborn (1905, 1906), included errors: most notably, the tail was too long (when later mounted, the tail of AMNH 5027 was too long by approximately 3 m, as pointed out by Newman in 1970). Also, the then unknown manus was reconstructed as in *Allosaurus*, with 3 functional digits rather than the correct 2 (e.g., Lambe 1914; Carpenter and Smith 2001). Later, Osborn (1916), following Lambe (1914) and also the verbal opinion of Charles Whitney Gilmore, agreed that the *Tyrannosaurus* manus could eventually prove to be functionally didactyl with a vestigial D III (see Lipkin and Carpenter this volume). Although AMNH 5027 was first mounted with a 3-fingered hand, it was reduced to 2 when the skeleton was remounted in 1927 (R. A. Long, personal communication 2006; also photographs in Dingus 1996).

Figure 21.2. The first life restoration of **Tyrannosaurus rex**, painted by Charles R. Knight in 1906 to accompany the first mounted specimen of this dinosaur, based on Osborn's 1906 published skeletal reconstruction. Courtesy Department of Library Services, American Museum of Natural History.

Clearly, Knight based his 1906 painting of *T. rex* directly on the figure published by Osborn that same year, because the painting approximates a fleshed-out mirror image of the skeletal drawing, complete with tridactyl hands and overly long tail. As in the Sterling reconstruction (Osborn 1906), the tail is almost at ground level and the left leg is positioned slightly back. Furthermore, the animal is depicted in a rather stationary, almost lethargic position, appearing almost disinterested as it, significantly (see below), faces down a family of the horned dinosaur *Triceratops* (probably because parts of a *Triceratops* frill had been found with AMNH 5866).

Curiously, Knight painted the eye of *Tyrannosaurus* almost directly below the lacrimal horn. Why Knight, who worked closely with the scientists and knew animal anatomy well, would make such an error is puzzling. It seems unlikely that Knight, already firmly under the mentorship of Osborn and working closely with Brown, would have followed his former mentor Cope in misidentifying the theropod antorbital opening. One suggestion for Knight's error is that the artist had referred to Osborn's (1905) original statement regarding the lacrimal of *T. rex* being positioned above the orbit, and that no one—Knight, Osborn, or Brown—had noticed the error until the painting was on exhibit.

Figure 21.3. Skeleton (AMNH 5027) as it was mounted and photographed in 1915. Note the incorrect upright posture, 3-fingered hands, and overly long tail. Courtesy Department of Library Services, American Museum of Natural History. Photo by A. E. Anderson.

Nonetheless, both Knight's painting and the skeletal display had an immediate impact on the public when the exhibit opened in 1910. A *New York Times* headline announced "The Prize Fighter of Antiquity Discovered and Restored" (see Horner and Lessem 1993). Visitors to the exhibit were confronted by the partial remains of a carnivorous animal that had never before been seen, and the Knight painting conveyed the image of this fantastic and spectacular beast, with a huge head, muscular body, and eaglelike hind feet—an animal suggesting a dragon out of myth. Indeed, visitors to the museum found themselves gazing at the partial remains and life restoration of a gigantic, terrifying denizen of another age, yet one, by virtue of its being extinct, that was now a harmless attraction.

The experience was heightened immensely when, in 1915, the referred skeleton of *T. rex*, AMNH 5027, was mounted in an imposing vertical pose, and with the hands, as in Knight's accompanying painting, sporting 3 fingers (Fig. 21.3). The mounted skeleton used a cast of the skull because the original was too heavy to be mounted and so was displayed separately. The mount also used a cast of the limb elements from AMNH 973. As Osborn (1916, p. 762) stated, "*Tyrannosaurus* is the most superb carnivorous mechanism among the terrestrial Vertebrata, in which raptorial destructive power and speed are combined; it represents the climax in the evolution of a series which began with the relatively small and slender Triassic carnivore *Anchisaurus*." As far as the public was concerned, this skeleton, with its skull towering almost to the museum hall's ceiling, and with Knight's painting reinforcing the terror, *T. rex* was a real monster—albeit a safe one, as, being long dead, it was no longer a threat—and therein, I suggest, lay this dinosaur's appeal.

Figure 21.4. *Cover for the book* **America Before Man** *(1953) featuring C. B. Falls's version of Knight's 1906* **Tyrannosaurus**. *Courtesy of the Viking Press.*

However, it must be stressed that at the time of its completion, Knight's painting was unique: it constituted the only life restoration of *Tyrannosaurus rex* available to the scientist or the public. Indeed, to many people, especially those who would never stand in the presence of the skeletal mount, the painting would constitute their only visual impression of *T. rex*. As a work of art, the painting cannot be criticized, except for the already-stated errors. Nonetheless, this image—of the presumably slow-moving, 3-fingered, and long-tailed *Tyrannosaurus*, with its misplaced eyes—became, for decades, *the* image of this dinosaur.

Osborn (1917) included what seems to have been the first photographic reproduction, in black and white, of Knight's *Tyrannosaurus* painting in

The Origin and Evolution of Life, a book comprising a collection of that author's lectures (also including photographs, taken in 1915 by A. E. Anderson, of the mounted AMNH 5027; see Osborn 1916, fig. 17). Science Press is credited in the book as having published these lectures the previous year; however, they apparently did not include the Knight image (R. A. Long and J. S. McIntosh, personal communications 2006). Knight's painting was republished numerous times over the succeeding years in both technical and popular articles (e.g., Brown 1919), textbooks (e.g., Schuchert and Dunbar 1937), popular books (e.g., Colbert 1945), books targeted at juvenile readers (e.g., Reed 1930), encyclopedia entries (e.g., Anonymous 1954), and elsewhere.

Edwin H. Colbert (1980, 1989), a prolific author and media consultant as well as a vertebrate paleontologist, authored numerous technical papers and popular books about dinosaurs during his lengthy career. His first use of the 1906 Knight painting was in *The Dinosaur Book*, published in 1945, which, including a number of reprints, became a minor best seller, remaining the most popular and often-read book on the subject available to the public into the early 1960s. Colbert, in fact, continued to use that same 1906 image as representative of *T. rex* in a number of subsequent popular writings (e.g., see Colbert 1953, 1961). Even as late as 1983, many years after the errors in Knight's painting had been recognized and corrected in other restorations, Colbert included Knight's famous painting in one of his books. Particularly during the 1950s and 1960s, Colbert was well known and well regarded among dinosaur-informed laymen as *the* authority on dinosaurs. His popular publications were readily available in bookstores, museum shops, and the public library, and his name was well known as authoritative outside of scientific circles. Consequently, inaccuracies notwithstanding, Colbert's use of the Knight restoration gave it an implied stamp of approval.

Aside from Colbert, many other writers frequently included redrawn (and inferior) copies of Knight's *Tyrannosaurus* (e.g., C. B. Falls for Baity 1953), sometimes with only subtle changes (e.g., head or tail turned; Figs. 21.4 and 21.5). Not everyone, however, read books or articles about dinosaurs,

Figure 21.5. *The geographically displaced* **Tyrannosaurus** *based on Knight's 1906 painting, accompanying "Dinosaur Hunting in the Gobi Desert," a popular article written by Roy Chapman Andrews for the November 1954 issue of* **Boys' Life** *magazine.*

Figure 21.6. Knight's 1906 *Tyrannosaurus* painting comes to life, courtesy of sculptor Marcel Delgado and stop-motion animator Willis O'Brien, in the 1933 classic motion picture **King Kong**. Courtesy RKO Radio Pictures.

even those issued by the popular press. Nevertheless, Knight's imagery profoundly influenced ideas about *Tyrannosaurus rex* in other venues. Perhaps most influentially, the 1906 painting was the source material for sculptor Marcel Delgado when, around 1931, he was commissioned by RKO Radio Pictures to create a fleshed-out *Tyrannosaurus* model that was to be brought to life by special effects master Willis O'Brien by stop-motion animation. The animation was intended for a motion picture entitled *Creation*. Although *Creation* was prematurely aborted before Delgado's *T. rex* could go through its animated movements, the model was salvaged for another RKO project, *King Kong* (1933), a movie that became an immediate and colossal box-office success (e.g., see Goldner and Turner 1975; Glut 1978; Berry 2002). Thanks to numerous theatrical and television reissues of the film, and later through videotape and DVD releases, this seemingly living incarnation of Knight's painting of *T. rex*—misplaced eye, 3-fingered hand, and all—battling the imaginary giant prehistoric gorilla, Kong, has become a classic motion-picture sequence. Indeed, to many people, this dramatic image of *T. rex*—hissing, biting, and swishing its tail like some gigantic serpent—constitutes the very essence of this great theropod dinosaur (Fig. 21.6).

Although variations on Knight's 1906 work continued to appear and can yet be seen even today in various media- and merchandise-related incarnations, it was another painting, created some 2 decades later by the same artist, that had equal, if not more profound, effect on popular culture. Between 1926

and 1931, Knight was commissioned by the Field Museum of Natural History (now the Field Museum) to paint a series of large murals portraying life through time for the exhibit hall where some of its vertebrate fossil specimens were displayed. Among this series—regarded by both Knight and his admirers as his magnum opus—was a Late Cretaceous scene, completed by 1928 (N. Cummings, personal communication 2006) and perhaps the most famous dinosaur painting ever executed, depicting *Tyrannosaurus* confronting *Triceratops* (Czerkas and Glut 1982). This time, however, there were important differences from the 1906 version of *T. rex*. Knight portrayed *T. rex* with its eyes correctly placed, its hands bearing 2 rather than 3 fingers, and, prophetically, it seems, its body held in a horizontal posture with the still too long tail off the ground (a second *T. rex* individual in the background is propped up in the standard subvertical pose still fashionable at the time). Moreover, the scenario had also altered. No longer the rather lethargic animals depicted in 1906, this

Figure 21.7. *Late Cretaceous mural, painted by Charles R. Knight circa 1928 for the Field Museum of Natural History, featuring the classic **Tyrannosaurus** versus **Triceratops** scenario, possibly the best known and influential of any **T. rex** life restorations. Courtesy the Field Museum, negative CK9T. Photo by Ron Testa.*

Figure 21.8. ***Tyrannosaurus*** *based on Knight's Field Museum mural as it appeared in a comic book adaptation of the 1940 motion picture **One Million B.C. (1940)**, published in the **Reg'lar Fellers** comic book.*

Figure 21.9. *Another **Tyrannosaurus** based on Knight's Field Museum mural (along with an **Apatosaurus** and caveman) appeared in the 1960 science fiction motion picture, **Dinosaurus!** Courtesy Universal-International Pictures.*

Tyrannosaurus and *Triceratops* were painted actively charging toward one another in what promised to be a violent conflict (Fig. 21.7).

Like its earlier predecessor, Knight's *Tyrannosaurus* versus *Triceratops* mural had lasting impact on the public. The Field Museum sold postcards of the painting, reproduced first in black and white and, in later years, full color. Not only would reproductions of this painting become the norm in books about extinct animals (e.g., Knight's own book, *Before the Dawn of History*, 1935), largely supplanting the 1906 image in popularity, but it also would also become the image of *T. rex* most often copied by other artists and adapted to other media. Knight's new and more accurate *T. rex* was soon recognizable in a variety of publications, including popular dinosaur books, mainstream magazines, and comic books (Fig. 21.8), and also tyrannosaurs portrayed in motion pictures (Fig. 21.9).

Among such publications were the cheaply printed and purchased pulp magazines, which were mass-distributed periodicals displayed at newsstands with lurid covers during the 1930s and 1940s. Prolific pulp artist J. Allen St. John adapted Knight's Field Museum *Tyrannosaurus* to many of his magazine paintings and drawings during those decades, primarily illustrating the fantastic fiction scribed by adventure author Edgar Rice Burroughs, the creator of Tarzan and other larger-than-life heroes. This tradition of copying Knight was later inherited by such St. John successors as Frank Frazetta in the 1960s and beyond (Glut 1980, 2001). Some publications even created their own chimeras by melding elements from different Knight paintings to create a new whole image. W. W. Robinson's (1934) children's book, for example, combined Irene Robinson's redrawn (and reversed) head of Knight's Field Museum *Tyrannosaurus* with the body of the one painted for the American Museum, again with 3-fingered hands (Fig. 21.10).

Surely, the Field Museum mural affected popular culture tremendously, not only in its physical portrayal of *Tyrannosaurus*, but also in its

Figure 21.10. *Irene Robinson combined the head from Knight's Field Museum* **Tyrannosaurus** *mural with the body from his 1906* **Tyrannosaurus** *painting, resulting in this composite image for W. W. Robinson's children's book* **Ancient Animals of America** *(1934). Courtesy the Ward Ritchie Press.*

establishment in the public consciousness that this dinosaur and *Triceratops* were natural enemies that encountered, probably frequently, one another in battle (see Happ this volume). Inadvertently, the mural also led to a third depiction of *Tyrannosaurus*, which, like its Knight inspiration, would also have considerable influence on popular culture.

In 1933–1934, Chicago played host to the World's Fair. Dubbed A Century of Progress, the 2-year event was held on grounds located within walking distance of the Field Museum. Among the attractions was the Sinclair Dinosaur Exhibit, featuring life-size, mechanically moving, 3-dimensional recreations of the figures based directly on Knight's murals. Designed by P. G. Alen for the Sinclair Refining Company (Anonymous 1966) under the

Figure 21.11. *Based directly on Knight's Field Museum mural, this life-size mechanical* **Tyrannosaurus rex** *figure appeared in the Sinclair Dinosaur Exhibit at the 1933–1934 Chicago World's Fair (A Century of Progress).*

A Century of Celebrity   409

Figure 21.12. *The Sinclair Dinosaur Exhibit **Tyrannosaurus** as photographed in its highly influential three-quarter view.*

technical consultation of Barnum Brown, this suite of Mesozoic characters included, of course, the popular *Tyrannosaurus-Triceratops* confrontation (although the *Tyrannosaurus* again sported a 3-fingered hand).

In profile, the Sinclair *T. rex* resembled its 2-dimensional inspiration at the Field Museum (Fig. 21.11). However, it was a three-quarter-view photograph (Fig. 21.12) of the Sinclair *Tyrannosaurus* that soon began appearing in various venues, including a series of souvenir newspapers and postcards from the World's Fair. Seen in this aspect, however, the figure took on its own unique and odd appearance that deviated in various ways from Knight's profile view. Most notably, the dinosaur looked unusually wide in the hips and hind legs, whereas the head, seen from this angle, assumed a horselike shape. More significantly, that photograph was the basis for a painting executed by artist James E. Allen for *The Sinclair Dinosaur Book*, a free booklet (Fig. 21.13) promoting the oil company and providing information about dinosaurs, reviewed and perhaps authored by Barnum Brown in 1934 (see listing in Chure and McIntosh 1989). The book was sent out to schools and other targeted audiences, and eventually became a highly collectible item. Like the Knight mural that inspired it, numerous copies of Allen's painting (manus again correctly didactyl) were also re-

printed in numerous other publications. Again *Tyrannosaurus* was depicted about to engage its by now traditional foe, *Triceratops*, in combat. Also, as with Knight's version, Allen's rendition of *T. rex* became fodder for numerous artists, including St. John, who adapted it to their own needs (Glut 1980). Oddly, some artists like St. John, in later illustrating their own side-view portrayals of *T. rex*, based them on Allen's three-quarter-aspect painting, resulting in distorted depictions that only superficially resembled Knight's original lateral view (Fig. 21.14).

New artistic variations of *T. rex*, usually with 3 fingers and based either on the Allen painting or three-quarter-view photographs of the World's Fair model, subsequently debuted in other widely accessible venues. Among these was the *Tyrannosaurus* presented in the mid-1930s at Dinosaur Park in Rapid City, SD, one of a suite of such figures made by local sculptor Emmett A. Sullivan (Fig. 21.15). Celebrated as the first full-scale dinosaur figures ever made in the United States for permanent display, the Dinosaur Park *Tyrannosaurus* and kin impressed visitors if only for their size, while at the same time boosting the Black Hills tourist trade.

Collectors' stamps were also issued by the Sinclair Refining Company, various children's books, pulp magazines, and comic books (see Glut 1980) that boasted variations on Allen's work. Counted among the myriad items

Figure 21.13. *James E. Allen's painting of* **Tyrannosaurus rex**, *based on three-quarter-view photographs of the Sinclair Dinosaur Exhibit figure, prepared for* **The Sinclair Dinosaur Book**. *This image was widely seen during the 1930s through 1950s. Courtesy Sinclair Refining Company.*

A Century of Celebrity   411

Figure 21.14. J. Allen St. John based this ***Tyrannosaurus rex*** on James E. Allen's three-quarter-view painting, producing, for a pulp magazine (1942), a new side-view interpretation that only slightly resembled Knight's original. Courtesy Ziff-Davis Publishing Company.

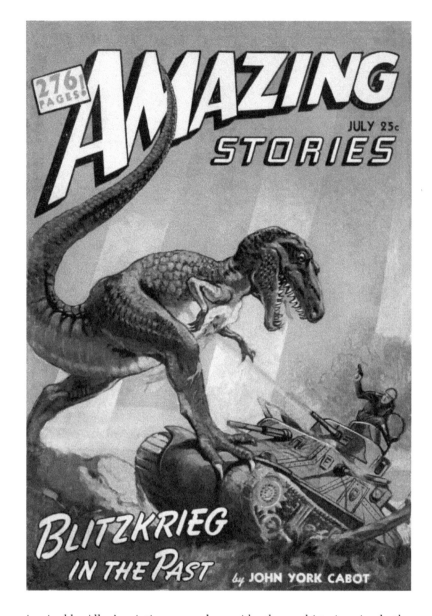

inspired by Allen's painting were, along with other prehistoric animals, the heavy, bronze-coated lead *T. rex* figurines (Fig. 21.16) first issued circa 1947 by Sell Rite Giftware (SRG) of New York (see Cain and Fredericks 1999). Primarily sold in museum gift shops and through mail-order advertisements printed in magazines, such as the American Museum's *Natural History*, these greenish, perennially popular *Tyrannosaurus* figures constituted, for nearly a decade, virtually the only 3-dimensional incarnations of the dinosaur available to the public. They were available in both large and small sizes, again possessing 3-fingered hands but assuming a rather squatty posture. These figures, which were also put out in ceramic and plastic versions, could still be purchased in recent years and are highly collectible today.

Into the early 1950s, those 3 most influential images of *Tyrannosaurus rex*—the 2 Knight paintings and the Allen painting—continued to pervade

Figure 21.15. *Full-scale statues of **Tyrannosaurus** and its traditional foe, **Triceratops**, based on the works of both Knight and Allen, sculpted by Emmett A. Sullivan for permanent display at Dinosaur Park in Rapid City, SD. Courtesy Travel Division, South Dakota Department of Highways.*

popular culture. By that time, Knight (1942, 1946), as both artist and author, had already published new depictions of *Tyrannosaurus* in the popular *National Geographic Magazine* and his book *Life through the Ages*. Knight's illustration in the latter, in which 2 *T. rex* individuals walk side by side, one crouched, both apparently ready to strike at each other, seemingly echoes Osborn's (1913) aborted plan to mount both AMNH 973 and AMNH 5027 as if about to fight for possession of a carcass. Although Knight did not significantly deviate from his mural in regards to the physical look of *Tyrannosaurus*, he now portrayed the animal in conflict with another of its kind rather than with the ubiquitous *Triceratops*. These wonderful pieces of art, portraying *T. rex* as a considerably more active animal than suggested by the 1906 painting, were widely seen and influential in their own right. However, these later images did not affect the public as strongly as had Knight's earlier works.

Indeed, the main influence on the public consciousness regarding *Tyrannosaurus rex* came neither from Knight nor anyone associated with Sinclair, but from Yale University. During the 1940s, artist Rudolph F. Zallinger, who had been taught about extinct life by paleontologists Richard Swann Lull and Alfred S. Romer, was persuaded to paint an expansive mural, now known as the Age of Reptiles, to adorn an empty wall of the university's Peabody Museum of Natural History. *Tyrannosaurus* would be featured prominently in the Cretaceous portion of the work. Zallinger worked on this mural for about 5 years, finishing it in 1947 and garnering a Pulitzer Prize for his visually stunning efforts (see Scully et al. 1990; Debus and Debus 2002).

Figure 21.16. *The large-size* **Tyrannosaurus rex** *figure put out during the late 1940s by SRG (Sell Rite Giftware) and sold mainly through museum gift shops. Courtesy Mike Fredericks and* **Prehistoric Times** *magazine.*

Zallinger's version of *T. rex* deviated significantly from Knight's versions in a number of ways. Most notably, it had a somewhat triangular head and a potbelly. The forelimbs, then known only from the humerus, were restored too small, as would be evidenced by future discoveries (Carpenter and Smith 2001; Lipkin and Carpenter this volume). Also, the giant carnivore was shown standing as if motionless, tail on the ground, disinterested in the numerous Cretaceous animals surrounding it (Fig. 21.17). Nevertheless, it was Zallinger's *T. rex* that imparted the strongest influence on popular culture during the late 1950s and early 1960s, mostly thanks to a periodical. In 1953, the extremely popular *Life Magazine* included a reverse-image (or flopped) full-color reproduction of the small preliminary study Zallinger painted of the Age of Reptiles in its September 7 issue. The actual mural could not be photographed because some of the museum's mounted dinosaur skeletons obstructed the camera's view. *Life Magazine* was prominently displayed on virtually every newsstand in the United States, so myriad readers brought Zallinger's work, *Tyrannosaurus* included, into their homes.

Again, a work of art rapidly shaped anew the public's awareness of *T. rex*, with reproductions of Zallinger's painting (or portions of it) being published elsewhere. Yale professor Carl O. Dunbar in 1955 included both the Zallinger *T. rex* and Knight's 1906 original in the new edition of his classic textbook, *Historical Geology*. More significantly, in terms of affecting public awareness, Zallinger's *Tyrannosaurus* was the template for countless other incarnations of this dinosaur during the 1950s and 1960s, many of them in comic books such as Dell Publishing Company's *Turok, Son of Stone* and ACG's (American Comics Group) *Forbidden Worlds* (Fig. 21.18), sometimes, as had happened with Allen's painting, reconstructing the original image as viewed from different perspectives. Zallinger's restoration is easily recognizable in the form of performer Tim Smyth wearing an unconvincing *Tyrannosaurus* costume (Fig. 21.19) made by the Bud West-

more makeup shop in the motion picture *The Land Unknown*, released by Universal-International in 1957. Other prehistoric creatures in the motion picture, as well as overall set design, were also based on the Yale mural (Glut 1980, 2001; Berry 2002). Perhaps most significantly, it was the model for the first commercially mass-produced dinosaur toys, which appeared in 1956, courtesy of the Miller (Fig. 21.20) and Marx (Fig. 21.21) companies (Glut 1980; Cain and Fredericks 1999). These inexpensive small plastic figures, which could be purchased in dime stores and other places that sold toys, allowed children to invent their own prehistoric scenarios while at the same time being educated.

Other popular, often-reproduced images of *Tyrannosaurus rex* have also influenced our culture (Debus and Debus 2002), although without the same impact as those of Knight, Allen, and Zallinger. One of these is the monochrome restoration by British artist Neave Parker during the 1950s, first made available as a series of postcards issued by the British Museum

Figure 21.17. *Tyrannosaurus rex* dominates the Late Cretaceous portion of the Age of Reptiles mural completed in 1947 by artist Rudolph F. Zallinger for Yale University's Peabody Museum of Natural History. Courtesy Peabody Museum of Natural History, Yale University.

Figure 21.18. Zallinger's *Tyrannosaurus* inspired many copycat drawings appearing in comic books during the 1950s through 1960s. This version is revised to a three-quarter view. Courtesy American Comics Group.

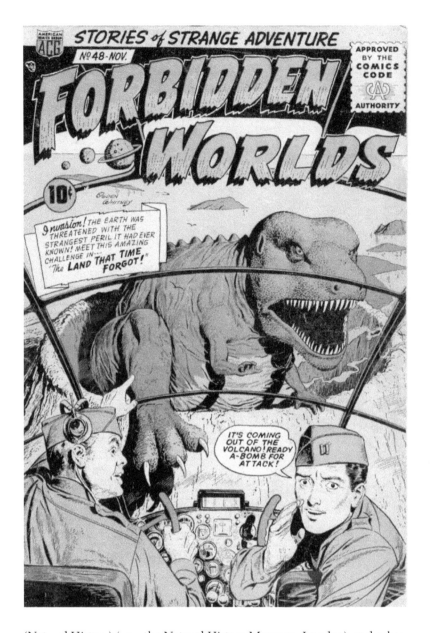

(Natural History) (now the Natural History Museum, London), and subsequently included by W. E. Swinton (1962) in his popular-science book *Dinosaurs*. Another well-known image is the painting made by Czech artist Zdenek Burian in 1938, first published decades later by Joseph Augusta (1960) and then published again, slightly revised (Augusta 1972).

Also receiving much media attention was the full-scale, totally upright, mechanized (the lower jaw moved) *Tyrannosaurus* (Fig. 21.22) built for the Sinclair Oil Company's Dinoland exhibit at the 1964 New York World's Fair by the Jonas Studios (Anonymous 1966; Glut 1980; Debus and Debus 2002). All of the Sinclair dinosaur figures, like their 1930s precursors, were made with the technical advice of Barnum Brown shortly before

Figure 21.19. *Tyrannosaurus rex* awkwardly demonstrates predacious behavior in this scene from **The Land Unknown**. The prehistoric animals in this 1957 motion picture, as well as the flora, were based on imagery from Zallinger's Age of Reptiles mural. Courtesy Universal-International Pictures.

his death in 1963, and with additional input from Peabody Museum paleontologist John H. Ostrom.

Scaled up from a sculpture made by company owner Louis Paul Jonas, this *T. rex* subsequently took on its own impressive life during the 1960s, creating another wave of *Tyrannosaurus* influence on popular culture. Constituting what today would be called an ideal photo op, the huge figure became a familiar prop, with photographers posing their human models before the towering giant. Advertising agencies used either photographs or artwork based on Jonas' World's Fair dinosaurs—most often, the *Tyrannosaurus rex*—to promote anything from the *Encyclopedia Britannica* (Fig. 21.23) to Taco Bell (Fig. 21.24). Sinclair prominently featured its new *T. rex* in *Sinclair and the Exciting World of Dinosaurs*, which was illustrated with paintings that were based on the Jonas statues. The pamphlet was taken home by countless visitors to the fair during 1964–1965 (Fig. 21.25). Plastic models of the *T. rex* sold by vending machines at the fair and later in stores countrywide soon replaced in the older Allen- and Zallinger-based figures made by SRG, Miller, and Marx (Fig. 21.26). Additionally, copies of the *T. rex* and other figures (also the small-size prototype figures sculpted by Louis Paul Jonas) were distributed among museums and parks even beyond U.S. borders (e.g., the Calgary Zoo Prehistoric Park; see Glut 1980). The original World's Fair *T. rex* statue eventually came to reside at the famous Lower Cretaceous dinosaur tracksite near Glen Rose, TX. For a brief time during the 1960s, the image of *T. rex* most seen by the public was that provided by the Jonas Studios.

Figure 21.20. *Waxy plastic Tyrannosaurus rex* based on Zallinger's painting. Made by the Miller Company and originally issued in 1956, this and other prehistoric animal figures in the Miller set constituted the first mass-produced, commercially sold toy dinosaurs. Courtesy Mike Fredericks and **Prehistoric Times** magazine.

Figure 21.21. Made by the Marx Company, this Zallinger-inspired, plastic **Tyrannosaurus** toy (known affectionately among collectors as the "potbellied **Tyrannosaurus rex**") debuted in dime stores in 1956. Courtesy Mike Fredericks and **Prehistoric Times** magazine.

Consequently, these various versions of T. rex inspired their share of new incarnations in the media and in merchandise. Although these T. rex restorations offered their own anatomical inaccuracies, as pointed out by Paul (1988), they were visually powerful, dramatic, and kinetic, and for those reasons strikingly distinct from the more stoic works preceding them. For such reasons, I (Glut 1972) selected the Parker T. rex as the dust jacket illustration for *The Dinosaur Dictionary*. The Parker, Burian, and Jonas versions of *Tyrannosaurus rex* were portrayed in the old-fashioned, impossibly upright-posed, tail-dragging modes.

Figure 21.22. The Jonas Studios' towering ***Tyrannosaurus rex*** model, scaled up from a sculpture by Louis Paul Jonas, in a photograph taken shortly before being transported from the Catskills to the 1964 New York World's Fair. Courtesy Sinclair Oil Corporation.

THE MAGIC DOOR/from Britannica
The Tyrant King

What is happening? Did the sun suddenly slip behind a huge cloud? No, it's something much stranger than that. It's a huge and hungry tyrannosaur—the biggest meat-eating dinosaur that ever walked the Earth! It was longer than a school bus. And when it stood on its hind legs, its head was so high above the ground that it could look over small hills.

When a tyrannosaur, sometimes called *Tyrannosaurus rex*, was hungry, it was hungry enough and big enough to eat even another dinosaur! Its jaws were huge, and its teeth were sharp with jagged edges.

And so, whenever dusk seemed to come suddenly or a cloud seemed to hide the warmth of the sun, all of the other dinosaurs probably started running. They knew that a huge tyrannosaur might be looming up nearby and shutting out the sunlight. And when they were close enough to *Tyrannosaurus rex* to see its long shadow, they were too close!

Figure 21.23. *Advertisement, which appeared in newspapers during the mid-1960s, incorporating the Jonas Studios' interpretation of* **Tyrannosaurus rex**.

Not until relatively recent years have life restorations of *T. rex* been more anatomically accurate, with the body posed horizontally and tail off the ground. Among the first such restorations, if not the first, was the sculpture by Sylvia Massey (now Czerkas) created during the late 1970s (Fig. 21.27) under the guidance of paleontologists Robert A. Long and Ralph E. Molnar (see Glut 1980, 1982). Subsequently, a new breed of modern paleoartist—e.g., Robert T. Bakker, Stephen A. Czerkas, Brian Franczak, John Gurche, Mark Hallett, Douglas Henderson, Gregory S. Paul—created more correctly restored versions of how *T. rex* appeared and possibly behaved. These versions gradually replaced the older and outdated models, not only in technical and popular publications, but also in merchandising (including a plethora of scientifically accurate model assembly kits) and other venues. Perhaps more than those of any other paleoartist, Paul's (Fig. 21.28) numerous life restorations of *T. rex*—dynamic as well as scientifically accurate—became the norm, influencing a new generation of paleoartists, whose drawings, paintings, and sculptures leave no question as to their main source of inspiration (see Paul 1987).

Without doubt, the single most impacting image on our culture by *Tyrannosaurus rex* was that portrayed in the motion picture *Jurassic Park* (1992) (Fig. 21.29), which was based on the novel by Michael Crichton. Producer-director Steven Spielberg raised the paradigm for dinosaurs on screen by bringing a consultant paleontologist, John R. Horner, onto the project. Horner worked closely with the special effects artists and technicians to ensure that the film's Mesozoic animals were presented with rea-

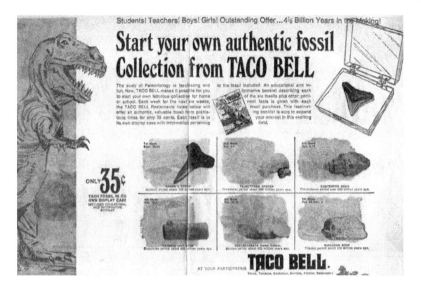

Figure 21.24. *Another advertisement from the mid-1960s based on the Jonas Studios'* **Tyrannosaurus rex**.

sonable accuracy. Thanks both to the lifelike animatronics designed and built by Stan Winston and his team and to the process of computer-generated imagery, the bar was significantly raised in movie special effects, with the *Jurassic Park* dinosaurs seemingly coming to life. Allowing for a just few minor discrepancies (mostly for dramatic effect), the movie's largely Paul-inspired *T. rex* and other dinosaurs appeared closer to what paleontologists now believe about dinosaurs than those of any motion picture made up until that time.

Jurassic Park was an enormous success from its first day of release and was eventually seen by millions of viewers in its original release. Inevitably, the film launched a series of hit sequels, as well as a plethora of authorized and imitation publications and merchandise. Moreover, its pioneering special-effects work led, either directly or indirectly, to such later well-received

Figure 21.25. **Tyrannosaurus rex** *as painted for the pamphlet* **Sinclair and the Exciting World of Dinosaurs**, *a souvenir item from the 1964 New York World's Fair. Courtesy Sinclair Refining Company.*

Figure 21.26. Miniature reproduction of the Jonas Studios' ***Tyrannosaurus rex***. These waxy plastic figures were originally sold in vending machines at the 1964–1965 New York World's Fair. Courtesy Mike Fredericks and ***Prehistoric Times*** magazine.

Figure 21.27. Sculpture by Sylvia Massey (now Czerkas) of ***Tyrannosaurus rex*** made during the late 1970s, possibly the first modern life restoration of this dinosaur. Courtesy Sylvia Czerkas. Photograph by Robin Robin.

Figure 21.28. *Life restoration of **Tyrannosaurus rex** by the highly influential paleoartist Gregory S. Paul, prepared for his book **Predatory Dinosaurs of the World: A Complete Illustrated Guide** (1988). Courtesy of Gregory S. Paul.*

Figure 21.29. ***Tyrannosaurus rex** (life-size animatronics model by the Stan Winston studios), the first of its kind to be based on modern interpretations, as it appeared in the groundbreaking motion picture **Jurassic Park** (1992). Courtesy Universal Pictures and MCA.*

Figure 21.30. The *Tyrannosaurus rex* skeleton (FMNH PR2081) known as Sue, as mounted in 2000 at the Field Museum. Courtesy the Field Museum, negative GN89677_47c. Photo by John Weinstein.

projects as the 1999 Imax motion picture *T-Rex: Back to the Cretaceous*, shown on enormous screens in 3-D, and the BBC's *Walking with Dinosaurs* (1999) and the Discovery Channel's *When Dinosaurs Roamed America* (2001) television programs (see Berry 2002).

As a result of *Jurassic Park*, viewers experienced, most of them for the first time, a truer-to-reality idea of how *T. rex* appeared in life (at least as far as it is understood today), how the animal may have moved, and how it could have behaved. Thus, the public was already primed to experience the spectacular remains of the authentic animal when, after several years of extensive legal wrangling and media coverage, Sue (e.g., see Larson and Donnan 2002)—the largest and best-preserved, and one of the most complete, *T. rex* skeletons (FMNH PR2081) yet discovered—was unveiled on May 17, 2000, in the Field Museum's grand Stanley Field Hall (Fig. 21.30). This was the first time that an individual *Tyrannosaurus* specimen arose to true celebrity status. Sue immediately became one of Chicago's main tourist attractions.

The world and its cultures are constantly changing. Also changing over the century after its discovery is *T. rex*. With the release of *Jurassic Park*, and also thanks to the subsequent imagery it inspired, coupled with the now global superstar notoriety of Sue, most former conceptions of *Tyrannosaurus rex* (once so powerful and seemingly immutable) have been erased. The public awareness of this tyrant-lizard king has changed forever.

Acknowledgments

I sincerely thank the following people who contributed to, or in some other way provided assistance, materials, or advice in, the writing of this chapter: Kenneth Carpenter, Denver Museum of Nature & Science: Luis M. Chiappe, Dinosaur Institute, Natural History Museum of Los Angeles County; Nina Cummings and Jerice Barrios, Photography Department, The Field Museum; Alan A. Debus, Hell Creek Creations; Mike Fredericks, *Prehistoric Times* magazine; Jacques Gauthier, Peabody Museum of Natural History, Yale University; Dean Hannotte; Robert A. Long; John S. McIntosh; Gregory S. Paul; and also the library, Natural History Museum of Los Angeles County.

References Cited

Anonymous. 1910. The *Tyrannosaurus*. *Journal of the American Museum of Natural History* 10: 3–8.
———. 1954. *Tyrannosaurus*. P. 8679–8680 in Morse, J. L. (ed.). *Universal Standard Encyclopedia*. Unicorn Publishers, New York.
——— 1966. *A Great Name in Oil: Sinclair through Fifty Years*. F. W. Dodge Company/McGraw Hill, New York.
Augusta, J. 1960. *Prehistoric Animals*. Paul Hamlyn, London.
———. 1972. *Life before Man*. American Heritage Press, New York.
Baity, E. C. 1953. *America before Man*. Viking Press, New York.
Berry, M. F. 2002. *The Dinosaur Filmography*. McFarland, Jefferson, NC.
Brown, B. 1919, Hunting big game of other days. *National Geographic Magazine* 37: 407–429.
———. 1934. *The Sinclair Dinosaur Book*. Sinclair Refining Company.
Cain, D., and Fredericks, M. 1999. *Dinosaur Collectibles*. Antique Trader Books, Norfolk, VA.
Carpenter, K., and Smith, M. 2001. Forelimb osteology and biomechanics of *Tyrannosaurus rex*. P. 90–116 in Tanke, D., and Carpenter, K. (eds.). *Mesozoic Vertebrate Life*. Indiana University Press, Bloomington.
Chure, D. J., and Mcintosh, J. S. 1989. *A Bibliography of the Dinosauria (Exclusive of the Aves), 1677–1986*. Museum of Western Colorado, Paleontology Series 1.
Colbert, E. H. 1945. *The Dinosaur Book*. American Museum of Natural History, New York.
———. 1953. *Dinosaurs*. Science Guide 70. American Museum of Natural History, New York.
———. 1961. *Dinosaurs: Their Discovery and Their World*. E. P. Dutton, New York.
———. 1980. *A Fossil-Hunter's Notebook*. E. P. Dutton, New York.
———. 1983. *Dinosaurs: An Illustrated History*. Hammond, Maplewood, NJ.

———. 1989. *Digging into the Past: An Autobiography.* Dembner Books, New York.

———. 1992. *William Diller Matthew, Paleontologist: The Splendid Drama Observed.* Columbia University Press, New York.

Cope, E. D. 1866. Remarks on dinosaur remains from New Jersey. *Proceedings of the Academy of Natural Sciences of Philadelphia* 18: 275–279.

Czerkas, S. M., and Glut, D. F. 1982. *Dinosaurs, Mammoths and Cavemen: The Art of Charles R. Knight.* E. P. Dutton, New York.

Debus, A. A., and Debus, D. E. 2002. *Paleoimagery: The Evolution of Dinosaurs in Art.* McFarland, Jefferson, NC.

Dingus, L., 1996. *Next of Kin.* Rizzoli, New York.

Dunbar, C. O. 1955. *Historical Geology.* John Wiley & Sons, New York.

Glut, D. F. 1972. *The Dinosaur Dictionary.* Citadel Press, Secaucus, NJ.

———. 1978. *Classic Movie Monsters.* Scarecrow Press, Metuchen, NJ.

———. 1980. *The Dinosaur Scrapbook.* The Citadel Press: Secaucus, NJ.

———. 1982. The New Dinosaur Dictionary. The Citadel Press, Secaucus, NJ.

———. 2001. *Jurassic Classics.* McFarland & Company, Jefferson, NC.

Goldner, O., and Turner, G. E. 1975. *The Making of King Kong.* A. S. Barnes, Cranbury, NJ.

Horner, J. R., and Lessem, D. 1993. *The Complete* T. Rex. Simon & Schuster, New York.

Knight, C. R. 1935. *Before the Dawn of History.* McGraw-Hill Book Company, New York.

———. 1942. Parade of life through the ages. *National Geographic Magazine* 81: 141–184.

———. 1946. *Life through the Ages.* A. A. Knopf, New York.

———. 1914. On the fore-limb of a carnivorous dinosaur from the Belly River Formation of Alberta, and a new genus of Ceratopsia from the same horizon, with remarks on the integument of some Cretaceous herbivorous dinosaurs. *Ottawa Naturalist* 27: 129–135.

Larson, P., and Donnan, K. 2002. *Rex Appeal: The Amazing Story of Sue, the Dinosaur that Changed Science, the Law, and My Life.* Invisible Cities Press, Montpelier, VT.

Madsen, J. H., Jr. 1976. *Allosaurus fragilis: A Revised Osteology.* Utah Geological and Mineral Survey Bulletin 109.

Marsh, O. C. 1877. Characters of the Odontornithes, with notice of a new allied genus. *American Journal of Sciences,* ser. 3, 14: 85–87.

Newman, B. H. 1970. Stance and gait in the flesh-eating *Tyrannosaurus. Biological Journal of the Linnean Society* 2: 119–123.

Osborn, H. F. 1905. *Tyrannosaurus* and other Cretaceous carnivorous dinosaurs. *Bulletin of the American Museum of Natural History* 21: 259–265.

———. 1906. *Tyrannosaurus,* Upper Cretaceous carnivorous dinosaur (second communication). *Bulletin of the American Museum of Natural History* 22: 150–165.

———. 1912. Crania of *Tyrannosaurus* and *Allosaurus. Memoirs of the American Museum of Natural History,* n.s., 1: 1–30.

———. 1913. *Tyrannosaurus,* restoration and model of the skeleton. *Bulletin of the American Museum of Natural History* 32: 91–92.

———. 1916. Skeletal adaptations of *Ornitholestes, Struthiomimus, Tyrannosaurus. Bulletin of the American Museum of Natural History* 35: 761–771.

———. 1917. *The Origin and Evolution of Life: On the Theory of Action, Reaction and Interaction of Energy.* Charles Scribner's Sons, New York.

Paul, G. S. 1987. The science and art of restoring the life appearance of dinosaurs and their relatives. P. 4–49 in Czerkas, S. J., and Olson, E. C. (eds.).

Dinosaurs Past and Present, vol. 2. Natural History Museum of Los Angeles County, Los Angeles.

———. 1988. *Predatory Dinosaurs of the World: A Complete Illustrated Guide.* Simon & Schuster, New York.

Reed, W. M. 1930. *The Earth for Sam.* Harcourt, Brace, New York.

Robinson, W. W. 1934. *Ancient Animals of America.* Ward Ritchie Press, Los Angeles.

Schuchert, C., and Dunbar, C. O. 1937. *Outlines of Historical Geology.* 3rd ed. John Wiley & Sons, New York.

Scully, V., Zallinger, R. F., Hickey, L. J., and Ostrom, J. H. 1990. *The Age of Reptiles: The Great Dinosaur Mural at Yale.* Peabody Museum of Natural History, New Haven, CT.

Swinton, W. E. 1962. *Dinosaurs.* Trustees of the British Museum (Natural History), London.

INDEX

Achelosaurus, 330
Acrocanthosaurus, 241, 382
Afrovenator, 382
Albertosaurus, 57, 103, 104, 109–111, 143, 145, 146, 149, 156, 235, 238, 239, 284–286, 289, 291, 293, 299, 310, 311, 316, 325, 326, 335–337; *A. libratus*, 110, 306, 314, 319, 326; *A. megragracilis*, 12; *A. sarcophagus*, 289, 294, 296, 300, 308, 321
Alectrosaurus, 288, 309; *A. olseni*, 287, 289
Alioramus, 309; *A. remotus*, 287
Allosaurus, 107, 144, 147, 153, 166, 310, 312, 314, 316, 319, 382, 400, 401; *A. fragilis*, 255, 259, 261, 289
Amargosaurus, 155
Anchiceratops, 361
Anchisaurus, 143
Ankylosaurus, 6, 337
Apatosaurus, 143, 144
Appalachiosaurus, 234, 287
Archaeopteryx, 242, 318
Aublysodon molnari, 2, 287
Aviatyrannis, 287

Bambiraptor, 148, 149
Brachiosaurus, 143
Brachytrachelopan, 155

Carcharodontosaurus, 384
Carnotaurus, 131, 133, 147, 149, 152, 158, 160, 315; *C. sastrei*, 255
Centrosaurus, 382
Ceratosaurus, 58, 142, 144, 149, 153, 316, 377
Chasmosaurus, 330, 358; *C. belli*, 358; *C. russelli*, 358; *C. irvenensis*, 358; *C. mariscalensis*, 358
Chingkankousaurus, 287

Coelophysis, 133, 142, 144, 152, 153, 154
Crocodylus porosus, 256

Daspletosaurus, 55, 109, 110, 233, 234, 234, 236, 237, 238, 239, 241, 287–289, 298, 310, 311, 314, 319, 323, 325–327, 335–337; *D. torosus*, 287, 292, 306, 314, 355
Deinocheirus, 382
Deinodon, 58; *D. horridus*, 57, 287
Deinonychus, 10, 148, 149, 166, 176, 382; *D. antirrhopus*, 272
Dilong, 143, 145, 146, 150, 287
Dilophosaurus, 143, 144, 382; *D. wetherilli*, 374, 375
Dinotyrannus, 287; *D. megragracilis*, 12
Diplodocus, 144, 373
Dromaeosaurus albertensis, 280
Dryptosaurus, 287, 387, 400
Dynamosaurus imperiosus, 1, 6, 57–59, 79, 233, 400

Edmontonia, 337
Edmontosaurus, 20, 22, 35, 42, 44, 382; *E. annectens*, 72, 104, 298, 355, 364, 384, 389
Einosaurus, 358
Eotyrannus, 148; *E. lengi*, 287

Gallimimus, 145, 149
Giganotosaurus, 147, 315, 373–375, 382
Gorgosaurus, 103, 104, 107–112, 114–116, 119, 122, 153, 168, 233–239, 241, 288, 291, 292, 294, 300, 310; *G. lancensis*, 108; *G. libratus* 255, 287, 288, 290, 355, 360, 370

Herrerasaurus ischigualastensis, 355, 377

429

Hesperornis, 247, 251

Iguanodon, 247

Laelaps incrassatus, 384, 400
Liliensternus, 144

Maleevosaurus, 287
Manospondylus gigas, 1, 13, 37, 50, 58

Nanotyrannus, 14, 20, 34, 103, 104, 107, 108, 115, 116, 234–239, 309, 326, 338; **N. lancensis,** 2, 83, 109, 110, 111, 112, 122, 287

Ornithomimus, 57, 133, 145, 147, 149, 152; **O. grandis,** 57

Pachycephalosaurus, 373
Pachyrhinosaurus, 330, 358
Parahesperornis, 242
Pentaceratops, 358
Piatnizkysaurus, 147
Plateosaurus, 143, 241
Protoceratops, 333

Ricardoestesia, 359

Saurolophus, 337
Scipionyx, 373
Shanshanosaurus, 287
Shantungosaurus, 337
Siamotyrannus isanensis, 287
Sinraptor dongi, 241, 319, 355
Stokesosaurus, 287
Struthiomimus, 145, 147, 149, 315
Stygivenator molnari, 2
Styracosaurus, 358

Tarbosaurus, 10, 55, 104, 107, 110, 237, 241, 382; **T. bataar** 103, 107, 111, 112, 233, 234, 235, 239, 287, 288, 370
Therizinosaurus, 382
Thescelosaurus, 20, 373, 379
Triceratops, 22, 44, 57, 63, 64, 78, 298, 327, 330, 331, 354–359, 361–365, 384, 389, 402, 407–411, 413
Troodon, 151

Tyrannosauripus pillmorei, 208, 209
Tyrannosauropus, 208
Tyrannosaurus, 1, 2, 11, 30, 78, 109, 131, 134, 143, 145–147, 149, 152, 166, 168, 170, 173, 175, 176, 178–182, 184–187, 208, 212, 234–236, 239–242, 286, 291–294, 296–298, 307, 373, 381, 399–414, 416, 418, 425; **T. bataar,** 261, 306, 314, 319, 325, 326, 335, 337; **T. rex,** 1–3, 6, 8, 9, 11, 13, 15, 17, 18, 20–24, 28, 29, 31, 32, 34–37, 40–44, 47, 50, 51, 55–60, 63, 69, 73, 83, 87, 93, 96, 98, 103–108, 110–117, 120–122, 124, 132, 133, 156–158, 160, 167, 169, 170, 192–195, 197–199, 201, 209, 210, 211, 213, 216, 227, 233, 238, 245, 247–250, 254, 255, 257–278, 281–285, 287, 288, 294, 295, 299, 306, 309–311, 314, 316, 319, 323–327, 329–331, 335, 337, 338, 340, 358–365, 371, 372, 376, 377, 383, 384, 386, 388–390, 398–400, 402–416, 417–425; "007," 29; B-rex, 5, 39, 122; Barney, 399; Barnum, 4, 31; Black Beauty, 3, 16, 17, 51, 122; Bob, 39, 40, 96; Bowman, 4, 24; Bucky, 5, 43, 44, 51, 55; C-rex, 5, 40; County rex, 32; Cowley **T. rex,** 16; Devil rex, 18; Duffy, 4, 26, 27, 38, 51; E. D. Cope, 5, 37; F-rex, 5, 44; Fox, 4, 32, 35; Foxy Lady, 32; Ivan, 5, 49, 51; Jane, 82, 83, 85, 86, 87, 88, 89, 110, 294; G-rex, 5, 44; Hager rex, 3, 17; Henry, 122; Huxley, 14, 15, 122; Monty, 4, 38; Mr. Zed, 23; Mr. Z, 23; Mud Butte **T. rex,** 13; N-rex, 5, 44; Ollie, 4, 35, 51; Otto, 5, 44; Peck's rex, 4, 33, 34, 51, 55, 74–76, 78, 79, 87, 122; Pete, 4, 30, 31, 67, 69; Rex A, 35; Rex B, 4, 35; Rex C, 4, 35; Samson, 4, 23, 51, 104, 105, 107, 108, 114, 115, 122, 295; Scotty, 4, 28, 51; Stan, 4, 9, 21, 22, 26, 36, 44,

46, 51, 105, 122, 192, 194, 195, 196, 198, 212, 247, 289, 295; Steven, 4, 29, 30, 38; Sue 3, 9, 11, 19, 20, 21, 44, 51, 55, 69, 110, 122, 196, 288, 294, 424, 425; Thomas, 5, 47, 48, 51; Tinker, 4, 34, 51; **Triceratops**-Alley **T. rex,** 35; Wankel **rex,** 3, 18, 122; Wayne, 49; Wyrex, 5, 46, 47, 51, 55, 122; Z-rex 23, 24, 51, 122; **T.** x, 103, 104, 105, 106, 107, 108, 110, 111, 112, 115, 116, 122

Velociraptor, 148, 149, 333, 374, 376

Wintonopus, 224

Yangchuanosaurus, 316

ABOUT THE EDITORS

Peter Larson is founder and president of the Black Hills Institute of Geological Research in Hill City, SD. He and his staff have amassed the largest research collection of *T. rex* specimens, including "Stan," casts of which are seen in museums around the world.

Kenneth Carpenter is the dinosaur paleontologist for the Denver Museum of Natural History. He is author of *Eggs, Nests, and Baby Dinosaurs* (2000), editor of *The Carnivorous Dinosaurs* (2005) and *The Armored Dinosaurs* (2001), and co-editor of *Thunder-Lizards: The Sauropodomorph Dinosaurs* (with Virginia Tidwell, 2005) and *Mesozoic Vertebrate Life* (with Darren H. Tanke, 2001), all with Indiana University Press. He is also co-editor of *Dinosaur Systematics; Dinosaur Eggs and Babies;* and *The Upper Jurassic Morrison Formation.*

COLOPHON

Tyrannosaurus rex, *the Tyrant King* was designed at Indiana University Press by Jamison Cockerham and set in type by Mike Kelsey at Inari Information Services. June Silay was the project editor and Robert Sloan was the sponsoring editor.

The text type is Electra, designed by William A. Dwiggins in 1935, and the display type is Frutiger, designed by Adrian Frutiger in 1975. This book was printed by Sheridan Books, Inc.

www.ingramcontent.com/pod-product-compliance
Ingram Content Group UK Ltd.
Pitfield, Milton Keynes, MK11 3LW, UK
UKHW052218180425
457603UK00007B/55